Developments in Food Science 3C

FOOD FLAVOURS
Part C. The Flavour of Fruits

DEVELOPMENTS IN FOOD SCIENCE

DEVELOPMENTS IN FOOD SCIENCE 3C

FOOD FLAVOURS
Part C. The Flavour of Fruits

Edited by

I.D. MORTON
Department of Food and Nutritional Sciences

A.J. MACLEOD
Department of Chemistry

*King's College London, University of London, Strand,
London WC2R 2LS, England*

ELSEVIER
Amsterdam — Oxford — New York — Tokyo   1990

ELSEVIER SCIENCE PUBLISHERS B.V.
Sara Burgerhartstraat 25
P.O. Box 211, 1000 AE Amsterdam, The Netherlands

Distributors for the United States and Canada:

ELSEVIER SCIENCE PUBLISHING COMPANY INC.
655, Avenue of the Americas
New York, NY 10010, U.S.A.

Library of Congress Cataloging-in-Publication Data
(Revised for vol. 3)

Food flavours.

    (Developments in food science ; 3)
    Includes bibliographies and indexes.
    Contents: pt. A. Introduction. -- pt. B. The flavour
of beverages. -- pt. C. The flavour of fruits.
    1. Flavor. 2. Food--Analysis. I. Morton, I. D.
(Ian Douglas) II. MacLeod, A. J. (Alexander Joseph)
III. Series.
TX541.F66 1982   664   80-17054
ISBN 0-444-41857-1 (v. 1)
ISBN 0-444-87362-7 (v. 3)

ISBN 0-444-87362-7

PREFACE

This is the third volume in the series *Food Flavours*. The first provided a general introduction to the subject, in which matters of basic importance and common to the majority of food and food products were considered. The second dealt with the flavour of a specific group of foods, namely beverages, and this third volume continues that approach, dealing with the flavour of fruit. Fruit is consumed principally because of its desirable flavour and texture. Since much fruit is eaten raw, flavour composition is often simpler than that of heated foods, e.g. roast meat, coffee. Nevertheless, studying and understanding the flavour of fruit is still a great challenge, as shown very clearly by the contributions in this book.

Any volume on food flavour chemistry must inevitably include large numbers of tables listing identified volatile constituents contributing to flavour. Undoubtedly these are of great value, and they supplement specific compilations of such data. However, throughout this volume, authors have also included detailed critical evaluation of these lists, as well as discussion of many additional aspects of flavour composition and qualities of various fruit. For example, in some cases, aroma biogenesis has been considered; in others, the effects of agricultural practices, storage conditions or processing on the development of fruit flavour have been reviewed. We hope that the reader will find this mixture both stimulating and informative.

It has not, of course, been possible to include consideration of all fruit, although discussion of a wide range of the less common and the more exotic is included, as well as the more popular types. Furthermore, there is still a number of fruit that has yet to be properly studied with regard to flavour composition. There is ample scope for more research within this area.

Taking into account the widespread consumption throughout the world of a large variety of fruit, it is very appropriate that the major groupings of fruit have been covered in this book by International authorities from Finland, France, Germany, New Zealand, the U.S.A. and the U.K. Each chapter reflects the individual approach of the authors and the different aspects of the fruit flavour under discussion. We are extremely grateful for the co-operation of the authors which has permitted citations to the early 1989 literature, as well as the older literature.

As before, the editors must, of course, accept full responsibility for the selection of topics covered and for any omissions. We should be pleased to hear from any readers offering comment or constructive criticism.

A.J. MacLeod
I.D. Morton
King's College London
University of London.

CONTENTS

VIII

Chapter I

THE FLAVOUR OF APPLES, PEARS AND QUINCES

NICOLE M.M. PAILLARD
Laboratoire de Physiologie Végétale Appliquée, Université Pierre et Marie Curie, Tour 53, 1er étage, 4, Place Jussieu, 75252 Paris Cedex 05 (France)

INTRODUCTION

The apple (*Pirus malus* L.), the pear (*Pirus communis* L.), and the quince (*Cydonia oblonga* Mill, *Cydonia vulgaris* Pers., or *Pirus cydonia* Lin.) are fruits produced by trees or bushes of the Rosaceae family, and, like the medlar, are often referred to as 'Pome Fruits'. The first two are eaten as table fruit or after processing, whereas the quince is only eaten after cooking and processing. The hardness, bitterness and astringency of the flesh renders the raw fruit uneatable. Table I.I summarises worldwide production of apples and pears. Quince production is much smaller and does not appear in FAO statistics. In France, for example, quince is grown in small orchards or on isolated trees.

Given the extent of its cultivation and its economic importance, the apple is the most studied of these three fruit: chemical and biochemical composition, pre- and post-harvest physiology, ripening, storage, etc. Aroma and flavour in particular have been the subject of a large number of studies over many years. Far fewer studies have been devoted to pear aroma and flavour, and have generally involved the Bartlett cultivar (= Williams). The case of quince is different. Its pectin-rich composition and characteristic aroma have resulted in traditional use in the manufacture of fruit pulp, jelly, syrup, and jam. The manufacture of juice or nectar could be developed to diversify the assortment of fruit-based drinks (Schobinger *et al.,* 1982). In France, the quince, like the apple and pear, is used to make pulp and fruit purée. The apple alone accounts for 50 and 63%, respectively, of tinned purée manufactured between 1980 and 1985, and in 1986, whereas the pear represents only 6%. Deep-frozen quince and pear purées account for 6 and 12%, respectively, of deep-frozen fruit purée production (Les cahiers du CTIFL, 1987).

1. CHEMICAL COMPOSITION

1.1 History

The first analysis of volatile compounds in apples was published in 1920 by

Power and Chesnut. It indicated the presence of acetaldehyde, pentyl esters of formic, acetic, hexanoic and octanoic acids, traces of methanol and ethanol, and perhaps free formic, acetic and hexanoic acids. Gas chromatography was not used in the dozen or so studies which followed. Some studies dealt with total volatile organic compounds (Fidler, 1948, 1950; Gerhardt, 1954), while others involved analyses using a variety of techniques: combination of volatiles, separation by paper chromatography, spectrophotometry. The esters were generally hydrolysed. Meigh (1956, 1957) thus identified seven aldehydes and six alcohols, and the esters were converted into hydroxamic acids.

Analytical progress was made possible by the use of gas chromatography, and the number of identified compounds increased regularly with technical advances, particularly with capillary column separation and the combined use of gas chromatography and mass spectrometry (GC – MS). Nineteen apple aroma constituents were reported in 1961 by Strackenbrock (1961b), 56 were identified in apple essence in 1967 (Flath et al., 1967), 96 volatiles emitted from apples were separated in 1968, and about 40 were identified (Paillard, 1968). One hundren and seventeen compounds in apple extracts were listed by Drawert et al. (1969a), including 39 acids, 34 esters, 30 alcohols and 12 aldehydes. Dimick and Hoskin (1983) have listed 266 volatiles from apple. In recent publications, far fewer apple volatiles are usually considered, and studies of apple aromas and

TABLE I.I

Apple and pear production (100 MT[a]) (FAO, 1985)

	Apples				Pears			
	1979 – 81	1983	1984	1985	1979 – 81	1983	1984	1985
World	34466	39289	39598	37922	8558	9551	9372	9212
Africa	488	472	624	606	221	275	244	285
N. America	4491	4576	4501	4454	871	777	717	747
S. America	1379	1379	1562	1593	256	283	276	248
Asia	7971	10120	9551	9642	2733	3212	3483	3634
Europe	13324	14409	15762	14041	3753	4178	3936	3628
Oceania	528	534	498	586	149	130	135	149

	1985			1985	
	Apples	Pears		Apples	Pears
Australia	340	136	Italy	2092	874
Argentina	943	145	Japan	907	470
Canada	504	25	Poland	1343	100
Chile	401	46	Turkey	1772	370
France	2315	450	U.K.	318	52
Germany F.R.	1410	335	U.S.A.	3542	670

[a] MT = metric tonnes.

flavours have focused on other aspects: comparison of cultivars, changes during ripening and storage, organoleptic significance of constituents, biosynthesis, and metabolic origin, etc. Table I.II shows a chronological list of volatile analyses, together with information regarding the cultivars studied and the methods used. Not all authors, however, have provided all details or have analysed commercially manufactured apple essences. The list relates principally to analytical work and is not exhaustive; further results will be cited in various sections of this chapter, particularly in terms of sensory evaluation.

Far fewer studies have been devoted to the aroma and flavour of the pear. The first chromatograms were published in 1961 (Buttery and Teranishi), and the recent analyses carried out for pear essences have involved HPLC (high-performance liquid chromatography) (Russell *et al.*, 1981; Quamme, 1984). Studies of quince aroma and flavour are recent, and the latest have used gas chromatography combined with mass spectrometry (GC – MS), infrared spectroscopy, nuclear magnetic resonance and preparative HPLC, with the most refined methods being HRGC (high-resolution gas chromatography) and HRMS (high-resolution mass spectroscopy) (Tsuneya *et al.*, 1983; Ishihara *et al.*, 1986; Umano *et al.*, 1986; Winterhalter *et al.*, 1987).

1.2 Volatile Compounds Identified in Apple Aroma

Table I.III lists the compounds identified in apple aromas, as well as some details of the methods used and the authors. The preparative techniques have been classified in two groups. On the one hand, those that allow trapping and analysis of volatiles emitted by whole, intact fruit: headspace, condensation in a cold trap, trapping on an adsorbent followed by elution or desorption, etc. These compounds comprise the odour from the fruit and the first impression given by the smell. In the second group of techniques, fruit are fragmented to a greater or lesser degree, or are processed by various distillations, extractions and concentrations of fruit juice. The volatiles play rôles in odour but also in the aroma perceived in the mouth, via the retronasal pathway, and the overall flavour and taste. It is necessary to distinguish the secondary aromas formed during these treatments, notably by enzymic reaction (oxidation, hydrolysis). Destruction of cellular structures calls for inhibition of enzymic reactions to avoid or limit release of secondary compounds (Drawert *et al.*, 1966; Drawert, 1975).

1.2.1 Esters
Ninety-two esters form the largest chemical category, including 44 straight-chain esters, combinations of the first in the alcohol series with the first in the fatty acid series. These constitute the major chromatogram peaks. Esters are particularly well represented in analyses of volatiles emitted by fruit. In various cultivars they account for 78 – 92% of the total emission adsorbed by activated charcoal (Paillard, 1967b) and 81 – 96% in headspace vapour (Kakiuchi *et al.*, 1986), but only 11 – 33% in distillates of the same materials. In Golden

TABLE I.II

Analysis of volatile compounds of apples, pears and quinces. History

	Authors	Sample preparation	Analysis	Cultivars	
Apples					
1920	Power and Chesnut	Distillation	Reactions	Several cultivars	1
1922	Power and Chesnut		Reactions	McIntosh	2
1942	Walls	Trapping	Total volatiles	Several cultivars	3
1948	Fidler	Trapping		Several cultivars	4
1950	Fidler	Filters		King Edward VII	5
1950	White	Distillation	Reactions	MacIntosh	6
				Stayman Wines	
1951	Thompson	Cold trap	Tests, paper chromatography		7
1952	Huelin	Absorption	Paper chromatography	Granny Smith	8
1953	Henze et al.	Adsorption, extraction, elution	Partition chromatography		9
1954	Gerhardt	Absorption, adsorption	Column chromatography	Several cultivars	10
1954	Henze et al.	Adsorption			11
1956	Meigh	Cold trapping	Paper chromatography	Several cultivars	12
1957	Meigh				
1961a	Strackenbrock	Adsorption	GC	Golden Delicious	13
1961b	Strackenbrock			Several cultivars	
1962	Grevers and Doesburg	Adsorption	GC	Several cultivars	14
1962	Von Wucherpfennig and Bretthauer	Distillation	Paper chromatography	Several cultivars	15
1964	Koch and Schiller	Concentration	GC		16
1964	MacGregor et al.	Juice concentrate	GC	McIntosh	17
1964	Nishimura and Hirose	Extraction, distillation	GC	Kogyoku	18
1965	Osman	Essence	GC		19
1965	Paillard	Adsorption	GC	Calville blanc	20
1967	Angelini and Pflug	Adsorption	GC, MS	McIntosh	21
1967	Flath et al.	Commercial essence	GC	Delicious	22
1967a	Paillard	Adsorption	GC, IR	Calville blanc	23
1967	Schultz et al.	Commercial essence	GC	Delicious	24
1968	Brown et al.	Headspace	GC	Royal Red Delicious, Starkrimson	25

Year	Author	Method	Analysis	Cultivar	Ref.
1968	Drawert et al.	Extraction	GC	Granny Smith	26
1968	Huelin and Coggiola	Coating extract	Column, spectroscopy	Calville blanc	27
1968	Paillard	Adsorption	Capillary GC	Industrial aroma	28
1968	Pompéi	Concentration, extraction	GC		29
1969a	Drawert et al.	Extraction	GC		30
1969b	Drawert et al.	Headspace			
1969	Flath et al.	Essence	GC–MS	Gravenstein	31
1971a	Guadagni et al.	Peel and flesh		Red Delicious	32
1971b	Guadagni et al.	Headspace, essence	GC	Gravenstein	
1973	Brulé	Distillation	GC–MS	Apple juice	33
1977	Dirinck et al.	Adsorption	GC–MS	Golden Delicious	34
1977	Williams et al.	Headspace	GC–MS, IR	Several cultivars	35
1982	Schamp and Dirinck	Headspace-adsorption	GC–MS	Golden Delicious	36
1984	Berger et al.	Liquid–liquid extraction	GC–MS	Red Delicious	37
1984	Yajima et al.	Distillation	GC–MS, NMR, IR	Kogyoku	38
1986	Kakiuchi et al.	Headspace, distillation	GC–MS	Several cultivars	39

Pears

Year	Author	Method	Analysis	Cultivar	Ref.
1961	Buttery and Teranishi	Vapours	GC	Comice d'Anjou	40
1966	Heinz and Jennings	Essence	GC, IR	Bartlett	41
1966	Romani and Ku	Headspace	GC	Bartlett	42
1969	Gascó et al.	Juice concentration	GC		43
1970	Paillard et al.	Cold trap	GC	Passe-Crassane	44
1974	Jennings and Tressl	Porapak trap	GC	Bartlett	45
1981	Russel et al.	Distillation–extraction	GC, HPLC	Several cultivars	46
1982	Strandjev	Essence	GC	Beurré Will., Passe-Crassane	47
1984	Quamme	Distillation–extraction	HPLC	Several cultivars	48

Quinces

Year	Author	Method	Analysis	Cultivar	Ref.
1979	Schreyen et al.	Distillation, headspace	GC–MS, IR, NMR		49
1982	Schobinger et al.	Distillation–extraction	GC	Several cultivars	50
1983	Tsuneya et al.	Distillation	GC–MS, IR, NMR		51
1986	Ishihara et al.	Distillation	GC, HPLC, NMR, MS		52
1986	Umano et al.	Fruit peel, headspace, Distillation–extraction	GC–MS		53
1987	Winterhalter et al.	Distillation–extraction	HRGC, GC–MS		54

TABLE I.III

Volatile flavour compounds identified from apples

V (vapours): volatiles emitted from fruits, sample preparation by headspace, cold trap, adsorption, etc.
E (essence, extract): volatile constituents of fruits obtained by distillation, extraction, concentration, etc.
Number: reference in Table I. II

Esters

Methyl formate   V: 14, 23, 28   E: 17
Methyl acetate   V: 13, 14, 23, 25, 28, 39
   E: 19, 26, 29, 30, 39
Methyl propanoate   V: 23, 28
Methyl butanoate   V: 25, 28   E: 22, 29
Methyl hexanoate   V: 23, 28   E: 17
Methyl methylpropanoate   V: 14
Methyl 3-methylbutanoate   V: 13
Methyl 2-methylbutanoate   E: 22
Ethyl formate   V: 28   E: 6, 17, 33
Ethyl acetate   V: 13, 14, 20, 23, 25, 28, 32, 34, 36, 39
   E: 6, 17, 19, 22, 24, 26, 29, 30, 31, 33, 39
Ethyl propanoate   V: 13, 20, 25, 28, 32
   E: 6, 22, 24, 29, 31
Ethyl butanoate   V: 14, 20, 21, 23, 25, 28, 32, 39
   E: 6, 17, 22, 24, 26, 29, 30, 31, 33, 39
Ethyl pentanoate   V: 13, 20, 23, 28,   E: 22, 29, 31
Ethyl hexanoate   V: 13, 14, 20, 25, 28, 32, 39
   E: 6, 17, 18, 22, 26, 29, 30, 31, 39
Ethyl octanoate   V: 14, 28   E: 26, 30, 31
Ethyl nonanoate   E: 26, 30
Ethyl decanoate   E: 26, 30
Ethyl dodecanoate   E: 26, 30
Ethyl methylpropanoate   V: 7, 28, 32
Ethyl 3-methylbutanoate   V: 7
Ethyl 2-methylbutanoate   V: 32, 39   E: 22, 31, 33, 39
Ethyl (E)-but-2-enoate   E: 31
Ethyl (E)-oct-2-enoate   E: 37
Ethyl (Z)-dec-4-enoate   E: 37
Ethyl (E, Z)-deca-2,4-dienoate   E: 37
Propyl formate   V: 23, 28
Propyl acetate   V: 20, 23, 25, 28, 34, 36, 39
   E: 18, 22, 24, 26, 29, 30, 31, 33, 38, 39
Propyl propanoate   V: 25, 28, 34   E: 22, 38
Propyl butanoate   V: 14, 20, 23, 25, 28, 32, 34, 36
   E: 22, 26, 30, 31
Propyl pentanoate   V: 28   E: 29
Propyl hexanoate   V: 39   E: 39
Propyl methylpropanoate   V: 14
Propyl 2-methylbutanoate   V: 34   E: 22
Butyl formate   V: 21, 23, 28
Butyl acetate V: 7, 13, 14, 20, 21, 23, 25, 28, 32, 34, 36, 39
   E: 17, 18, 19, 22, 24, 26, 29, 30, 31, 38, 39
Butyl propanoate   V: 14, 20, 23, 28, 32, 34, 36, 39
   E: 17, 22, 31, 38, 39
Butyl butanoate   V: 13, 14, 21, 23, 25, 28, 34, 36, 39
   E: 19, 22, 26, 29, 30, 31, 38, 39
Butyl pentanoate   V: 23, 28, 34
Butyl hexanoate   V: 14, 28, 34, 36, 39
   E: 22, 26, 29, 30, 39
Butyl octanoate   E: 26, 30
Butyl decanoate   E: 26, 30
Butyl methylpropanoate   V: 14
Butyl 2-methylbutanoate   V: 34, 36, 39   E: 38, 39
Butyl 3-hydroxybutanoate   E: 38
Pentyl formate   E: 1, 19
Pentyl acetate   V: 13, 14, 23, 25, 28, 32, 34, 36, 39
   E: 1, 18, 19, 22, 26, 30, 38, 39
Pentyl propanoate   V: 28   E: 29

Pentyl butanoate   V: 28, 34, 36   E: 19, 26, 30, 34, 36
Pentyl hexanoate   V: 13, 34, 39   E: 1, 26, 30, 39
Pentyl octanoate   E: 1, 26, 30
Pentyl decanoate   E: 26, 30
Pentyl 2-methylbutanoate   V: 34   E: 22
Hexyl formate   V: 28
Hexyl acetate   V: 13, 14, 20, 23, 28, 32, 34, 36, 39
   E: 18, 19, 22, 24, 26, 29, 30, 31, 38, 39
Hexyl propanoate   V: 13, 34, 36, 39   E: 22, 28, 29, 38, 39
Hexyl butanoate   V: 21, 28, 34, 36, 39
   E: 22, 26, 30, 31, 38, 39
Hexyl pentanoate   V: 21, 34
Hexyl hexanoate   V: 34, 39   E: 26, 30, 38, 39
Hexyl octanoate   V: 39   E: 39
Hexyl methylpropanoate   V: 34
Hexyl 2-methylbutanoate   V: 34, 36, 39   E: 38, 39
Heptyl acetate   V: 34
Heptyl 2-methylbutanoate   V: 34
Octyl acetate   V: 21, 34
Isopropyl butanoate   V: 14
2-Methylpropyl formate   V: 21, 28
2-Methylpropyl acetate   V: 23, 25, 28, 32, 34, 36
   E: 22, 26, 29, 30, 31, 38
2-Methylpropyl propanoate   V: 25
2-Methylpropyl butanoate   V: 28   E: 22
1-Methylpropyl acetate   V: 28
3-Methylbutyl acetate   V: 13, 14, 23, 25, 28, 34
   E: 17, 26, 29, 30, 31
3-Methylbutyl propanoate   V: 14, 25, 28
3-Methylbutyl butanoate   V: 28   E: 26, 30
3-Methylbutyl pentanoate   E: 26, 30
3-Methylbutyl hexanoate   V: 34
3-Methylbutyl octanoate   E: 26, 30
3-Methylbutyl decanoate   E: 26, 30
3-Methylbutyl 3-methylbutanoate   V: 14
2-Methylbutyl acetate   V: 32, 39
   E: 22, 24, 38, 39

2-Methylbutyl propanoate   V: 39   E: 39
2-Methylbutyl butanoate   V: 39   E: 39
2-Methylbutyl hexanoate   V: 39   E: 39
2-Methylbutyl 2-methylbutanoate   V: 39   E: 39
(E)-Hex-2-enyl acetate   E: 24, 38
(Z)-Hex-3-enyl acetate   E: 38
(E, Z)-Octa-3,5-dien-1-yl acetate   E: 38
(Z, Z)-Octa-3,5-dien-1-yl acetate   E: 38
3-Hydroxyoctyl acetate   E: 38
3-Hydroxy-(Z)-oct-5-en-1-yl acetate   E: 38
Benzyl acetate   E: 38
2-Phenethyl acetate   E: 22, 31, 38

Alcohols

Methanol   V: 11, 12, 28   E: 6, 15, 16, 17, 19, 26, 29, 30
Ethanol   V: 9, 12, 20, 23, 25, 28
   E: 6, 15, 16, 17, 18, 19, 22, 24, 26, 29, 30, 31, 33, 38
Propan-1-ol   V: 9, 13, 20, 25, 28, 36
   E: 6, 15, 16, 17, 18, 19, 22, 24, 26, 29, 30, 31, 33, 38
Butan-1-ol   V: 9, 13, 14, 20, 23, 25, 28, 34, 36, 39
   E: 6, 15, 16, 17, 18, 19, 22, 24, 26, 29, 30, 31, 33, 38, 39
Pentan-1-ol   V: 3, 9, 13, 23, 25, 28, 39
   E: 15, 16, 17, 18, 19, 26, 29, 30, 33, 38, 39
Hexan-1-ol   V: 8, 9, 12, 13, 21, 23, 28, 32, 34, 39
   E: 6, 15, 16, 17, 18, 22, 24, 26, 29, 30, 33, 38, 39
Heptan-1-ol   E: 26, 30, 38
Octan-1-ol   E: 26, 30, 38
Nonan-1-ol   E: 26, 30
Decan-1-ol   E: 26, 30, 38
Undecan-1-ol   E: 26, 30
Dodecan-1-ol   E: 26, 30
Propan-2-ol (or isopropanol)   V: 12, 21   E: 6, 16, 17, 19
2-Methylpropan-1-ol   V: 12, 14, 20, 21, 28, 34, 39
   E: 16, 17, 18, 19, 22, 24, 26, 30, 31, 33, 38, 39
Butan-2-ol   V: 28
2-Methylpropan-2-ol   V: 21

(continued)

TABLE 1.III (*continued*)

3-Methylbutan-1-ol V: 3, 14, 21, 22, 28, 34
  E: 16, 17, 22, 26, 29, 30, 33
2-Methylbutan-1-ol V: 12, 21, 39
  E: 16, 17, 18, 22, 24, 26, 30, 31, 38, 39

Pentan-2-ol E: 26, 30
3-Methylbutan-2-ol V: 21
2-Methylbutan-2-ol V: 21 E: 31
2-Methylpentan-1-ol V: 34
Hexan-2-ol E: 26, 30
2-Methylpentan-2-ol E: 31
Secondary alcohols, $C_7 - C_{12}$ E: 26, 30
(*E*)-Hex-2-en-1-ol V: 39 E: 16, 22, 24, 26, 29, 30, 33, 38, 39
(*Z*)-Hex-3-en-1-ol E: 18, 22, 26, 30, 33, 38
(*Z*)-Oct-5-en-1-ol E: 38
6-Methylhept-5-en-2-ol E: 38
(*E, Z*)-Octa-3,5-dien-1-ol E: 38
(*Z, Z*)-Octa-3,5-dien-1-ol E: 38
Octane-1,3-diol E: 38
(*E*)-Oct-5-ene-1,3-diol E: 38
2-Phenylethanol E: 26, 30, 33, 38

Aldehydes

Formaldehyde E: 16
Acetaldehyde V: 8, 12, 14 E: 1, 16, 18, 19, 22, 24, 26, 29, 30, 31, 38
Propanal V: 8, 12, 14, 28 E: 16, 18, 19, 26, 29, 30
Butanal V: 12, 14 E: 16, 18, 19, 22, 26, 29, 30, 31, 38
Pentanal V: 14 E: 16, 22, 26, 30, 31
Hexanal V: 8, 13, 14, 28, 32, 36, 39 E: 6, 16, 22, 24, 26, 29, 30, 31, 38, 39
Heptanal V: 21 E: 26, 30
Octanal V: 21
Nonanal V: 21 E: 18, 26, 30, 38
Decanal V: 21 E: 26, 30, 38
Undecanal E: 26, 30

Dodecanal E: 26, 30
Methylpropanal E: 29
3-Methylbutanal V: 12, 21 E: 31
2-Methylbutanal E: 22, 31
(*E*)-Hex-2-enal V: 8, 28, 32, 34, 36, 39
  E: 6, 16, 18, 19, 22, 24, 26, 29, 30, 31, 38, 39
(*Z*)-Hex-3-enal V: 34 E: 26, 30, 38
(*E*)-Hex-3-enal E: 26, 30
(*E*)-Hept-2-enal E: 26, 30
Methylpropenal V: 21
2-Methylbut-2-enal V: 34
Benzaldehyde V: 34
Phenylacetaldehyde E: 33
Furfural (2-furaldehyde) E: 6, 31, 33

Ketones

Acetone V: 8, 12, 13, 28 E: 6, 16, 17, 19, 29
Butanone V: 12 E: 16, 31
Pentan-2-one V: 12 E: 31
Pentan-3-one E: 22
Heptan-2-one E: 31
7-Methyloctan-4-one E: 31
2-Methylhept-2-en-6-one V: 34
6-Methylhept-5-en-2-one E: 38
Acetoin (3-hydroxybutanone) E: 26, 30
Butanedione E: 26, 30, 31
Acetophenone E: 38

Free acids

Formic V: 3, 9 E: 6, 15, 16, 17
Acetic V: 3, 7, 9, 14, 23 E: 6, 15, 16, 17, 26, 30, 33
Propanoic V: 7, 9 E: 6, 16, 17, 26, 30
Butanoic V: 7 E: 6, 16, 26, 30
Pentanoic V: 7 E: 6, 16, 26, 30
Hexanoic V: 8, 9 E: 15, 16, 17, 18, 26, 30

Heptanoic E: 26, 30
Octanoic V: 9 E: 16, 26, 30
Nonanoic E: 26, 30
Decanoic E: 26, 30
Undecanoic E: 26, 30
Dodecanoic E: 26, 30
Saturated fatty acids, $C_{13} - C_{20}$ E: 26, 30
Methylpropanoic E: 16, 26, 30
3-Methylbutanoic E: 18, 26, 30
2-Methylbutanoic E: 18, 26, 30
(E)-Hex-2-enoic E: 16, 26, 30
Unsaturated fatty acids $C_{5:1} - C_{19:1}$ E: 26, 30
Benzoic E: 18

Ethers

1-Ethoxy-1-methoxyethane E: 22
1,1-Diethoxyethane E: 22, 31
1-Ethoxy-1-propoxyethane E: 22, 31
1,1-Diethoxypropane E: 22
1-Butoxy-1-ethoxyethane E: 22
1-Ethoxy-1-(2-methylbutoxy)ethane E: 22, 31
2,4,5-Trimethyl-1,3-dioxolane E: 22, 31

Hydrocarbons

Ethane E: 26
Hexane V: 21
Heptane V: 21
Octane V: 21
Nonane V: 21, 34

Decane V: 21
Undecane V: 21, 36
Tridecane V: 36 E: 38
Tetradecane E: 38
Ethylene V: 3, 5 E: 26
Hept-1-ene V: 21
Oct-1-ene V: 21
α-Farnesene V: 34, 39 E: 27, 38, 39
Benzene V: 28, 34
Toluene V: 21
Ethylbenzene V: 21, 28, 34
1-Methylnaphthalene E: 22
2-Methylnaphthalene E: 22

Terpenes

α-Pinene V: 34
Camphor E: 38
Geraniol E: 2
Linalool E: 38
Linalool oxide E: 38
α-Terpineol E: 38

Miscellaneous

γ-Hexalactone E: 38
4-Methoxyallylbenzene V: 34, 35, 36
Methyleugenol E: 38
Methylchavicol V: 39 E: 38, 39
Benzothiazole E: 38

Delicious, for example, esters represent 78% of total volatiles adsorbed by activated charcoal, 96% in headspace vapour, and 29% in the distillate.

The vapour pressures of the lower esters (up to nine carbon atoms) at room temperature are relatively high: 1 mm of mercury for pentyl butanoate, 3 mm for pentyl acetate, 8 mm for butyl acetate, 78 mm for ethyl acetate, and 204 mm for methyl acetate. However, vapour pressure is not the only factor in volatility. Solubility in lipids and skin waxes increases for longer aliphatic chains, whereas in dilute aqueous solution, ester volatility increases for chains of C_9 and longer. The same is true for aldehydes and ketones (Buttery *et al.,* 1969, 1971). C_{10} esters, such as butyl hexanoate (boiling point 204.3°C) and ethyl octanoate (b.p. 208°C), and C_{11} esters, such as pentyl hexanoate (b.p. 222.2°C), have vapour pressures close to 1 mm of mercury. These esters are found both in the volatile emissions and in extracts, and appear to constitute a limit, since higher esters ($> C_{11}$) are only identified in extracts. This is perhaps due to reduced volatility, which prevents liberation from the fruit in detectable quantities. Only hexyl hexanoate and hexyl octanoate are detected, in trace amounts, in the emission from ester-rich cultivars (Kakiuchi *et al.,* 1986).

Some analyses suggest a lower proportion of esters, no doubt due to enzymic hydrolysis. Short-chain aliphatic esters seem especially susceptible, and are easily degraded during treatment of apple juices or essences. They are, for example, hydrolysed by commercial preparations of pectolytic enzymes (Jakob *et al.,* 1973).

1.2.2 Alcohols

Alcohols constitute the second largest category of compounds, and give large chromatographic peaks, especially straight-chain butan-1-ol and hexan-1-ol, and branched-chain 3-methylbutan-1-ol (isopentanol). Depending on the cultivar, alcohols represent 6 – 16% of the total volatile emission analysed after adsorption on activated charcoal (Paillard, 1967b). Kakiuchi *et al.* (1986) have reported markedly lower amounts in the headspace vapours. However, the proportion of alcohols in distillates ranges from 48.3 to 75.5%.

Primary alcohols, particularly straight-chain aliphatic ones, are present both in volatile emissions from fruit and in extracted essences. Their vapour pressures at room temperature are at least equal to 1 mm of mercury, a value corresponding to hexan-1-ol, and greater than those of lower molecular weight alcohols (8 mm for butan-1-ol, 16 mm for propan-1-ol). However, compared with other classes of compounds, such as methyl esters, aliphatic aldehydes, and even ketones, alcohols are less volatile in very dilute aqueous solutions, and higher molecular weight alcohols, such as octan-1-ol, are therefore more volatile than methanol (Buttery *et al.,* 1971). Higher alcohols have only been found in extracts by some authors (Drawert *et al.,* 1969a; Yajima *et al.,* 1984). The former authors used liquid – liquid extraction to obtain higher molecular weight compounds contained in the lipids and wax coating of apple skin. Some terpene alcohols have been identified in the extracts (Power and Chesnut, 1922; Yajima *et al.,* 1984). Yajima *et al.* reported the presence of alcohols with double bonds in the chain. Numerous authors have identified (*E*)-hex-2-en-1-ol and (*Z*)-hex-3-en-1-ol.

1.2.3 Aldehydes

Straight-chain or branched aliphatic aldehydes identified among apple volatiles generally accompany the corresponding alcohol. Acetaldehyde, often considered as a fermentation product, is found when physiological alteration of the fruit occurs, for example under unfavourable storage conditions. It is also produced by damage or cutting, and the quantities emitted are then proportional to the area of skin removed (Paillard, 1975a). Numerous authors have identified the C_6 aldehydes, hexanal, (*E*)-hex-2-enal and (*Z*)-hex-3-enal, produced by enzymic reaction when cellular structures are disrupted in green vegetables (Drawert *et al.*, 1966). These aldehydes are present in large amounts in extracts, but also exist, in much smaller quantities, in emissions from whole, intact fruit.

1.2.4 Ketones and ethers

Apart from acetone, ketones have only been reported by a few authors, and ethers have only been noted by Flath *et al.* (1967) in apple essences.

1.2.5 Fatty acids

Frequent in esterified form, free fatty acids have also been identified. $C_1 - C_6$ fatty acids appear in extracts and volatile emissions, but some may derive from hydrolysis of esters. The appearance of large amounts of free fatty acids in extracts is often linked to a decreased proportion of esters, indicating enzymic hydrolysis. Drawert *et al.* (1968) have identified aliphatic fatty acids, saturated or with one double bond, up to C_{20} (the higher fatty acids are not given in Table I.III). However, these analyses were performed with concentrates obtained after liquid – liquid extraction with a pentane/dichloromethane mixture, and these long-chain fatty acids doubtlessly derive from the lipids of the fruit, in which they have been found by some other authors (Meigh and Hulme, 1965; Mazliak, 1969).

1.2.6 Hydrocarbons

Certain hydrocarbons, such as nonane, benzene and ethylbenzene, are considered to be pollutants by Dirinck *et al.* (1977), and should be viewed with caution (other aromatic compounds have, nonetheless, been identified: benzyl acetate, benzoic acid, benzaldehyde and acetophenone). Of the alkanes, alkenes and benzene compounds detected by Angelini and Pflug (1967), not all of which are listed in Table I.III, some could be formed by microorganisms present in controlled-atmosphere storage areas.

The ripening hormone ethylene (ethene), present in large amounts in gaseous emissions from fruit, should be considered separately (McGlasson, 1970; Pratt, 1974; Rhodes and Reid, 1974).

Because of its involvement in scald, α-farnesene has been studied in detail (Huelin and Coggiola, 1968, 1970a), and has been identified in apple skin extracts (Huelin and Coggiola, 1968), apple juice distillates (Yajima *et al.*, 1984), and in headspace vapour (Kakiuchi *et al.*, 1986).

12

1.2.7 Terpenes
Terpenes are rare and are represented by a few hydrocarbons, and in oxygenated form by some alcohols, such as geraniol (Power and Chesnut, 1922).

1.2.8 General characteristics of apple volatiles
Compounds with even-numbered carbon chains are frequently found in large quantities, and for esters these include combinations of acetic, butanoic and hexanoic acids with ethyl, butyl and hexyl alcohols. Butyl acetate may account for more than 50% of the total volatile emission in the Calville blanc cultivar (Paillard, 1967b).

The main differences noted between volatiles emitted by fruit and those extracted can be explained by differences in volatility. Several factors are involved: vapour pressure, solubility in water and in certain fruit structures. High molecular weight volatiles, often with one or two hydrophobic aliphatic chains, are likely to be trapped by skin waxes. Nonetheless, distribution of the main compounds between the pulp and serum of apple juice shows that they are contained principally in the serum; 12.2% of hexyl acetate, for example, is found in the pulp, and 87.8% in the serum (Radford et al., 1974).

1.3 Compositional Variations of Apple Aromas From Different Cultivars

The results given above summarise analyses of a large number of cultivars, but some studies have compared composition of volatile emissions or of essences from different cultivars.

Observations of emission from nine cultivars grown in France show that most identified components are always present, but in variable proportions, and no component seems to be characteristic of a single variety (Paillard, 1967b). Cultivars can be categorised according to the type of esters predominant in emission: acetates (Calville blanc, Golden Delicious), butanoates (Richared, Belle de Boskoop, Canada blanc), propanoates (Reinette du Mans, Richared, Starking), and ethanol esters (Starking). Emission from the Canada gris apple is weak, and ester-poor (the fruits of the cultivar have a cork coating).

Drawert (1978) has compared the formation of C_6 aldehydes, hexanal, (E)-hex-2-enal, (Z)-hex-2-enal, and the corresponding alcohols, secondary aromas, in six apple cultivars. C_6 aldehyde formation was noted in all six cultivars, but was always less pronounced in Golden Delicious; hexanal formation was four- to five-fold greater, and (E)-hex-2-enal about 100-fold greater in Cox Orange and Jonathan cultivars.

Volatiles from five apple cultivars grown in Japan (Hatsuaki, Kogyoku (Jonathan), Golden Delicious, Mutsu, Fuji) have been analysed in headspace vapours and after vacuum distillation (Kakiuchi et al., 1986). Thirty-nine compounds were identified both in headspace vapours and in distillates, and were present in all varieties, although sometimes in trace amounts or at concentrations of less than 0.001 ppm. Esters account for 81% of volatiles in headspace vapours in the Fuji cultivar, and 96% in Golden Delicious. These values naturally decrease in the distillates, where alcohols predominate (45% in Golden Delicious, 75% in Fuji). The volatiles of the Hatsuaki cultivar resemble those of its parents (Golden Delicious and Kogyoku), and similarity is also noted for volatiles of Mutsu and Golden Delicious.

Williams *et al.* (1977) have identified 4-methoxyallylbenzene, which accounts for 0.270% of the aroma volatiles in the concentrated headspace vapour of Ellison's Orange. It is present in the aroma emission of 13 other cultivars: 0.265% in Spartan, 0.060% in Golden and Red Delicious, 0.002% in Bramley's Seedling.

1.4 Estimation of Production of Aromas and Volatiles by Apples

Strackenbrock (1961a) has reported a production of 0.9 ml/1000 kg/24 h for the Golden Delicious cultivar stored at 4°C, and 0.001 ml/1000 kg/24 h for the Boskoop cultivar, i.e. respective mean values of 0.9 and 0.001 µl/kg/24 h.

We have estimated daily emission from one Calville blanc apple ripening at 15°C as between 0.66 and 1.32 mg; the respective range at 4°C is 0.5 – 1.0 mg (Paillard, 1975a). Various experiments have given a value of 0.9 mg for mean daily production by one ripe Calville blanc apple.

Kakiuchi *et al.* (1986) have reported amounts (measured by the headspace method) of 1.889 and 1.531 ppm, respectively, for the Hatsuaki and Kogyoku cultivars, and respective values of 0.196 and 0.055 ppm for Golden Delicious and Fuji. The corresponding distillate values were 8.936, 9.415, 5.964 and 2.273 ppm, respectively.

These latter results show that it is difficult to compare the values cited above, since they relate to different cultivars, were obtained by different methods, and are expressed differently. Nonetheless, the value of 5.964 ppm for Golden Delicious (i.e. 5.9 mg/kg fruit) is of the same order of magnitude as the mean daily production of almost 1 mg in the Calville blanc apple. The emissions and chromatograms of these two cultivars are very similar.

1.5 Volatiles Identified in Pear Aroma

Table I.IV shows the results of the main analytical studies of pear aroma. Esters account for most identified components, principally methyl and ethyl esters, and $C_1 - C_8$ alcohols are present. Esterified fatty acids present have an even number of carbon atoms, and are saturated or unsaturated with one, two or even three double bonds.

1.6 Volatiles Identified in Quince Aroma

The list of volatiles identified in quince aromas and flavours is shown in Table I.V. Esters and alcohols are the most common compounds, and saturated esters and low molecular weight alcohols are the major components in headspace vapours (Umano *et al.*, 1986). Ethyl esters are especially numerous:

TABLE I.IV

Volatile flavour compounds identified from pears
V (vapours): volatiles emitted from fruits, sample preparation by headspace, cold trap, adsorption, etc.
E (essence, extract): volatile constituents of fruits obtained by distillation extraction, concentration, etc.
Number: reference in Table I.II

Esters

Methyl formate E: 43
Methyl acetate V: 44, 45 E: 41
Methyl octanoate V: 45 E: 41
Methyl decanoate V: 45 E: 41
Methyl hexadecanoate V: 45
Methyl octadecanoate V: 45
Methyl methylpropanoate E: 43
Methyl 4-oxy-(*E*)-butenoate V: 45 E: 41
Methyl (*E*)-oct-2-enoate V: 45 E: 41
A methyl 3-hydroxyoctenoate V: 45 E: 41
Methyl (*Z*)-dec-4-enoate V: 45 E: 41, 47
Methyl (*E*)-dec-2-enoate E: 41, 47
Methyl (*Z*)-tetradec-8-enoate V: 45
Methyl (*Z*)-tetradec-5-enoate V: 45
A methyl hexadecenoate V: 45
Methyl octadecanoate V: 45
Methyl (*Z*, *E*)-deca-2,4-dienoate V: 45
Methyl (*E*, *E*)-deca-2,4-dienoate V: 45 E: 41, 47
Methyl (*E*, *Z*)-deca-2,4-dienoate V: 45 E: 41, 47
Methyl (*E*, *Z*)-dodeca-2,6-dienoate V: 45
Methyl (*Z*, *Z*)-tetradeca-5,8-dienoate V: 45
Methyl (*E*, *E*, *Z*)-tetradeca-2,4,8-trienoate V: 45
Ethyl acetate V: 44, 45 E: 41, 43, 47
Ethyl propanoate E: 43
Ethyl butanoate V: 44 E: 43
Ethyl octanoate V: 45 E: 41
Ethyl decanoate V: 45 E: 41, 47
Ethyl dodecanoate V: 45
Ethyl tetradecanoate V: 45

Ethyl 4-oxy-(*E*)-butenoate V: 45 E: 41
Ethyl (*E*)-oct-2-enoate V: 45 E: 41
An ethyl 3-hydroxyoctenoate V: 45 E: 41
Ethyl (*Z*)-dec-4-enoate V: 45 E: 41, 47
Ethyl (*E*)-dec-2-enoate V: 45 E: 41
Ethyl (*Z*)-dodec-6-enoate V: 45
Ethyl (*E*)-dodec-2-enoate V: 45
Ethyl (*Z*)-tetradec-8-enoate V: 45
Ethyl (*Z*)-tetradec-5-enoate V: 45
An ethyl hexadecenoate V: 45
Ethyl (*E*, *E*, *Z*)-deca-2,4-dienoate V: 45 E: 41, 47
Ethyl (*E*, *Z*)-deca-2,4-dienoate E: 41
Ethyl (*E*, *Z*)-dodeca-2,6-dienoate V: 45
Ethyl (*Z*, *Z*)-tetradeca-5,8-dienoate V: 45
An ethyl hexadecadienoate V: 45
Ethyl (*E*, *E*, *Z*)-deca-2,4,7-trienoate V: 45
Ethyl (*E*, *Z*, *Z*)-deca-2,4,7-trienoate V: 45
Ethyl (*E*, *Z*, *Z*)-dodeca-2,6,9-trienoate V: 45
Ethyl (*E*, *E*, *Z*)-tetradeca-2,4,8-trienoate V: 45
Propyl acetate V: 44, 45 E: 41, 43, 47
Propyl propanoate E: 43, 47
Propyl (*E*, *Z*)-deca-2,4-dienoate V: 45
Butyl acetate V: 44, 45 E: 41, 43, 47
Butyl (*E*, *Z*)-deca-2,4-dienoate V: 45
Pentyl acetate V: 44, 45 E: 41, 47
Hexyl acetate V: 44, 45 E: 41, 47
Hexyl (*E*, *Z*)-deca-2,4-dienoate V: 45
Heptyl acetate V: 45 E: 41
Octyl acetate V: 45 E: 41
Isopropyl formate E: 43

2-Methylpropyl acetate   E: 47
2-Methylpropyl methylpropanoate   E: 43
2-Methyl-2-propyl acetate   E: 43
3-Methylbutyl acetate   V: 44   E: 43
(Z)-Hex-2-enyl acetate   V: 45

Alcohols
Methanol   E: 47
Ethanol   V: 44, 45   E: 41, 43, 47
Propan-1-ol   V: 45   E: 41, 43
Butan-1-ol   V: 44, 45   E: 41, 47
Pentan-1-ol   V: 45   E: 41, 43, 47
Hexan-1-ol   V: 44, 45   E: 41, 43

Heptan-1-ol   V: 45   E: 41
Octan-1-ol   V: 45   E: 41
2-Methylpropan-1-ol   E: 43
Butan-2-ol   E: 43

Aldehydes
Acetaldehyde   E: 43, 47
Propanal   E: 43
Hexanal   E: 43
Hex-2-enal   E: 43

Hydrocarbons
α-Farnesene   V: 45

TABLE I.V

Volatile flavour compounds identified from quince

V (vapours): volatiles emitted from fruits, sample preparation by headspace
E (essence, extract): volatile constituents of fruits obtained by distillation, extraction
Number: reference in Table I.II

Esters

Methyl hexanoate   E: 49
Methyl octanoate   E: 49
Ethyl acetate   V: 49, 53   E: 49, 53
Ethyl propanoate   V: 49, 53   E: 53
Ethyl butanoate   V: 49, 53   E: 49, 53
Ethyl pentanoate   V: 53
Ethyl hexanoate   V: 49, 53   E: 49, 51, 53
Ethyl heptanoate   V: 49   E: 49, 53
Ethyl octanoate   V: 49, 53   E: 49, 51, 53
Ethyl nonanoate   E: 49, 53
Ethyl decanoate   E: 49, 51, 53
Ethyl dodecanoate   E: 49, 51, 53
Ethyl tetradecanoate   E: 53
Ethyl hexadecanoate   E: 53
Ethyl methylpropanoate   V: 53   E: 49
Ethyl 3-methylbutanoate   V: 49   E: 49
Ethyl 2-methylbutanoate   V: 49, 53   E: 49, 53
Ethyl 3-hydroxybutanoate   E: 53
Ethyl but-2-enoate   V: 49, 53   E: 49
Ethyl but-3-enoate   V: 53
Ethyl hex-2-enoate   E: 49
Ethyl hex-3-enoate   V: 53   E: 49
Ethyl oct-6-enoate   E: 53
Ethyl non-6-enoate   E: 53
Ethyl dec-4-enoate   E: 51, 53
Ethyl dodec-7-enoate   E: 51, 53
Ethyl tetradec-9-enoate   E: 53
Ethyl methylpropenoate   V: 49   E: 53
Ethyl 2-methylbut-2-enoate (or ethyl tiglate)   V: 49   E: 49

Ethyl benzoate   E: 49
Ethyl cinnamate   E: 49
Propyl acetate   V: 49
Propyl octanoate   E: 53
Propyl 2-methylbut-2-enoate   E: 49
Butyl acetate   V: 53   E: 49
Butyl butanoate   V: 53
Butyl octanoate   E: 53
Butyl methylpropanoate   V: 53
Hexyl acetate   V: 49, 53   E: 49, 53
Hexyl butanoate   V: 53
Hexyl hexanoate   E: 51
Hexyl octanoate   E: 51
Hexyl methylpropanoate   E: 53
Hexyl 2-methylbutanoate   E: 53
2-Methylpropyl acetate   V: 53
2-Methylpropyl propanoate   V: 53
2-Methylpropyl butanoate   V: 53
2-Methylpropyl hexanoate   E: 49, 53
2-Methylpropyl octanoate   E: 53
2-Methylpropyl methylpropanoate   V: 53
3-Methylbutyl formate   E: 49
3-Methylbutyl acetate   E: 49
3-Methylbutyl 3-methylbutanoate   E: 49
2-Methylbutyl 3-methylbutanoate   E: 49
Hex-2-enyl acetate   V: 53
Hex-3-enyl acetate   V: 49, 53
Hex-3-enyl butanoate   E: 53
3-Methylbutyl benzoate   E: 49

(continued)

Alcohols
Methanol   E: 49
Ethanol   V: 49, 53   E: 49
Propan-1-ol   E: 49
Butan-1-ol   V: 49, 53   E: 49, 53
Pentan-1-ol   V: 53   E: 49, 53
Hexan-1-ol   V: 49, 53   E: 49, 51, 53
Octan-1-ol   E: 49
2-Methylpropan-1-ol   E: 49
Butan-2-ol   V: 53
3-Methylbutan-1-ol   E: 49   E: 49, 53
2-Methylbutan-1-ol   V: 49   E: 49
2-Methylbut-2-en-1-ol   V: 49   E: 49
(Z)-Hex-3-en-1-ol   V: 49, 53   E: 51
(E)-Hex-2-en-1-ol   V: 53   E: 51
Benzyl alcohol   V: 49   E: 49, 51
2-Phenylethanol   V: 49   E: 51
Furfuryl alcohol   E: 51

Aldehydes
Acetaldehyde   V: 53   E: 49
Hexanal   E: 49, 53
Heptanal   E: 53
Octanal   E: 51
Nonanal   E: 51
3-Methylbutanal   E: 49
2-Methylbutanal   E: 49
(E)-Hex-2-enal   V: 53   E: 53
(E)-Oct-2-enal   E: 53
2-Methylbut-2-enal   V: 49   E: 53
2,6-Dimethylhept-2-enal   E: 49
(E, E)-Deca-2,4-dienal   E: 51
(E, Z)-Deca-2,4-dienal   E: 51
Benzaldehyde   V: 49   E: 49
p-Tolualdehyde   E: 49
Phenylacetaldehyde   E: 49
Furfural (2-furaldehyde)   E: 49, 51, 53
5-Methylfurfural   E: 49, 51

Ketones
Acetone   V: 49   E: 53
Butanone   V: 49   E: 49
Heptan-2-one   E: 49
Octan-2-one   E: 49
Nonan-2-one   E: 49
Nonan-5-one   E: 49
Decan-2-one   E: 49
Pentadecan-2-one   E: 49
2-Methylhept-2-en-6-one   V: 49   E: 49
6-Methylhept-5-en-2-one   E: 51
Megastigma-4,6,8-trien-3-one (four stereoisomers)   E: 51
Theaspirone (two stereoisomers)   E: 51
β-Ionone   E: 51, 53
3,4-Dehydro-β-ionone   E: 51
Furfuryl methyl ketone   E: 51
Methyl 5-methylfurfuryl ketone   E: 51

Free acids
Acetic   V: 53
Decanoic   E: 53
Methylpropanoic   E: 53

Lactones
γ-Butyrolactone   E: 51
γ-Pentalactone   E: 51
γ-Hexalactone   V: 49   E: 49, 51
γ-Decalactone   E: 53
γ-Dodecalactone   E: 53
β-Decalone   E: 49
2,7-Dimethyl-4-hydroxyocta-5(E),7-dienoic acid lactone
   (stereoisomers)   E: 53
   (or *trans*-Marmelo lactone)   E: 51
   (and *cis*-Marmelo lactone)   E: 51

Hydrocarbons
trans-α-Farnesene   E: 51, 53, 54
trans-β-Farnesene   E: 49, 51

TABLE I.V (*continued*)

Benzene   E: 49
Toluene   E: 49
A trimethylbenzene   E: 49
1,2,3-Trimethyl-5-(prop-2-enyl)benzene   E: 53

Terpenes
Limonene   E: 49
Car-3-ene   E: 49
Linalool   E: 49, 51
trans-Linalool oxide   E: 51
cis-Linalool oxide   E: 51
α–Terpineol   E: 49, 51
Terpinen-4-ol   E: 51
trans-Farnesol   E: 51
Citral   E: 49
Geranial   E: 51
Neral   E: 51
trans-Farnesal   E: 51

Miscellaneous
Theaspirane (two stereoisomers)   E: 51, 53, 54
Vitispirane   E: 51
Eugenol   E: 51, 53
Methyleugenol   E: 51
p-Vinylguiacol   E: 51
3,4-Didehydro-β-ionol   E: 51, 52, 54
Anisyl propanoate (*p*-methoxybenzyl propanoate)   E: 53
Dibutyl phthalate   E: 51
Hydroquinone monoacetate   E: 49
2-Acetylfuran   E: 49
Benzothiazole   E: 49
2,2,6,7-Tetramethylbicyclo(4.3.0)nona-4,7,9(1)-triene   E: 52, 54
(+)-2,2,6,7-Tetramethylbicyclo(4.3.0)nona-4,9(1)-dien-8-one   E: 52, 54
(−)-2,2,6,7-Tetramethylbicyclo(4.3.0)nona-4,9(1)-dien-8-ol   E: 52, 54
(−)-2,2,6,7-Tetramethylbicyclo(4.3.0)nona-4,9(1)-diene-7,8-diol   E: 52, 54

ethyl esters of saturated fatty acids with an odd or even number of carbon atoms up to C_{16}, and ethyl esters of mono-unsaturated fatty acids up to C_{14}. Double bonds are noted in aldehydes and ketones. Free fatty acids are rare and lactones are relatively numerous.

1.7 Analogies in the Composition of the Three Aromas

The apple, the odour of which is strong, largely produces and emits relatively low molecular weight constituents. Pear and quince odours are more discrete, and contain high proportions of higher molecular weight constituents. This observation is confirmed by simple comparison of aroma chromatograms of apple and quince (Schobinger et al., 1982).

Alcohols and low molecular weight esters are noted in the three fruit, with acetates predominating in the pear and ethyl esters in the quince. These constituents are also found in emissions from numerous ripe fruit, and could be viewed as ripening products. Most constituents in the three species are aliphatic chain compounds, or aliphatic chain derivatives (esters, alcohols, acids, aldehydes, ketones, lactones), saturated or unsaturated. The latter are infrequent in the apple, but form the main constituents of pear aroma. This constant presence of aliphatic chain derivatives, often with even-numbered carbon chains, indicates a common, lipid-related metabolic origin.

Low levels of benzene and terpene compounds have been identified. All three fruit contain trans-α-farnesene, which seems to be characteristic of 'Pome Fruits'.

2. ORGANOLEPTIC VALUE OF IDENTIFIED VOLATILES

The results mentioned above only provide indications of the chemical composition of aromas, but not all the identified volatiles are of equal organoleptic importance.

2.1 Characteristics of the Odour of Analysed Volatiles

2.1.1 Olfactory thresholds

Compounds present even in trace amounts can have an effect, as indicated by their olfactory detection thresholds (minimum concentration triggering the stimulus) or recognition thresholds (usually higher). Table I.VI summarises the values found for some of the above-mentioned compounds. Aldehydes are noted in fat oxidation products, and have been the subject of numerous studies. Widely varying values have been reported for hexanal and hex-2-enal, although variability appears less marked for the taste threshold (Pangborn, 1981).

2.1.2 Odour characterisation for some volatiles

Odour descriptions for volatiles have been established using dilute solutions

TABLE I.VI

Odour thresholds of volatiles identified from pome fruits

Esters		
Methyl hexanoate	0.087 ppm in water	Pyysalo *et al.*, 1977
Ethyl acetate	5 ppm v/v	Teranishi *et al.*, 1971
Ethyl butanoate	0.001 ppm v/v	Teranishi *et al.*, 1971
Ethyl pentanoate	0.005 ppm v/v	Flath *et al.*, 1967
Ethyl hexanoate	0.003 ppm in water	Pyysalo *et al.*, 1977
Ethyl 2-methylbutanoate	0.0001 ppm v/v	Teranishi *et al.*, 1971
Propyl propanoate	0.057 ppm v/v	Teranishi *et al.*, 1971
Butyl acetate	0.066 ppm v/v	Teranishi *et al.*, 1971
Butyl propanoate	0.025 ppm v/v	Flath *et al.*, 1967
Pentyl acetate	0.005 ppm v/v	Teranishi, 1967
Hexyl acetate	0.002 ppm v/v	Flath *et al.*, 1967
Alcohols		
Ethanol	100 ppm v/v	Teranishi *et al.*, 1971
Propan-1-ol	9 ppm v/v	Flath *et al.*, 1967
Butan-1-ol	0.5 ppm v/v	Flath *et al.*, 1967
	1.2×10^3 ppb v/v	Hall and Andersson, 1983
Hexan-1-ol	0.5 ppm v/v	Flath *et al.*, 1967
	150 ppb in water	Aoki and Koizumi, 1986
Heptan-1-ol	94 ppb in water	Aoki and Koizumi, 1986
Octan-1-ol	77.5 ppb in water	Aoki and Koizumi, 1986
Nonan-1-ol	34 ppb in water	Aoki and Koizumi, 1986
Aldehydes		
Hexanal	$4.5/10^9$ in water	Guadagni *et al.*, 1963
	0.005 ppm v/v	Teranishi *et al.*, 1971
	$120/10^9$ in vegetable oil	Buttery *et al.*, 1973
	0.02 ppm in water	Pyysalo *et al.*, 1977
	9.18 ppb in water	Ahmed *et al.*, 1978
	16 ppb v/v	Hall and Andersson, 1983
	20 ppb in water	Aoki and Koizumi, 1986
Hex-2-enal	$17/10^9$ in water	Guadagni *et al.*, 1963
	0.017 ppm v/v	Teranishi *et al.*, 1971
	$850/10^9$ in vegetable oil	Buttery *et al.*, 1973
	85.3 ppb in water	Ahmed *et al.*, 1978
	1.7×10^2 ppb v/v	Hall and Andersson, 1983
Ketones		
Butanone	3.7×10^4 ppb v/v	Hall and Andersson, 1983
Pentan-2-one	1.1×10^4 ppb v/v	Hall and Andersson, 1983
Heptan-2-one	5.0×10^3 ppb v/v	Hall and Andersson, 1983
Acids		
Butanoic	6.8 ppm in water	Pyysalo *et al.*, 1977
Hexanoic	3.7 ppm in water	Pyysalo *et al.*, 1977

of pure compounds. These have often been developed by comparison with fruits, but attempts have been made to draw up a classified vocabulary (Dürr, 1979). Table I.VII shows that all esters have essentially fruity odours, and the odour of ethyl 2-methylbutanoate is apple-like.

2.2 Identification of Compounds Responsible for the Characteristic Aroma of Apple

Panel evaluation of compounds responsible for certain aromas has been carried out by sniffing gas chromatography effluents. Guadagni *et al.* (1966) and Flath *et al.* (1967) have reported that ethyl 2-methylbutanoate, which has a very low olfactory threshold, has an intense apple odour giving an impression of ripeness. Also reported are hex-2-enal and hexanal, the latter is associated with a green apple odour. Hexyl 2-methylbutanoate also has a typical apple-like aroma (Jakob *et al.*, 1969), and increased levels of this volatile improve quality in Golden Delicious (Williams, 1983). Other esters, such as ethyl butanoate, which has a low olfactory threshold, also have a rather typical fruity odour (Nursten and Woolfe, 1972).

Four minor components of Kogyoku apple essence, (*Z, Z*)-octa-3,5-dien-1-ol,

TABLE I.VII

Odour characterization of some volatiles identified from pome fruits

Esters	
Methyl hexanoate	Fruity (2)
Ethyl acetate	Fruity (1), solvent-like (3)
Ethyl butanoate	Fruity (4), banana, pineapple (5)
Ethyl hexanoate	Fruity, fresh, sweet (2)
Propyl butanoate	Pineapple, apricot (5)
Butyl acetate	Pungent, pear (5)
Butyl propanoate	Fruity, pineapple (5)
Butyl butanoate	Pear, pineapple (5)
Butyl pentanoate	Apple, raspberry (5)
Butyl hexanoate	Pineapple (5)
Pentyl acetate	Banana oil (1), fruity pineapple (5)
Hexyl acetate	Fruity, floral (5)
Ethyl 2-methylbutanoate	Apple-like (3), green, fruity (5)
Alcohols	
Ethanol	Alcoholic (3)
2-Methylbutan-1-ol	Butter, straw-like (2) fusel oil-like (3)
Aldehydes	
Hexanal	Grass-like (4), fatty, green, grassy (5)
Hex-2-enal	Grass-like (3), fruity (4), green, fruity (5)

(1) Amerine *et al.*, 1965, p. 186; (2) Pyysalo *et al.*, 1977; (3) Dürr, 1979; (4) Dürr and Schobinger, 1981; (5) Chua *et al.*, 1987.

(*E,Z*)-octa-3,5-dien-1-ol and their two acetates, contribute sweet and fruity notes to apple flavour. These compounds have been synthesized, and their isomers have been reported to have green and oily notes (Iwamoto *et al.*, 1983).

4-Methoxyallylbenzene has been identified in the aromas of several cultivars, and is particularly prevalent in the aroma of Ellison's Orange, giving it a recognisable aniseed-like character (Williams *et al.*, 1977).

Comparison of the chromatograms of four cultivars with the differences in olfactory impressions recorded by an evaluation panel has shown that all components contribute to the aromatic subtleties of the apple, particularly with regard to varietal differences (Paillard, 1975b). Panasiuk *et al.* (1980) have compared aroma, determined by sensory evaluation, and the composition of volatiles in McIntosh apples. Ethyl esters (ethyl propanoate, ethyl butanoate, ethyl 2-methylbutanoate) were correlated with overripe and cheesy notes, and hexanal with aromatic and fruity notes. The intensity of the aroma, the impression of ripeness, and fruity and aromatic notes, were correlated with C_6 aldehydes, and overripeness and a cheesy note with esters.

Odour profiles defined on the basis of sensory evaluation by a panel of experts have been used to develop a synthetic apple juice odour (Dürr, 1979; Dürr and Röthlin, 1981). After dilution, the synthetic odour could not be distinguished from the odour of natural apple juice on the basis of a series of sensory tests. The formulation of this synthetic juice odour was: 5 g of water, 30 mg of butan-1-ol, 50 mg of 2-methylbutan-1-ol, 30 mg of hexan-1-ol, 5 mg of ethyl butanoate, 50 mg of pentyl acetate, 3 mg of hexanal, 100 mg of (*E*)-hex-2-enal, 0.1 mg of benzaldehyde, and 5 mg of ethyl 2-methylbutanoate, made up to 50 g with ethanol.

2.3 Composition and Hedonic Value of Apple Aroma

Chua *et al.* (1987) and Rao *et al.* (1987) have described lists of odour-active volatiles comprising C_6 aldehydes, hexanal and hex-2-enal, and the esters ethyl butanoate, butyl acetate, ethyl 2-methylbutanoate, 3-methylbutyl acetate and hexyl acetate.

Dürr and Schobinger (1981) have evaluated the contribution of 30 volatiles to the sensory quality of commercial apple essences. They found that unsaturated C_6 aldehydes ((*E*)-hex-2-enal and (*Z*)-hex-3-enal), 2-methylpropan-1-ol and 2-methylpropyl acetate contribute strongly to aroma intensity. The authors correlated hedonic values with the relative amounts of three chemical groups (alcohols, esters and carbonyl compounds). Bad essences were characterised by high levels of alcohols and low levels of carbonyl compounds. Good essences contained a low proportion of alcohols and a high proportion of carbonyl compounds, and, in some cases, of esters. The volatiles were classified into three groups depending on their contribution to sensory quality: important constituents ((*E*)-hex-2-enal, (*Z*)-hex-3-enal, (*E*)-hex-2-en-1-ol, (*Z*)-hex-3-en-1-ol, ethyl butanoate, ethyl 2-methylbutanoate), desirable constituents (hexanal, benzaldehyde, propyl butanoate, pentyl acetate, pentan-2-one, 2-

methylpropyl acetate), and undesirable constituents (ethanol, 2-methylpropan-1-ol, 2-methylbutan-1-ol, 3-methylbutan-1-ol, 2-phenylethanol) (Dürr and Schobinger, 1981).

Adulteration of apple juice liquors with (E)-hex-2-enal was sometimes apparent, through comparison of levels of (E)-hex-2-enal, hexan-1-ol and butan-1-ol. This control method can also be applied to foodstuffs to which apple aromas have been added (Von Reinhard and Eichholz-Fecher, 1984).

Petró-Turza et al. (1986) have used gas chromatography and sensory evaluation to correlate sensory quality and chemical composition for 10 industrial natural apple aroma condensates, in terms of determination of the recognition threshold value and description of odour quality. The authors correlated recognition threshold with hexanal content, and with the overall concentration of measured esters and carbonyl compounds. The curves of sensory intensity as a function of aroma condensate concentration indicated that the part of the graph corresponding to a pleasant apple odour was broadened by increases in levels of butan-1-ol, butyl acetate, 3-methylbutyl acetate, hexyl acetate, hexanal, and hex-2-enal, but was narrowed by increases in levels of ethanol, hexan-1-ol and ethyl acetate. Williams (1983) has reported that odour quality for Golden Delicious apples improved at higher levels of hexyl 2-methylbutanoate.

It is ironic to note that esters (hexyl acetate, butyl 2-methylbutanoate, propyl hexanoate, hexyl propanoate, butyl hexanoate and hexyl butanoate) are attractive to insects such as the apple maggot (Diptera tephririadae) (Fein et al., 1982; Reissig et al., 1982).

2.4 Volatile Compounds Responsible for Pear and Quince Aromas

Early studies reported identification by gas chromatography of the volatiles responsible for Bartlett pear aroma in essences collected without loss of aroma characteristics (Jennings and Sevenants, 1964; Heinz and Jennings, 1966). Hexyl acetate and methyl (E, Z)-deca-2,4-dienoate were identified by infrared spectroscopy. The former was described as a contributory flavour compound, and the latter as a character impact compound. Ethyl (E, Z)-deca-2,4-dienoate was identified as a constituent typical of the flavour of the Bartlett pear. Synthesis showed that the esters methyl, ethyl, propyl, butyl, pentyl and hexyl (E, Z)-deca-2,4-dienoate all possessed a powerful pear aroma.

Subsequent studies of several cultivars employing gas chromatography, HPLC and sensory evaluation have shown that high concentrations of decadienoates are characteristic of the Bartlett pear and of cultivars with Bartlett-like flavours (Quamme, 1984). Essences from cultivars such as Magness and Kieffer are distinguished by a predominance of constituents with low boiling points, and the lack of significant amounts of high boiling point compounds. The three varieties Bartlett, Magness and Kieffer have been classified respectively as highly, intermediate and poorly flavoured (Russel et al., 1981). Brief, low-temperature sterilisation of Beurré Williams and Passe-Crassane pears causes loss of volatiles with low boiling points, but no loss of aroma quality. The methyl and ethyl esters of deca-2,4-dienoic acid have high boiling points, and are retained in stored pears (Strandjev, 1982).

As in the apple and pear, a mixture of volatile esters forms the basis of the fruity impression in the quince. Essential oils obtained by condensation of headspace vapours have a pleasant and natural quince flavour, whereas distillates have the flavour of cooked quince. An important contribution to the typical quince flavour is made by ethyl 2-methylbut-2-enoate (or ethyl tiglate) (Schreyen *et al.*, 1979). The principal aroma of the fruit is highly dependent on the sweet and powerful aromatic contribution of the two isomers (2*R*, 4*S*)-(+)- and (2*R*, 4*R*)-(−)-2,7-dimethyl-4-hydroxyocta-5(*E*),7-dienoic acid lactones, named respectively (+)-*trans*- and (−)-*cis*-marmelo lactones. Theaspiranes have been identified. Tsuneya *et al.* (1983) report that the characteristic odour of quince is dependent on contributions from *trans*- and *cis*- marmelo oxides and a series of derivatives of β-ionone bearing various functional groups, notably 3,4-didehydro-β-ionol, which has recently been identified by Ishihara *et al.* (1986). The presence of appreciable quantities of these volatiles depends on the method of sample preparation (pH, enzyme inhibition) and the distillation techniques (high vacuum distillation (HVD) or simultaneous distillation extraction (SDE)) (Winterhalter *et al.*, 1987). HVD reveals theaspirane diastereoisomers, and SDE a series of volatiles including 3,4-didehydro-β-ionol, although the latter is probably a thermally formed volatile. Traces of acetic acid in headspace vapours from quince peel give it a slightly pungent odour (Umano *et al.*, 1986).

2.5   Flavour and Sensory Quality of Fruit

Aroma is one criterion of fruit quality. Alston (1981) has distinguished eating quality and appearance in assessment of apple quality, as well as storability, suitability for processing, resistance to parasites and disease, and crop yield. Odorous compounds associated with sugars and acids form the main characteristics of the flavour. The overall sensory quality of the fruit also depends on texture. Appearance, colour and smoothness, without surface defects, also play rôles in organoleptic quality.

2.5.1   Quality tests
Quality tests involve a certain number of measurements and sensory evaluations. Redalen (1984, 1986), for example, has compared different apple cultivars, and considered flavour, size, greenness, redness, firmness, soluble solids, titratable acids, and the soluble solid/titratable acid ratio. Some tests involve measurements (weight, acidity, firmness), others relate to notes attributed by a panel. Redalen (1986) has also defined the optimal period for eating, and the starch content. Gormley (1975) investigated the processing suitability of Golden Delicious and Bramley's Seedling apples by determining soluble solids, acidity, firmness (Shear Press) and flesh colour (Hunter 'b' value), and by comparison of the quality of apple products by taste panels. The acidity of

Bramley's Seedlings is largely responsible for the excellent flavour of apple tarts and the quality of the cooked fruit. Subsequent comparative sensory evaluation of Bramley's Seedlings with Spartan and East Malling A 3022 apples has demonstrated the superior acceptability of these two new cultivars (Williams *et al.*, 1982).

2.5.2 Contribution of non-volatile constituents: sugars, acids and phenolic compounds

Taste quality is directly linked to certain constituents present in variable amounts. Hulme and Rhodes (1971a) consider that taste in pome fruits is principally based on acid – sugar balance. However, phenolic compounds contribute to flavour, since the astringency of several acts on gustation. Odour and taste also interact, and have been examined by von Sydow *et al.* (1974) for fruit juices, some observations also being applicable to pome fruits. These authors distinguish between the odour perceived by the nose, the taste recorded by the mouth, and the odour perceived by the mouth, and note that the addition of sucrose decreases bitter tastes (acid, tart, astringent). Pleasant aromatic notes (fragrant, sweet odour) develop in the presence of sucrose, whereas unpleasant notes decrease (vinegar, green).

Acceptable internal quality for calibrated Golden Delicious apples evaluated by sensory analysis and chemical measurements corresponds to an L-malate content equal to, or greater than, 0.55%, and a sucrose concentration equal to, or greater than, 2% of the fresh weight (Gorin and Klop, 1982). Sensory evaluation has shown, however, that L-malate is not correlated with an impression of sourness, but with juiciness and texture. Sucrose is correlated with juiciness and texture, but also with sugariness and flavour. To sum up, all sensory characteristics are represented by the concentrations of L-malate and sucrose (Gorin and Klop, 1982). Slightly different observations have been made with other cultivars of apple (Mast and Feldheim, 1983a). Quality assessment involves analysis of a sensory profile established with organoleptic tests (external appearance, feel, external and internal odour, internal appearance, taste with and without peel, structure of flesh). Fair correlation has been noted between taste and malate and sucrose contents, and good correlation between sensory evaluation of the flesh and firmness of $5-5.3$ kg/cm^2 (Mast and Feldheim, 1983b).

Table I.VIII shows comparative sugar and acid contents for apple, pear and quince. Fructose content is always markedly higher than sucrose content, and its sweetening power is higher (if sucrose is taken as unity, respective values for fructose and glucose are $1.2-1.5$ and $0.5-0.7$) (Solms, 1971). The acid contents of apple and quince are characterised by high proportions of malic acid (hydroxybutanedioic acid).

Phenolic compounds contribute to flesh taste, and measurements on Passe-Crassane pears and Calville blanc apples have shown that the concentrations of these compounds are generally higher in the peel than in the parenchyma (Macheix and Delaporte, 1973; Billot, 1983). Chlorogenic acid content of the

Calville blanc apple, per 100 g fresh weight, is 19.1 mg for the peel and 15.1 mg for the flesh, and the respective values for catechins are 45.8 and 22.6 mg. A taste perception threshold of 46.1 mg/l (i.e. 4.61 mg/100 ml) has been estimated by a panel for (+)-catechin in aqueous solution (Delcour *et al.*, 1984).

3. FORMATION OF AROMAS BY FRUITS

Fruit aromas develop during ripening, and gas chromatography has defined changes in levels of the main constituents. Total volatile emission, measured for example in the Calville blanc apple cultivar ripening at 15°C, reaches a maximum which coincides with loss of greenness and the climacteric (Paillard, 1967c). Numerous constituents contribute to this maximum, notably alcohols (butan-1-ol, hexan-1-ol) and especially esters (propyl, butyl and hexyl acetate, butyl butanoate). Changes in total volatiles (Fidler and North, 1969) and in esters (Brown *et al.*, 1968) follow similar patterns in different cultivars. Increases in ethanol levels at the end of ripening are often accompanied by rises in levels of low molecular weight ethyl esters. Aroma development depends on the date of harvest (Dirinck *et al.*, 1984); the sum of the headspace esters of Golden Delicious apples increases very rapidly for late harvested fruits, and the maximum reached is considerable by comparison with that for early harvested apples. Romani and Ku (1966) have shown in the Bartlett pear that development of total volatiles follows the emission peak for ethylene and the maximum of CO_2 emission. A maximum is also noted for emission of esters, such as butyl

TABLE I.VIII

Sugar and acid contents of apple, pear and quince

Sugars	% of fresh weight (1)			% of fresh weight (2)	% contributions (2)			
	Total sugars	Min value	Max value		Fructose	Glucose	Sucrose	Sorbitol
Apple dessert	11.57	6.01	16.60	9.28	49.8	28.3	21.9	tr
Apple culinary	9.64	5.34	13.05					
Pear	9.95	6.51	13.15	11.16	44.6	22.1	4.8	28.5
Quince	8.10	6.50	9.96	7.18	61.6	22.4	16	

Acids (2)	Total acidity % of fresh weight	% contributions		
		Malic	Citric	Quinic
Apple				
Antonowka	1.03	98.3	1.0	0.4
Bancroft	0.80	94.2	1.8	2.0
Starking	0.36	85.6	3.3	8.7
Quince	0.87	81.2	0.5	18.3

(1) Whiting, 1970; (2) Lesińska, 1987.

and hexyl acetate, whereas ethyl acetate levels increase towards the end of ripening.

3.1 Effects of Storage Conditions and Treatments on Development of Aromas and Volatiles

After harvesting, ripening of fruit is characterised by changes in colour, texture and aroma development, in relation to physiological processes (biosynthesis and ethylene emission, respiration climacteric) and metabolic processes. Refrigeration and controlled atmosphere storage are used to slow ripening and ageing, and to lengthen shelf life.

3.1.1 Effects of temperature and refrigeration of fruits

Volatile emission from the Calville blanc apple stored at 4°C has been compared with that from fruit ripening at 15°C. Changes in the fruit are greatly retarded, but total emission is still marked and reaches a peak corresponding to loss of greenness (Paillard, 1967c). Chromatograms obtained for apples stored at 4°C and ripened at 15°C are very similar. At the end of storage at 4°C, there is a relative decrease in high molecular weight compounds (butyl and hexyl acetate, for example), and an increase in low molecular weight compounds (ethyl acetate). After 32 weeks at 4°C, from October to May, volatile emission from yellow fruits of healthy appearance was weak, and there was pronounced loss of flavour. During long-term storage, the apples probably exhaust their substrates, and their capacity to synthesise aromas decreases.

Ageing phenomena which appear after prolonged storage are the consequence of normal physiological processes. But during cold storage in air, abnormal non-parasitic manifestations are due to physiological disorders. Some cold-related diseases appear to involve metabolism of volatiles (ester emission) and some particular volatiles (α-farnesene, geraniol).

3.1.2 Controlled atmosphere storage

Metabolism is slowed and changes blocked in fruit stored under controlled atmosphere (CA) conditions (CO_2 and O_2 concentrations increased and decreased respectively, humidity control, and usually refrigeration).

Adsorption on activated charcoal was used in the first analyses of volatiles emitted from McIntosh apples in CA storage (Angelini and Pflug, 1967). Alcohols, esters and aldehydes were identified, as well as numerous hydrocarbons produced by microorganisms or derived from contamination by material on the walls of the storage chamber.

All studies indicate that CA storage should be of limited duration if fruit is to retain its aromatic qualities during subsequent ripening. Hatfield and Patterson (1974) have shown that apples stored in controlled atmospheres (2% O_2 or 2% O_2 + 5% CO_2) often produce very few volatiles, and are incapable of synthesising normal amounts of esters during subsequent ripening. Willaert et al. (1983) stored Golden Delicious apples for various times (at 0.5°C, 1 – 2% CO_2,

93% relative humidity) and then allowed them to ripen at 20°C and 70% relative humidity. A significant decrease in apple flavour and deficient ester production were noted after long-term CA atmosphere storage. Short-term CA storage must be used in order to retain fruit quality (Dirinck *et al.*, 1984). Lidster *et al.* (1983a, 1983b) have shown that short-term storage of McIntosh apples in an atmosphere of 1.5% CO_2 and 1% O_2 at 2.8°C did not inhibit regeneration of ethyl butanoate and hexanal after subsequent return of the fruit to air. Long-term storage (320 days) under these conditions, however, resulted in complete loss of the main headspace volatiles, and blocked their formation after return of the fruit to air. However, Lidster *et al.* (1984) have shown that decreased loss of titrable acids allows production of high-quality apple juice over a longer processing period.

Short-term anaerobiosis (400 h), low-oxygen storage (0.2 – 0.4 – 1.0%) causes decreases in production of volatiles in four cultivars of apple (Jonathan, Golden Delicious, Idared, Starkrimson) (Golias, 1984). The ethanol that accumulates during anaerobiosis is subsequently oxidised, and ethyl acetate is formed (Golias, 1983). Nichols and Patterson (1987) have demonstrated that ethanol accumulation in Golden Delicious apples stored at 0.0% O_2 for 14 weeks is 10 times that noted for apples stored in low-oxygen atmospheres (0.5 – 1.5%). Half of the ethanol is lost after return of the apples to air for 7 days.

3.1.3 Pre-harvest AVG treatment
Pre-harvest treatment of Bartlett pears with the ethylene synthesis inhibitor, aminoethoxyvinylglycine (AVG), combined with storage treatment, reduces subsequent production of the volatiles methyl, ethyl and hexyl acetate (Romani *et al.*, 1983).

3.1.4 Effects of treatments
Analyses of volatiles have generally involved small numbers of alcohols and esters, but these compounds nonetheless provide useful information. Alcohols, such as butan-1-ol and hexan-1-ol, and the esters butyl and hexyl acetate and ethyl butanoate, indicate biosynthetic capacity. Aroma formation, and esterification reactions in particular, seem associated with ripening and respiration-linked processes. Increased levels of ethanol and acetaldehyde denote fermentation-directed metabolism. The esterification of ethanol to ethyl acetate is oxygen-dependent.

3.2 Biosynthesis of Primary Aromas

3.2.1 Biosynthesis of alcohols and esters from fatty acids in living apple tissue
Given the richness of volatile emission from apples in compounds with even-numbered carbon chains, an early study examined apple segments supplied with radioactive acetate as precursor (Paillard, 1969). Chromatographic analysis in-

dicated that virtually all compounds were labelled, including acetic esters, alcohols and numerous other constituents. The biosynthetic steps were therefore studied in discs of living apple tissue using other substrates (Paillard, 1979a). Straight chain alcohols are formed from aliphatic fatty acids with the same number of carbon atoms, or from higher fatty acids after β-oxidation. Even-carbon-numbered fatty acids give rise to butan-1-ol and hexan-1-ol, and odd-carbon-numbered fatty acids to propan-1-ol and pentan-1-ol. The low levels of odd-carbon-numbered alcohols is explained by the very small amounts of odd-carbon-numbered fatty acids in apple tissue. During ripening of apples, lipid metabolism releases fatty acids that may serve as substrates. Marked aldehyde synthesis was not seen during these studies; the aldehyde intermediate between the acid and alcohol is no doubt formed, but is not released.

The various enzymic systems called into play lack specificity, and the presence of substrates conditions ester biosynthesis. Varietal differences in composition often relate to proportions of esters, and depend on the reaction rates in the β-oxidation of fatty acids. The rate of transformation of butanoate into acetate, for example, is much higher in Golden Delicious apples, which emit high levels of acetic esters, than in red cultivars rich in butanoate esters (Starking, Richared). Biosynthesis from C_8 fatty acids has also been investigated in a cork-coated cultivar, the Canada gris, for which volatile emission is weak and ester-poor. Fatty acids are transformed into alcohols and ketones with one fewer carbon atom (pentan-2-one and heptan-2-one from C_6 and C_8). Ester synthesis is extremely limited in the parenchyma and nil in the skin (Paillard, 1979b).

3.2.2 Bioconversion of volatiles in cultured apple cells

Low-level transformation of fatty acids predominantly into ethanol occurs in Golden Delicious apple cells cultured *in vitro* in an open system and under conditions of slowed division. Aldehyde transformation is relatively pronounced; butan-1-ol, ethyl butanoate and 2-methylpropan-1-ol, for example, are produced from butanal (Ambid and Fallot, 1980). Bioconversion of sodium butanoate, hexanoate and octanoate fatty acids is possible in a closed system with a CO_2-rich atmosphere (Ambid and Fallot, 1981).

3.2.3 Bioconversion of volatiles in whole fruit

Treatment of whole, intact Golden Delicious apples with vapours of aldehydes ($C_3 - C_6$) or carboxylic acids ($C_2 - C_6$) induces increases in headspace alcohols and esters. But these effects are short-term, and do not result in important changes in organoleptic quality (De Pooter *et al.*, 1983). The concentrations of all ethyl esters increase during storage of Red Delicious apples in an atmosphere containing ethanol vapour (Berger and Drawert, 1984). Ester synthesis occurs in Cox's Orange Pippin apples kept at 3°C in an oxygen-poor atmosphere (2% O_2, 98% N_2) following treatment with alcohol ($C_2 - C_8$) vapour, or with vapours of methyl esters of fatty acids ($C_4 - C_8$). The alcohols are converted into the corresponding acetic esters (pentan-1-ol to pentyl acetate), and the methyl esters into shorter-chain fatty acid esters (two to four fewer carbon atoms). For instance, the supply of methyl heptanoate or methyl octanoate produces respective increases in pentyl acetate and butyl acetate, which indicates β-oxidation of the fatty acids (Bartley et al., 1985).

3.2.4 Other synthetic pathways

Synthesis of mono-, di- and tri-unsaturated esters identified in the aromas of quince and Bartlett pear can be accounted for by β-oxidation of polyunsaturated linoleic and linolenic fatty acids (Jennings and Tressl, 1974). Synthesis of branched chain alcohols and esters can be described by the scheme involving amino acids, which has been demonstrated in other fruit such as the banana. Methylpropanoic acid, and 2-methylpropan-1-ol and its esters, are synthesised from valine, 3-methylbutanoic acid, and 3-methylbutan-1-ol and its esters, from L-leucine, and 2-methylbutanoic acid and its esters (especially important in apple aroma), from L-isoleucine (Myers et al., 1970; Tressl et al., 1970; Drawert, 1975; Drawert and Berger, 1981).

The apple synthesises a great variety of compounds, including aromatic compounds, terpenoids, and compounds containing the allyl group (Williams et al., 1977). The biosynthetic pathways of these compounds must be investigated by studies of other plant material.

3.2.5 Importance of the outer part of the fruit and the skin

There is a difference in production of aromas, principally esters, between apples with waxy cuticles and those with cork skins. Production of volatiles, notably esters, is much greater in the skin than in the flesh or the whole fruit, and the use of Delicious apple peel for the manufacture of essence increases the yield without loss of organoleptic quality (Guadagni et al., 1971a,b). Lipid metabolism is pronounced in the epidermis and subepidermal layers of waxy skinned fruits, and may generate fatty acid substrates.

Volatiles seem to be more or less integrated in all the lipid constituents, notably waxes, that form the cuticle. The transition to the vapour phase of these compounds is certainly due to their volatility, but may also be due to physiological mechanisms. The presence of pores and transcuticular canals could allow diffusion of volatiles into the external atmosphere (Miller, 1982). The same is true for the lenticels and stomata which are more numerous in the quince than in the apple (Blanke, 1986), but are they open in the ripe fruit?

3.3 Formation of Secondary Aromas

Grinding destroys cellular structure and promotes a series of enzymic reactions. Esters, principally acetates, are hydrolysed, and C_6 aldehydes are produced (Schreier et al., 1978). Hexanal and (E)-hex-2-enal are formed by the action of lipoxygenase systems on polyunsaturated fatty acids, and are accompanied by lesser amounts of hexan-1-ol, (E)-hex-2-en-1-ol, (Z)-hex-3-enal and (Z)-hex-3-en-1-ol. The degree of synthesis depends on the cultivar (Drawert,

1978). Production of C_6 aldehydes by ground tissue from Golden Delicious apples is particularly abundant in the peel, and less so in the central parenchyma, although levels increase in the core. The distribution of linoleic and linolenic acids is identical (Paillard and Rouri, 1984). Feys *et al.* (1980) have also noted increased lipoxygenase activity in the skin and core of Schone von Boskoop apples. A lipoxygenase (EC.1.13.11.12) has been extracted from Golden Delicious apples and partially purified; it is partially associated with the membranes of the microsomal fraction (Kim and Grosch, 1979). During prolonged cold storage (very slow ripening), there is a relation between decreased capacity to produce hex-2-enal after grinding and loss of greenness of the fruit measured by the colour or by the disappearance of chlorophylls. Destruction of chloroplasts and their lamellae releases galactolipids rich in linolenic acid (Paillard, 1985).

3.4 Effects of Volatiles on Fruit Physiology and Quality

3.4.1 Possibility of induced ripening

Endogenous or exogenous ethylene (ripening hormone) can be considered separately, since it is not strictly one of the aroma-forming volatiles. But vapours of acetic, propanoic and butanoic acids could have an analogous effect. The treatment of whole Golden Delicious apples in the unripe pre-climacteric state with these acid vapours triggers ripening: respiration climacteric, yellowing, and aroma formation, with, in particular, production of esters (De Pooter *et al.,* 1982). Experiments with [2-^{14}C]-propanoic acid have shown that a small proportion of the added acid was transformed into [^{14}C]-ethylene, and that the observed triggering of ripening probably resulted from increased levels of endogenous ethylene (De Pooter *et al.,* 1984).

3.4.2 Implication of certain volatiles in physiological diseases

(a) Rôle of ester detoxification.   Soft scald can be provoked in Jonathan apples by injection of hexan-1-ol, and to a lesser extent by injection of hexanal, hexyl acetate and hexyl esters, but hexan-1-ol is the main promoter (Wills and Scott, 1970; Wills, 1972). High levels of acetic acid (or injection of this acid) are associated with low-temperature breakdown (Wills *et al.,* 1970). This disease and accumulation of acetic acid are only seen during cold storage; at higher temperatures, acetic acid is eliminated by transformation into acetate and release of esters (Wills and McGlasson, 1969, 1971). Accumulation of small molecules is favoured by low temperatures: after 6 months of storage at 0°C, acetaldehyde was detected in mitochondria of Starking Delicious apples (Kimura *et al.,* 1985). The outer tissues of the fruit play an active rôle in the transport of an ester from the biosynthetic site to the external atmosphere (Paillard, 1981). The esterification reaction seems to be a mechanism for elimination of acetic acid and C_2 molecules, but also of short-chain fatty acids and some alcohols in the form of esters, compounds which are more volatile and emitted in abundance.

(b) Geraniol and low-temperature breakdown. Geraniol and several compounds associated with isoprenoid metabolism can induce disease. Injection of geraniol into the core of Jonathan apples increases breakdown in the flesh and decreases levels of all other terpene volatiles. However, geraniol is quickly converted into non-volatile geranyl β-D-glucoside (Wills and Scriven, 1979).

(c) Superficial scald and α-farnesene. The emission of α-farnesene has been identified in three fruit, and has been studied in particular in the apple, because of its involvement in superficial scald (Huelin and Coggiola, 1968, 1970a). This unsaturated hydrocarbon (molecular weight 204, boiling point 125°C) is insoluble in water and is volatile; its evaporation is increased by ventilation (Huelin and Coggiola, 1970b; Anet, 1972b). α-Farnesene is subject to autoxidation (Meigh and Filmer, 1969; Anet, 1972a), and the oxidation products may be implicated in superficial scald (Huelin and Coggiola, 1970a; Anet, 1974; Anet and Coggiola, 1974).

3.5 Ripening Biochemical Phenomena Implicated in Production of Aroma and Volatiles

3.5.1 Presence of lipid substrates

Biosynthesis of aromas, alcohols and esters is directly associated with metabolism of fatty acids and lipids. The apple has carbohydrate reserves but is lipid-poor. However, Galliard (1968) has measured 880 and 947 mg of lipids respectively in 1000 g of pulp in pre- and post-climacteric Cox's Orange Pippin apples. Phospholipids constitute the largest category, and the respective amounts are 405 and 396 mg/1000 g fresh weight in pre- and post-climacteric apples (principally phosphatidylcholine and phosphatidylethanolamine). Ninety percent of phospholipids are found in the mitochondria of the parenchyma (Mazliak, 1970).

Fatty acids are generally esterified, and are found in very small amounts in the free state (Meigh and Hulme, 1965). The results of several authors for various cultivars are in agreement: the most abundant saturated and unsaturated fatty acids are, respectively, hexadecanoic acid and linoleic acid ($C_{18:2}$), which represent respective percentages of 31 and 53% of total acids (Mazliak, 1969). Short-chain saturated fatty acids do exist in small amounts, including even smaller quantities of odd-carbon-numbered fatty acids (C_9, C_{11}, C_{13}) in the peel (Meigh and Hulme, 1965; Mazliak, 1969). Fatty acid composition in mitochondria differs slightly, with $C_{14:0}$ as the most abundant saturated acid, $C_{18:2}$ as the most abundant unsaturated acid, and $C_{18:3}$ in notable quantities (Kimura *et al.*, 1982).

The parenchyma incorporates unsaturated fatty acids (notably linoleic acid) into triglyceride molecules during ripening (Mazliak, 1967a). Changes in microsomal membrane viscosity are related to variations in composition of lipids, sterols, phospholipids and fatty acids of phospholipids (Lurie and Ben Arie, 1983b). Fatty acids and acetate fragments derived from phospholipid catabolism enter the metabolic pool of available fatty acids or acetate, and are then used in various syntheses (Mazliak, 1967b). The peel of Jonagold apples acquires an oily appearance during ripening changes, which are characterised by an increase in fatty acid levels and a relative rise in low molecular weight acids (Noro et al., 1985). The fatty acid composition of the different lipid fractions of Ralls Janet and Starking Delicious apples has been monitored during complementary ripening after different periods of storage at 0°C (Kimura et al., 1979). During storage at 0°C there is an overall decrease in levels of the $C_{18\,:\,3}$ fatty acid, corresponding to the previous observation of decreased capacity for production of hex-2-enal.

3.5.2 Increases in enzymic activity during the respiration climacteric

Ethylene triggers the respiration climacteric in the apple by stimulating synthesis of RNA and proteins (Hulme et al., 1971b,c). The onset of ripening of Bartlett pears is accompanied by increased synthesis of enzymes required for normal ripening (Frenkel et al., 1968). Increases in large polyribosome levels in Passe-Crassane pears induced by ethylene treatment accompany the onset of ripening (Drouet and Hartmann, 1979).

The activities of hydrolytic enzymes, lipases and acid phosphatase increase with the respiration climacteric, and then decrease as respiration drops. This increased lipase activity in the apple could explain increased levels of free fatty acids (Rhodes and Wooltorton, 1967). Chlorophylls are degraded as a result of the increased activity of chlorophyllase which accompanies the disruption of chloroplast lamellae, and consequent possible release of lipid substrates. A sixfold increase in transaminase activity is noted in mitochondria prepared from apple skin at different stages of development of the climacteric. This reaction is involved in biosynthesis of acids converted into branched-chain alcohols and esters (Hulme et al., 1967). Plasmalemma ATPase activity in apples doubles between the pre- and post-climacteric phases; this rise is to be anticipated as a reflection of the increased energy requirements of biosynthesis during ripening (Lurie and Ben-Arie, 1983a).

CONCLUSIONS

Apple, pear and quince aromas are formed by mixtures of a great number of volatile constituents. Approximately 212 for the apple, 79 for the pear, and 156 for the quince have been identified in the volatile emission of the fruit, or in distillates and extracts. However, these figures reflect the fact that far more studies have been devoted to the apple than to the pear and quince.

A few constituents have been identified as more or less characteristic of the aroma of each fruit. Typical apple aroma, in particular, seems to be formed by ethyl 2-methylbutanoate, hexyl 2-methylbutanoate and hex-2-enal. Bartlett pear flavour derives from esters of (E, Z)-deca-2,4-dienoic acid. Ethyl 2-methylbut-2-enoate and the (+)-trans and (−)-cis marmelo lactones are especially implicated in quince aroma. Some constituents provide particular notes, such as the 4-methoxyallylbenzene in the aniseed-like character of some apple cultivars. Aroma quality is influenced favourably or otherwise by the different contents of some constituents (carbonyl compounds or alcohols). A range of esters gives an impression of ripeness in the three fruit, and the lower molecular weight compounds an impression of natural freshness.

The three fruit have analogous metabolic features. Numerous volatiles seem to have the same origin and lipids and fatty acids as precursors, with the formation of shorter aliphatic chains due to β-oxidation, reduction of fatty acids to aldehydes and alcohols, esterification, α-oxidation of fatty acids and formation of ketones, oxidation of aliphatic chains and formation of double bonds, formation of lactones, etc. Amino acids are no doubt the precursors of some branched-chain volatiles, esters of 2-methylbutanoic acid, 2-methylpropan-1-ol, 3-methylbutan-1-ol and their esters. These various reactions result in the biosynthesis of volatile constituents and aromas, and are associated with the physiological and metabolic processes accompanying ripening of the fruit.

REFERENCES

Ahmed, E.M., Dennison, R.A., Dougherty, R.H. and Shaw, P.E., 1978. Flavor and odor thresholds in water of selected orange juice components. *J. Agric. Food Chem.,* 26: 187 – 191.

Alston, F.H., 1981. Breeding high quality high yielding apples. In: P.W. Goodenough and R.K. Atkin (Editors), Quality in Stored and Processed Vegetables and Fruit. Academic Press, London, New York, pp. 93 – 102.

Ambid, C. and Fallot, J., 1980. Bioconversion d'acides gras et d'aldéhydes par des cellules de pommes cultivées in vitro. *Bull. Soc. Chim. Fr.,* 1 – 2: 104 – 107.

Ambid, C. and Fallot, J., 1981. Rôle of the gaseous environment on volatile compound production by fruit cell suspensions cultured in vitro. In: P. Schreier (Editor), Flavour '81. Walter de Gruyter, Berlin, New York, pp. 529 – 538.

Amerine, M.A., Pangborn, R.M. and Roessler, E.B., 1965. Principles of Sensory Evaluation of Food. Academic Press, London, New York, 602 pp.

Anet, E.F.I.J., 1972a. Superficial scald, a functional disorder of stored apples. VIII. Volatile products from the autoxidation of α-farnesene. *J. Sci. Food Agric.,* 23: 605 – 608.

Anet, E.F.I.J., 1972b. Superficial scald, a functional disorder of stored apples. IX. Effect of maturity and ventilation. *J. Sci. Food Agric.,* 23: 763 – 769.

Anet, E.F.I.J., 1974. Superficial scald, a functional disorder of stored apples. XI. Apple autoxidants. *J. Sci. Food Agric.,* 25: 299 – 304.

Anet, E.F.I.J. and Coggiola, I.M., 1974. Superficial scald, a functional disorder of stored apples. X. Control of α-farnesene autoxidation. *J. Sci. Food Agric.,* 25: 293 – 298.

Angelini, P. and Pflug, I.J., 1967. Volatiles in controlled-atmosphere apple storage: Evaluation by gas chromatography and mass spectrometry. *Food Technol.,* 21: 99 – 102.

Aoki, M. and Koizumi, N., 1986. Organoleptic properties of the volatile components of buckwheat flour and their changes during storage after milling. *Nippon Shokuhin Kogyo Gakkaishi,* 33: 769 – 772.

Bartley, I.M., Stoker, P.G., Martin, A.D.E., Hatfield, S.G.S. and Knee, M., 1985. Synthesis of aroma compounds by apples supplied with alcohols and methyl esters of fatty acids. *J. Sci. Food Agric.,* 36: 567 – 574.

Berger, R.G. and Drawert, F., 1984. Changes in the composition of volatiles by post-harvest application of alcohols to Red Delicious apples. *J. Sci. Food Agric.,* 35: 1318 – 1325.

Berger, R.G., Drawert, F. and Schraufstetter, B., 1984. Uber das natürliche Vorkommen der Ethylester von Octen-, Decen-und Decadiensaüre in Apfeln der Sorte Red Delicious. *Z. Lebensm.-Unters.-Forsch.*, 178: 104–105.

Billot, J., 1983. Evolution des composés phénoliques au cours de la maturation de la poire Passe-Crassane. *Physiol. Vég.*, 21: 527–535.

Blanke, M.M., 1986. Comparative SEM study of stomata on developing quince, apple, grape and tomato fruit. *Angew. Bot.*, 60: 209–214.

Brown, D.S., Buchanan, J.R. and Hicks, J.R., 1968. Volatiles from apples as related to variety, season, maturity, and storage. *Hilgardia*, 39: 37–67.

Brulé, G., 1973. Etude des produits volatils du jus et du concentré de pomme chauffé. *Ann. Technol. Agric.*, 22: 45–48.

Buttery, R.G. and Teranishi, R., 1961. Gas- liquid chromatography of aroma of vegetables and fruit. *Anal. Chem.*, 33: 1439–1441.

Buttery, R.G., Ling, L.C. and Guadagni, D.G., 1969. Volatilities of aldehydes, ketones and esters in dilute water solution. *J. Agric. Food Chem.*, 17: 385–389.

Buttery, R.G., Bomben, J.L., Guadagni, D.G. and Ling, L.C., 1971. Some considerations of the volatilities of organic flavor compounds in foods. *J. Agric. Food Chem.*, 19: 1045–1048.

Buttery, R.G., Guadagni, D.G. and Ling, L.C., 1973. Flavor compounds: Volatilities in vegetable oil and oil – water mixtures. Estimation of odor thresholds. *J. Agric. Food Chem.*, 21: 198–201.

Chua, H.T., Rao, M.A., Acree, T.E. and Cunningham, D.G., 1987. Reverse osmosis concentration of apple juice: flux and flavor retention by cellulose acetate and polyamide membranes. *J. Food Proc. Eng.*, 9: 231–245.

Delcour, J.A., Vandenberghe, M.M., Corten, P.F. and Dondeyne, P., 1984. Flavor thresholds of polyphenolics in water. *Am. J. Enol. Vitic.*, 35: 134–136.

De Pooter, H.L., Montens, J.P., Dirinck, P.J., Willaert, G.A. and Schamp, N.M., 1982. Ripening in-duced in preclimacteric immature Golden Delicious apples by propionic and butyric acids. *Phytochemistry*, 21: 1015–1016.

De Pooter, H.L., Montens, J.P., Willaert, G.A. , Dirinck, P.J. and Schamp, N.M., 1983. Treatment of Golden Delicious apples with aldehydes and carboxylic acids: Effect on the headspace composition. *J. Agric. Food Chem.*, 31: 813–818.

De Pooter, H.L., D'Ydewalle, Y.E., Willaert, G.A., Dirinck, P.J. and Schamp, N.M., 1984. Acetic and propionic acids, inducers of ripening in pre-climacteric Golden Delicious apples. *Phytochemistry*, 23: 23–26.

Dimick, P.S. and Hoskin, J.C. 1983. Review of apple flavor. State of the art. *CRC Crit. Rev. Food Sci. Nutr.*, 18: 387–409.

Dirinck, P., Schreyen, L. and Schamp, N., 1977. Aroma quality evaluation of tomatoes, apples and strawberries. *J. Agric. Food Chem.*, 25: 759–763.

Dirinck, P., De Pooter, H., Willaert, G. and Schamp, N., 1984. Application of a dynamic headspace procedure in fruit flavour analysis. In: P. Schreier (Editor), Analysis of Volatiles. Walter de Gruyter, Berlin, New York, pp. 381–400.

Drawert, F., 1975. Biochemical formation of aroma components. In: H. Maarse and P.J. Groenen (Editors), Aroma Research. Pudoc, Wageningen, pp. 13–39.

Drawert, F., 1978. Primaer-und sekundaeraromastoffe sowie deren analytik. In: Symposium, Aroma-stoffe in Früchten und Fruchtsäften, Internationale Fruchtsaft-union, Bern. Juris Druck-Verlag, Zurich, pp. 1–23.

Drawert, F. and Berger, R., 1981. Possibilities of the biotechnological production of aroma substances by plant tissues cultures. In: P. Schreier (Editor), Flavour '81. Walter de Gruyter, Berlin, New York, pp. 509–527.

Drawert, F., Heiman, W., Emberger, R. and Tressl, R., 1966. Uber die Biogenese vor Aromastoffen bei Pflanzen und Fruchten. II. Enzymatische Bildung von Hexen (2) al (1), Hexanal und deren Vorstufen. *Ann. Chem.*, 694: 200–208.

Drawert, F., Heimann, W., Emberger, R. and Tressl, R., 1968. Uber die biogenese von aromastoffen bei pflanzen und früchten. III. Gaschromatographische bestandsaufnanme von apfel-aromastoffen. *Phytochemistry*, 7: 881–883.

Drawert, F., Heimann, W., Emberger, R. and Tressl, R., 1969a. Gas-chromatographische Untersuchung pflanzlicher Aromen II. Anreicherung Trennung und Identifizierung von Apfelaromastoffen. *Chro-*

matographia, 2: 57 – 66.

Drawert, F., Heimann, W., Emberger, R. and Tressl, R., 1969b. Gas-chromatographische Untersuchung pflanzlicher Aromen. III. Vergleich der 'Heads Space' – Method mit der Flüssig-Flüssig-Extraktion zur Quantitativen Bestimmung von Apfelaromastoffen. *Chromatographia*, 2: 77 – 78.

Drouet, A. and Hartmann, C., 1979. Polyribosomes from pear fruit. *Plant Physiol.*, 64: 1104 – 1108.

Dürr, P., 1979. Development of an odour profile to describe apple juice essences. *Lebensm.-Wiss. Technol.*, 12: 23 – 26.

Dürr, P., and Röthlin, 1981. Development of a synthetic apple juice odour. *Lebensm.-Wiss. Technol.*, 14: 313 – 314.

Dürr, P. and Schobinger, U., 1981. The contribution of some volatiles to the sensory quality of apple and orange juice odour. In: P. Schreier (Editor), Flavour '81. Walter de Gruyter, Berlin, New York, pp. 179 – 193.

FAO, 1985. Production Yearbook, 39. FAO Statistics Series No. 70, pp. 191 – 193.

Fein, B.L., Reissig, W.H. and Roelofs, W.L., 1982. Identification of apple volatiles attractive to the apple maggot *Rhagoletis pomonella*. *J. Chem. Ecol.*, 8: 1473 – 1487.

Feys, M., Naessens, W., Tobback, P. and Maes, E., 1980. Lipoxygenase activity in apples in relation to storage and physiological disorders. *Phytochemistry*, 19: 1009 – 1011.

Fidler, J.C., 1948. Studies of the physiologically-active volatile organic compounds produced by fruits. I. The concentrations of volatile organic compounds occurring in gas stores containing apples. *J. Hortic. Sci.*, 24: 178 – 188.

Fidler, J.C. 1950. Studies of the physiologically-active volatile organic compounds produced by fruits. II. The rate of production of carbon dioxide and volatile organic compounds by King Edward VII apples in gas storage, and the effect of removal of the volatiles from the atmosphere of the store on the incidence of superficial scald. *J. Hortic. Sci.*, 25: 81 – 110.

Fidler, J.C. and North, C.J., 1969. Production of volatile organic compounds by apples. *J. Sci. Food Agric.*, 20: 521 – 526.

Flath, R.A., Black, D.R., Guadagni, D.G., McFadden, W.H. and Schultz, T.H., 1967. Identification and organoleptic evaluation of compounds in Delicious apple essence. *J. Agric. Food Chem.*, 15: 29 – 35.

Flath, R.A., Black, D.R., Forrey, R.R., McDonald, G.M., Mon, T.R. and Teranishi, R., 1969. Volatiles in Gravenstein apple essence identified by GC – mass spectrometry. *J. Chromatogr. Sci.*, 7: 508 – 512.

Frenkel, C., Klein, I. and Dilley, D.R., 1968. Protein synthesis in relation to ripening of pome fruits. *Plant Physiol.*, 43: 1146 – 1153.

Galliard, T., 1968. Aspects of lipid metabolism in higher plants – II. The identification and quantitative analysis of lipids from the pulp of pre and post-climacteric apples. *Phytochemistry*, 7: 1915 – 1922.

Gascó, L., Barrera, R. and De La Cruz, F., 1969. Gas chromatographic investigations of the volatile constituents of fruits aroma. *J. Chromatogr. Sci.*, 7: 228 – 238.

Gerhardt, F., 1954. Rates of emanation of volatiles from pears and apples. *Proc. Am. Soc. Hortic. Sci.*, 64: 248 – 254.

Golias, J., 1983. Reaktion von Apfelfrüchten auf eine gesteuerte Lageranaerobiose. *Arch. Gartenbau, Berlin*, 31: 355 – 365.

Golias, J., 1984. Biogenese Flüchtiger Aromastoffe von Äpfeln in Sauerstoffarmer Atmosphäre. *Acta Univ. Agric., Brno, Fac. Agron.*, XXXII, 2: 95 – 100.

Gorin, N. and Klop, W., 1982. L-Malate and sucrose as criteria for internal quality of Golden Delicious apples size 70 – 80 mm. *Z. Lebensm.-Unters.-Forsch.*, 174: 1 – 4.

Gormley, T.R., 1975. Effect of processing on the comparative quality of Golden Delicious and Bramley's Seedling apples. *Lebensm.-Wiss. Technol.*, 8: 168 – 171.

Grevers, G. and Doesburg, J.J., 1962. Gas chromatographic determination of some volatiles emanated by stored apples. In: Volatile Fruit Flavours. Julius-Verlag, Zürich, pp. 319 – 335.

Guadagni, D.G., Buttery, R.G. and Okano, S., 1963. Odour thresholds of some organic compounds associated with food flavors. *J. Sci. Food Agric.*, 14: 761 – 765.

Guadagni, D.G., Okano, S., Buttery, R.G. and Burr, H.K., 1966. Correlation of sensory and gas – liquid chromatographic measurements of apple volatiles. *Food Technol.*, 20: 166 – 169.

Guadagni, D.G., Bomben, J.L. and Hudson,J.S., 1971a. Factors influencing the development of aroma in apple peels. *J. Sci. Food Agric.*, 22: 110 – 115.

Guadagni, D.G., Bomben, J.L. and Harris, J.G., 1971b. Recovery and evaluation of aroma development in apple peels. *J. Sci. Food Agric.*, 22: 115 – 119.

Hall, G. and Andersson, J., 1983. Volatile fat oxidation products. I. Determination of odour thresholds and odour intensity functions by dynamic olfactometry. *Lebensm.-Wiss. Technol.*, 16: 354 – 361.

Hatfield, S.G.S. and Patterson, B.D., 1974. Abnormal volatile production by apples during ripening after controlled atmosphere storage. In: Facteurs et Régulation de la Maturation des Fruits. Colloques Internationaux CNRS, Paris, pp. 57 – 62.

Heinz, D.E. and Jennings, W.G., 1966. Volatile components of Bartlett pear. V. *J. Food Sci.*, 31: 69 – 80.

Henze, R.E., Baker, C.E. and Quackenbush, F.W., 1953. The chemical composition of apple storage volatiles. I. Acids, alcohols and esters. *Proc. Am. Soc. Hortic. Sci.*, 61: 237 – 245.

Henze, R.E., Baker, C.E. and Quackenbush, F.W., 1954. Fruit storage effects. Carbonyl compounds in apple storage volatiles. *J. Agric. Food Chem.*, 2: 1118 – 1120.

Huelin, F.E., 1952. Volatile products of apples. II. Identification of aldehydes and ketones. *Aust. J. Sci. Res.*, B5: 328 – 334.

Huelin, F.E. and Coggiola, I.M., 1968. Superficial scald, a functional disorder of stored apples. IV. Effect of variety, maturity, oiled wraps and diphenylamine on the concentration of α-farnesene in the fruit. *J. Sci. Food Agric.*, 19: 297 – 301.

Huelin, F.E. and Coggiola, I.M., 1970a. Superficial scald, a functional disorder of stored apples. V. Oxidation of α-farnesene and its inhibition by diphenylamine. *J. Sci. Food Agric.*, 21: 44 – 48.

Huelin, F.E. and Coggiola, I.M., 1970b. Superficial scald, a functional disorder of stored apples. VI. Evaporation of α-farnesene from the fruit. *J. Sci. Food Agric.*, 21: 82 – 86.

Hulme, A.C., Rhodes, M.J.C. and Wooltorton, L.S.C., 1967. The respiration climacteric in apple fruits: some possible regulatory mechanisms. *Phytochemistry*, 6: 1343 – 1351.

Hulme, A.C. and Rhodes, M.J.C., 1971a. Pome fruits. In: A.C. Hulme (Editor), The Biochemistry of Fruits and their Products, Vol. 2. Academic Press, London, New York, pp. 333 – 436.

Hulme, A.C., Rhodes, M.J.C. and Wooltorton, L.S.C., 1971b. The relationship between ethylene and the synthesis of RNA and protein in ripening apples. *Phytochemistry*, 10: 749 – 756.

Hulme, A.C., Rhodes, M.J.C. and Wooltorton, L.S.C., 1971c. The effect of ethylene on the respiration, ethylene production, RNA and protein synthesis for apples stored in low oxygen and in air. *Phytochemistry*, 10: 1315 – 1323.

Ishihara, M., Tsuneya, T., Shiota, H. and Shiga, M., 1986. Identification of new constituents of quince fruit flavor (*Cydonia oblonga* Mill = *C. vulgaris* Pers.). *J. Org. Chem.*, 51: 491 – 495.

Iwamoto, M., Takagi, Y., Kogami, K. and Hayashi, K., 1983. Synthesis of characteristic flavor constituents of Kogyoku apple (Jonathan) essential oil. *Agric. Biol. Chem.*, 47: 117 – 119.

Jakob, M.A., Hipler, R. and Lüthi, H.R., 1969. Ueber das Vorkommen von Hexyl 2-methylbutyrat in Apfelaroma. *Mitt. Lebensm. Unters. Hyg.*, 60: 223 – 227.

Jakob, M.A., Hipler, R. and Lüthi, H.R., 1973. Über den Einfluss pektolytischer enzym präparate auf das Aroma von Apfelsaft. *Lebensm.-Wiss. Technol.*, 6: 138 – 141.

Jennings, W.G. and Sevenants, M.R., 1964. Volatile esters of Bartlett pear III. *J. Food Sci.*, 29: 158 – 163.

Jennings, W.G. and Tressl, R., 1974. Production of volatile compounds in the ripening Bartlett pear. *Chem. Mickrobiol. Technol. Lebensm.*, 3: 52 – 55.

Kakiuchi, N., Moriguchi, S., Fukuda, H., Ichimura, N., Kato, Y. and Banba, Y., 1986. Composition of volatile compounds of apple fruits in relation to cultivars. *J. Jpn Soc. Hortic. Sci.*, 55: 280 – 289.

Kim, In-Sook, and Grosch, W., 1979. Partial purification of a lipoxygenase from apples. *J. Agric. Food Chem.*, 27: 243 – 245.

Kimura, S., Okamoto, T., Harada, J. and Meguro, M., 1979. Fatty acids composition of apples in storage. *Nippon Shokuhin Kogyo Gakkaishi*, 26: 162 – 167.

Kimura, S., Kanno, M., Yamada, Y., Takahashi, K., Murashige, H. and Okamoto, T., 1982. The contents of conjugated lipids and fatty acids and the cold tolerance in the mitochondria of Starking Delicious and Ralls Janet apples. *Agric. Biol. Chem.*, 46: 2895 – 2902.

Kimura, S., Yasuda, T. and Okamoto, T., 1985. Studies on chilling injury of apples. VII. Existence of acetaldehyde in mitochondria from pulp of Starking Delicious after cold storage. *Bull. Fac. Agric. Hirasaki Univ.*, 43: 1 – 6.

38

Koch, J. and Schiller, H., 1964. Beitrag zur kenntnis des Apfelaromas. *Z. Lebensm.-Unters.-Forsch.,* 125: 364 – 368.

Les cahiers du CTIFL (Centre Technique Interprofessionnel des Fruits et Légumes) 1987, Pulpes et purées de fruits, Productions et échanges, 30, 77 pp.

Lesínska E., 1987. Characteristics of sugars and acids in the fruit of East Asian quince. *Nahrung,* 31: 763 – 765.

Lidster, P.D., Lightfoot, H.J. and McRae, K.B., 1983a. Fruit quality and respiration of 'McIntosh' apples in response to ethylene, very low oxygen and carbon dioxide storage atmospheres. *Sci. Hortic.,* 20: 71 – 83.

Lidster, P.D., Lightfoot, H.J. and McRae, K.B., 1983b. Production and regeneration of principal volatiles in apples stored in modified atmospheres and air. *J. Food Sci.,* 48: 400 – 402.

Lidster, P.D., Sanford, K.A., McRae, K.B. and Stark, R., 1984. Apple juice quality and recovery from McIntosh apples stored in controlled atmospheres and air. *Can. Inst. Food Sci. Technol. J.,* 17: 86 – 91.

Lurie, S. and Ben-Arie, R., 1983a. Microsomal membrane changes during the ripening of apple fruit. *Plant Physiol.,* 73: 636 – 638.

Lurie, S. and Ben-Arie, R., 1983b. Characterization of plasmalemma ATPase from apple fruit. *Phytochemistry,* 22: 49 – 52.

MacGregor, D.R., Sugisawa, H. and Matthews, J.S., 1964. Apple juice volatiles. *J. Food Sci.,* 29: 448 – 455.

Macheix, J.J. and Delaporte, N., 1973. Etude de la répartition dans le fruit de *Pirus malus* L., des principaux O-diphénols par dosage oxydimétrique en milieu non aqueux. *Lebensm.-Wiss Technol.,* 6: 19 – 22.

Mast, G. and Feldheim, W., 1983a. Sensoriche Analyse, Malat- und Saccharosegehalt und Fruchtfleischfestigkert verschiedmer Apfelsorten. Teil I. Methode zur sensorichen Bewertung von Apfeln; dargestellt am Beispiel der Sorte Holsteiner Cox. *Z. Lebensm.-Unters.-Forsch.,* 176: 350 – 355.

Mast, G. and Feldheim, W., 1983b. Sensoriche Analyse Malat- und Saccharosegehalt und Fruchtfleischfestigkeit verschiedmer Apfelsorten. II. Beziehungen zwischen der sensorichen Bewertung und chemisch-physikolischen Parametern. *Z. Lebensm.-Unters.-Forsch.,* 177: 30 – 33.

Mazliak, P., 1967a. Recherches sur le métabolisme des glycérides et des phospholipides dans le parenchyme de pomme. II. Incorpoⅰ ⅰion dans les lipides de divers précurseurs radioactifs. *Phytochemistry,* 6: 941 – 956.

Mazliak, P., 1967b. Recherches sur le métabolisme des glycérides et des phospholipides dans le parenchyme de pomme. III. Mise en évidence directe de certaines transformations enzymatiques. *Phytochemistry,* 6: 957 – 961.

Mazliak, P., 1969. Le métabolisme des lipides au cours de la maturation des pommes. *Qual. Plant. Mater. Vég.,* XIX: 19 – 53.

Mazliak, P., 1970. Lipids. In: A.C. Hulme (Editor), The Biochemistry of Fruits and their Products, Vol. 1. Academic Press, London, New York, pp. 209 – 238.

McGlasson, W.B., 1970. The ethylene factor. In: A.C. Hulme (Editor), The Biochemistry of Fruits and their Products, Vol. 1. Academic Press, London, New York, pp. 475 – 519.

Meigh, D.F., 1956. Volatile compounds produced by apples. I. Aldehydes and ketones. *J. Sci. Food Agric.,* 7: 396 – 410.

Meigh, D.F., 1957. Volatile compounds produced by apples. II. Alcohols and esters. *J. Sci. Food Agric.,* 8: 313 – 325.

Meigh, D.F. and Filmer, A.A.E., 1969. Natural skin coating of the apple and its influence on scald in storage. III. α-Farnesene. *J. Sci. Food Agric.,* 20: 139 – 143.

Meigh, D.F. and Hulme, A.C., 1965. Fatty acid metabolism in the apple fruit during the respiration climacteric. *Phytochemistry,* 4: 863 – 871.

Miller, R.H., 1982. Apple fruit cuticles and the occurrence of pores and transcuticular canals. *Ann. Bot.,* 50: 355 – 371.

Myers, M.J., Issenberg, P. and Wick, E., 1970. ʟ-Leucine as a precursor of isoamyl alcohol and isoamyl acetate volatile aroma constituents of banana fruit discs. *Phytochemistry,* 9: 1693 – 1701.

Nichols, W.C. and Patterson, M.E., 1987. Ethanol accumulation and post storage quality of 'Delicious' apples during short-term, low-O_2, CA storage. *HortScience,* 22: 89 – 92.

Nishimura, K. and Hirose, Y., 1964. The aroma constituents of 'Kogyoku' apple. *Agric. Biol. Chem.*, 28: 1 – 4.

Noro, S., Kudo, N. and Kitsuma, T., 1985. Changes of lipids of 'Jonagold' apple peel in the harvest time. *J. Jpn Soc. Hortic. Sci.*, 54: 116 – 120.

Nursten, H.E. and Woolfe, M.L., 1972. An examination of the volatile compounds present in cooked Bramley's seedling apples and the changes they undergo on processing. *J. Sci. Food Agric.*, 23: 803 – 822.

Osman, A.E., 1965. Entwicklung von flüchtigen Geschmacksstoffen im Apfel. *Z. Lebensm.-Unters.-Forsch.*, 128: 333 – 337.

Paillard, N., 1965. Analyse des produits volatils émis par les pommes. *Fruits*, 20: 189 – 197.

Paillard, N., 1967a. Analyse des produits organiques volatiles émis par les pommes, par chromatographie en phase gazeuse et spectrophotométrie infra-rouge. *Physiol. Vég.*, 5: 95 – 107.

Paillard, N., 1967b. Analyse des produits organiques volatils émis par quelques variétés de pommes. *Fruits*, 22: 141 – 151.

Paillard, N., 1967c. Evolution des produits organiques volatils émis par les pommes au cours de leur maturation. *C.R. Acad. Sci. Paris*, 264: 3006 – 3009.

Paillard, N., 1968. Analyse de l'arôme de pommes de la variété 'Calville blanc' par chromatographie sur colonne capillaire. *Fruits*, 23: 383 – 387.

Paillard, N., 1969. Etude des produits organiques volatils émis par des pommes ayant reçu de l'acétate radioactif. *C.R. Acad. Sci. Paris*, 269: 51 – 54.

Paillard, N., 1975a. Nature et origine de l'émission organique volatile des pommes (*Pirus malus* L.). Thèse de Doctorat d'Etat, Université Pierre et Marie Curie, Paris, 118 pp.

Paillard, N., 1975b. Comparaison de l'arome de différentes variétés de pommes: Relation entre les différences d'impressions olfactives et les aromagrammes. *Lebensm.-Wiss. Technol.*, 8: 34 – 37.

Paillard, N., 1979a. Biosynthèse des produits volatils de la pomme: Formation des alcools et des esters à partir des acides gras. *Phytochemistry*, 18: 1165 – 1171.

Paillard, N., 1979b. Biosynthèse des produits organiques volatils de la pomme (*Pirus malus* L.): Formation des cétones et différences variétales de métabolisme. *C.R. Acad. Sci. Paris*, 288: 77 – 80.

Paillard, N., 1981. Biosynthèse et émission d'un ester accompagnant le transfert d'un alcool à travers un segment sphérique de pomme (*Pirus malus* L.). *C.R. Acad. Sci. Paris*, 292: 465 – 468.

Paillard, N., 1985. Evolution of the capacity of aldehyde production by crushed apple tissues, during an extended storage of fruits. In: G. Charalambous (Editor), The Shelf Life of Foods and Beverages. Elsevier, Amsterdam, pp. 369 – 378.

Paillard, N. and Rouri, O., 1984. Production des aldéhydes en C_6, hexanal et 2-hexènal, par des tissus de pommes broyés. *Lebensm.-Wiss. Technol.*, 17: 345 – 350.

Paillard, N., Pitoulis, S. and Mattei, A., 1970. Techniques de préparation et analyse de l'arome de quelques fruits. *Lebensm.-Wiss. Technol.*, 3: 107 – 114.

Panasiuk, O., Talley, F.B. and Sapers, G.M., 1980. Correlation between aroma and volatile composition of McIntosh apples. *J. Food Sci.*, 45: 989 – 991.

Pangborn, R.M., 1981. A critical review of threshold, intensity and descriptive analyses in flavor research. In: P. Schreier (Editor), Flavour '81. Walter de Gruyter, Berlin, New York, pp. 1 – 32.

Petró-Turza, M., Szarföldi-Szalma, I., Madarassy-Mersich, E., Teleky-Vamossy, G. and Füzesi-Kardos, K., 1986. Correlation between chemical composition and sensory quality of natural apple aroma condensates. *Nahrung*, 30: 765 – 774.

Pompéi, C., 1968. Etude sur la composition des concentrés d'aromes de pommes à cidre. *Ann. Technol. Agric.*, 17: 5 – 51.

Power, F.B. and Chesnut, V.K., 1920. The odorous constituents of apples. Emanation of acetaldehyde from the ripe fruit. *J. Am. Chem. Soc.*, 42: 1509 – 1526.

Power, F.B. and Chesnut, V.K., 1922. The odorous constituents of apples. II. Evidence of the presence of geraniol. *J. Am. Chem. Soc.*, 44: 2938 – 2942.

Pratt, H.K., 1974. The role of ethylene in fruit ripening. In: Facteurs et Régulation de la Maturation des Fruits. Colloques Internationaux CNRS, Paris, pp. 153 – 159.

Pyysalo, T., Suikho, M. and Honkanen, E., 1977. Odour thresholds of the major volatiles identified in Cloudberry (*Rubus chamaemorus* L.) and Artic Bramble (*Rubus articus* L.). *Lebensm.-Wiss. Technol.*, 10: 36 – 39.

Quamme, H.A., 1984. Decadienoate ester concentrations in pear cultivars and seedlings with Bartlett-like aroma. *HortScience,* 19: 822 – 824.

Radford, T., Kaoru, K., Friedel, P.K., Pope, L.E. and Gianturco, M.A., 1974. Distribution of volatile compounds between the pulp and serum of some fruit juices. *J. Agric. Food Chem.,* 22: 1066 – 1070.

Rao, M.A., Acree, T.E., Cooley, H.J. and Ennis, R.W., 1987. Clarification of apple juice by hollow fiber ultrafiltration: Fluxes and retention of odor-active volatiles. *J. Food Sci.,* 52: 375 – 377.

Redalen, G., 1984. Quality tests of early ripening cultivars. *Sci. Rep. Agric. Univ. Norway,* 63: 1 – 12.

Redalen, G., 1986. Quality tests of apple cultivars grown in Norway. *Gartenbauwissenschaft,* 51: 207 – 211.

Reissig, W.H., Fein, B.L. and Roelofs, W.L., 1982. Field tests of synthetic apple volatiles as apple maggot (*Diptera: Tephritidae*) attractants. *Environ. Entomol.,* 11: 1294 – 1298.

Rhodes, M.J.C. and Reid, M.S., 1974. The production of ethylene and its relationship with the onset of the respiration climacteric in the apple. In: Facteurs et Régulation de la Maturation des Fruits. Colloques internationaux CNRS, Paris, pp. 189 – 192.

Rhodes, M.J.C. and Wooltorton, L.S.C., 1967. The respiration climacteric in apple fruits. The action of hydrolytic enzymes in peel tissue during the climacteric period in fruit detached from the tree. *Phytochemistry,* 6: 1 – 12.

Romani, R.J. and Ku Lily (Lim), 1966. Direct gas chromatographic analysis of volatiles produced by ripening pears. *J. Food Sci.,* 31: 558 – 560.

Romani, R., Labavitch, J., Yamashita, T., Hess, B. and Rae, H., 1983. Preharvest AVG treatment of 'Bartlett' pear fruits: Effects on ripening, color change, and volatiles. *J. Am. Soc. Hortic. Sci.,* 108: 1046 – 1049.

Russell, L.F., Quamme, H.A. and Gray, J.I., 1981. Qualitative aspects of pear flavor. *J. Food Sci.,* 46: 1152 – 1158.

Schamp, N. and Dirinck, P. 1982. The use of headspace concentration on Tenax for objective flavor quality evaluation of fresh fruits (strawberry and apple). In: Chemistry of Foods and Beverages: Recent Developments. Academic Press, London, New York, pp. 25 – 47.

Schobinger, U., Dürr, P. and Aeppli, A., 1982. Quittensaft-eine interessante Möglichkeit zur Diversifikation des Getränkesortiments. *Flüss. Obst.,* 1: 10 – 14, 18 – 19.

Schreier, P., Drawert, F. and Schmid M., 1978. Changes in the composition of neutral volatile components during the production of apple brandy. *J. Sci. Food Agric.,* 29: 728 – 736.

Schreyen, L., Dirinck, P., Sandra, P. and Schamp, N., 1979. Flavor analysis of quince. *J. Agric. Food Chem.,* 27: 872 – 876.

Schultz, T.H., Flath, R.A., Black, D.R., Guadagni, D.G., Schultz, W.G. and Teranishi, R., 1967. Volatiles from Delicious apple essence. Extraction methods. *J. Food Sci.,* 32: 279 – 283.

Solms, J., 1971. Nonvolatile compounds and the flavor of foods. In: G. Ohloff and A.F. Thomas (Editors), Gustation and Olfaction. Academic Press, London, New York, pp. 92 – 110.

Strackenbrock, K.H., 1961a. Untersuchungen über Apfelaromen. Inaugural-Dissertation, Doktorgrades der Rheinischen Friedrich Wilhelms-Universität, Bonn, 64 pp.

Strackenbrock, K.H., 1961b. Nachweis sortenspezifischer Apfelaromen. *Erwerbsobstbau,* 3: 147 – 148.

Strandjev, A., 1982. Influence du temps et de la température de stérilisation sur les substances aromatiques des poires. *C.R. Acad. Bulg. Sci.,* 38: 1129 – 1132.

Teranishi, R., 1967. High resolution gas chromatography in aroma research. *Perfum. Essent. Oil Rec.,* 58: 172 – 178.

Teranishi, R., Issenberg, P., Hornstein, I. and Wick, E.L., 1971. Correlation with sensory properties. In: Flavor Research, Principles and Techniques. Marcel Dekker, Inc, New York, pp. 259 – 294.

Thompson, A.R., 1951. Volatile products of apples. I. Identification of acids and alcohols. *Aust. J. Sci. Res.,* B4: 283 – 292.

Tressl, R., Emberger, R., Drawert, F. and Heimann, W., 1970. Uber die Biogenese von Aromastoffen bei Pflanzen und Früchten. XI. Mitt: Einbau von 14C Leucin und Valin in Bananenaromastoffe. *Z. Naturforsch.,* 25: 704 – 707.

Tsuneya, T., Ishihara, M., Shiota, H. and Shiga, M., 1983. Volatile components of quince fruit (*Cydonia oblonga* Mill.). *Agric. Biol. Chem.,* 47: 2495 – 2502.

Umano, K., Shoji, A., Hagi, Y. and Shibamoto, T., 1986. Volatile constituents of peel of quince fruit, *Cydonia oblonga* Miller. *J. Agric. Food Chem.,* 34: 593 – 596.

Von Reinhard, C. and Eichholz-Fecher, M., 1984. Aromastoffanalytik in Lebensmitteln. Aromatisie-
rung von Apfelfruchtsaftliкören. *Dtsch. Lebensm.-Rundsch.*, 80: 50 – 52.
Von Sydow, E., Moskowitz, H., Jacobs, H. and Meiselman, H., 1974. Odor – taste interaction in fruit
juices. *Lebensm.-Wiss. Technol.*, 7: 18 – 24.
Von Wücherpfennig, K. and Bretthauer, G., 1962. Beitrag über veränderungen in der Zusammensetzung
von Aromadestillaten bei der Verarbeitung von Apfeln in Abhängigkeit von der angewandten Verfah-
renstechnik. *Fruchtsaft Ind.*, 6: 359 – 369.
Walls, L.P., 1942. The nature of the volatile products from apples. *J. Pomol. Hortic. Sci.*, 20: 59 – 67.
White, J.W., 1950. Composition of volatile fraction of apples. *J. Food Res.*, 15: 68 – 78.
Whiting, G.C., 1970. Sugars. In: A.C. Hulme (Editor), The Biochemistry of Fruit and their Products,
Vol. 1. Academic Press, London, New York, pp. 1 – 31.
Willaert, G.A., Dirinck, P., De Pooter, H. and Schamp, N., 1983. Objective measurement of aroma
quality of Golden Delicious apples as a function of controlled-atmosphere storage time. *J. Agric. Food
Chem.*, 31: 809 – 813.
Williams, A.A., 1983. The flavour of non-citrus fruits and their products. In: J. Adda and H. Richard
(Editors), International Symposium on Food Flavors, Paris, December 8 – 10, 1982. Technique et
Documentation-Lavoisier, pp. 46 – 81.
Williams, A.A., Tucknott, O.G. and Lewis, M.J., 1977. 4-Methoxyallylbenzene an important aroma
component of apples. *J. Sci. Food Agric.*, 28: 185 – 190.
Williams, A.A., Warrington, M. and Arnold, G.M., 1982. The evaluation of two new apple cultivars
(Suntan and East Malling A 3022) for culinary purposes. *Lebensm.-Wiss. Technol.*, 15: 80 – 82.
Wills, R.B.H., 1972. Effect of hexyl compounds on soft scald of apples. *Phytochemistry*, 11:
1945 – 1946.
Wills, R.B.H. and McGlasson, W.B., 1969. Association between loss of volatiles and reduced incidence
of breakdown in Jonathan apples achieved by warming during storage. *J. Sci. Food Agric.*, 20:
446 – 447.
Wills, R.B.H. and McGlasson, W.B., 1971. Effect of storage temperature on apple volatiles associated
with low temperature breakdown. *J. Hortic. Sci.*, 46: 115 – 120.
Wills, R.B.H. and Scott, K.J., 1970. Hexanol and hexylacetate and soft scald of apples. *Phytochemistry*,
9: 1035 – 1036.
Wills, R.B.H. and Scriven, E.M., 1979. Metabolism of geraniol by apples in relation to the development
of storage breakdown. *Phytochemistry*, 18: 785 – 786.
Wills, R.B.H., Scott, K.J. and McGlasson, W.B., 1970. A role for acetate in the development of low
temperature breakdown in apples. *J. Sci. Food Agric.*, 21: 42 – 44.
Winterhalter, P., Lander, V. and Schreier, P., 1987. Influence of sample preparation on the composition
of quince (*Cydonia oblonga*, Mill) flavor. *J. Agric. Food Chem.*, 35: 335 – 337.
Yajima, I., Yanai, T., Nakamura, M., Sakakibara, H. and Hayashi, K., 1984. Volatile flavor components of Kogyoku apples. *Agric. Biol. Chem.*, 48: 849 – 855.

Chapter II

STONED FRUIT: APRICOT, PLUM, PEACH, CHERRY

J. CROUZET, P. ETIEVANT and C. BAYONOVE
Centre de Génie et Technologie Alimentaires, U.S.T.L., Montpellier (France), Laboratoire de Recherche sur les Arômes, I.N.R.A., Dijon (France), Institut des Produits de la Vigne, I.N.R.A., Montpellier (France)

1. INTRODUCTION

According to the structure of their flowers and their fruits, apricot, cherry, peach and plum are members of the family Rosaceae and of the subfamily Prunoideae, genus *Prunus* (see Fig. II.1). They belong to the subgenus *Amygdalus: Prunus persica* L. (peach), *Prunus: Prunus armeniaca* L. (apricot), *Prunus domestica* L., *Prunus insititia* L. and *Prunus salicina* Lindl (plum), and *Cerasus: Prunus avium* L. and *Prunus cerasus* L. (cherry).

Fruits known as stoned fruits are appreciated by consumers for their colour, their palatability and their aromatic characteristics. From a morphological standpoint, stoned fruits are drupes characterized by highly lignified endocarp, fleshy mesocarp and thin epicarp (Romani and Jennings, 1971).

2. APRICOT

It is generally agreed that apricot originated in China, and was introduced into Europe at the beginning of the Christian era. It is presently cultivated in all Mediterranean countries, in South Africa, in South America, and in North America, especially in California. The more popular cultivars are Blenheim, Tilton and Derby in the United States, Rouge du Roussillon, Polonais and Bergeron in France, Canino, Moniqui, Cafona and Bulida in Spain, and Bebeco in Greece.

2.1 *The Volatile Constituents of Apricot*

2.1.1 *Qualitative data*

Studies related to the volatile components of apricot are relatively few (Kovacs and Wolf, 1964; Rhoades and Millar, 1965; Baldrati and Gianone, 1967; Tang and Jennings, 1967, 1968; Molina *et al.*, 1974; Rodriguez *et al.*, 1980; Chairote *et al.*, 1981; Guichard *et al.*, 1986; Guichard and Souty, 1988). However, 128 compounds have been identified: 14 hydrocarbons, 25 esters, 10

44

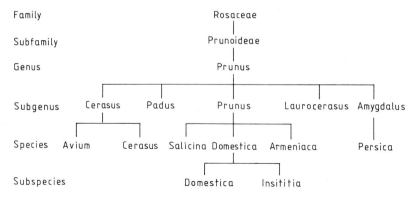

Fig. II.1. Botanical classification of stoned fruits.

TABLE II.I

Volatile constituents identified in apricot extracts

Volatile constituent	Reference[a]	Volatile constituent	Reference[a]
Hydrocarbons		Ethyl hexanoate	8
Toluene	8	cis-Hex-3-enyl acetate	8
Ethylbenzene	8	Hexyl acetate	8
Dimethylbenzene	8	trans-Hex-2-enyl acetate	8
Trimethylbenzene	8	Heptyl acetate	8
Naphthalene	8	Ethyl benzoate	8
p-Diacetylbenzene	8	Ethyl dodecanoate	6
Myrcene	3, 6	Ethyl tetradecanoate	6
Limonene	3, 6	Ethyl pentadecanoate	6
Terpinolene	3, 6	Methyl hexadecanoate	6
p-Cymene	3, 6	Ethyl hexadecanoate	6
Menthadiene	6	Geranyl acetate	6
Camphene	6		
β-Pinene	6	Aldehydes	
γ-Terpinene	6	Formaldehyde	2
		Acetaldehyde	2, 6
Esters		Butanal	8
Ethyl acetate	8	2-Methylbutanal	8
Propyl acetate	8	3-Methylbutanal	8
Isopropyl acetate	8	Pentanal	2, 6, 8
Ethyl propanoate	8	Hexanal	2, 6, 8
Ethyl methylpropanoate	8	trans-Hex-2-enal	6, 8
2-Methylpropyl acetate	8	Heptanal	8
Butyl acetate	8	Hexa-2,4-dienal	8
Ethyl 2-methylbutanoate	8	Benzaldehyde	6, 8
2-Methylpropyl propanoate	8	Hepta-2,4-dienal	8
3-Methylbutyl acetate	8	Octanal	8
Pentyl acetate	8	Phenylacetaldehyde	8
3-Methylbut-2-enyl acetate	8	Nonanal	6, 8

TABLE II.I (*continued*)

Volatile constituent	Reference[a]	Volatile constituent	Reference[a]
Decanal	8	Alcohols	
Geranial	3, 6	Butan-1-ol	8
p-Hydroxybenzaldehyde	7	3-Methylbut-3-en-1-ol	8
p-Methoxybenzaldehyde	7	3-Methylbut-2-en-2-ol	8
		Pentan-2-ol	8
Lactones		2-Methylpentan-2-ol	8
γ-Butyrolactone	6, 7, 8	Pentan-1-ol	6, 8
γ-Hexalactone	4, 6, 7	Hexan-2-ol	8
γ-Octalactone	3, 4, 5, 6, 7, 8	*cis*-Hex-3-en-1-ol	6, 8
δ-Octalactone	5, 8	*trans*-Hex-2-en-1-ol	4, 6, 8
γ-Nonalactone	8	Hexan-1-ol	6, 7, 8
γ-Decalactone	3, 4, 6, 7, 8	2-Ethylhexan-1-ol	6
δ-Decalactone	4, 6, 8	Benzyl alcohol	3, 6, 8
γ-Undecalactone	1	Hepta-1,5-diene-3,4-diol	8
γ-Dodecalactone	4, 5, 7	Geraniol	6, 7
Dihydroactinidiolide	7, 8	Linalool	3, 6, 7, 8
		α-Terpineol	3, 6, 7, 8
Ketones		4-Terpinenol	6, 7, 8
Acetone	2, 6	Citronellol	6
Butanone	2, 8	Nerol	6, 7
Pentan-2-one	8	2-Phenylethanol	6, 8
Hexan-2-one	8	Eucalyptol	8
Hexane-2,5-dione	8	α-Fenchol	8
Cyclopentanone	8	*p*-Cymen-8-ol	8
Heptan-2-one	8	*p*-Cymen-9-ol	8
5-Methylhexan-2-one	8	Farnesol	6, 7
Octan-2-one	8		
6-Methylhept-5-en-2-one	6, 8	Acids	
Non-3-en-2-one	7	Acetic acid	3
Acetophenone	8	2-Methylbutanoic acid	3
Fenchone	8	Hexanoic acid	6
Camphor	8		
Pinocamphone	8	Miscellaneous	
Isopinocamphone	8	*trans*-Rose oxide	7
Verbenone	8	Nerol oxide	7
β-Ionone	7, 8	Epoxydihydrolinalool I	4, 6, 7
Damascenone	7	Epoxydihydrolinalool II	4, 6, 7
		Epoxydihydrolinalool III	4, 6, 7
		Epoxydihydrolinalool IV	4, 6, 7

[a] (1) Kovacs and Wolf, 1964; (2) Baldrati and Gianone, 1967; (3) Tang and Jennings, 1967; (4) Tang and Jennings, 1968; (5) Molina *et al.*, 1973; (6) Rodriguez *et al.*, 1980; (7) Chairote *et al.*, 1980; (8) Guichard and Souty, 1988.

lactones, 19 ketones, 19 aldehydes, 25 alcohols, 3 acids and 6 miscellaneous compounds (see Table II.I).

2.1.2 Flavour impact information

According to Tang and Jennings (1967), none of the compounds isolated from essence concentrates obtained from the cv. Blenheim possessed by themselves an aroma suggestive of apricot. Several methods were used to obtain the concentrates: liquid – liquid extraction of the distillate collected from a vacuum deaerator; activated charcoal adsorption followed by ethyl ether extraction of the effluent of a vacuum deaerator; steam distillation at atmospheric and reduced pressure; direct extraction and molecular distillation. They concluded that apricot aroma is due to an integrated response to the proper ratio of the compounds present in the extract.

By silica gel separations of the aroma concentrate obtained after vacuum distillation of an industrial purée of apricot cv. Rouge du Roussillon, several fractions were isolated and pooled according to their odour (see Table II.II) (Chairote *et al.*, 1981). After mixing fractions 1, 2 and 5, a product possessing an apricot aroma was obtained. However, as previously stated by Tang and Jennings (1967), none of the compounds possessed separately an aroma characteristic of the fruit. The results of Table II.II suggest that this aroma is dependent upon several compounds, such as benzaldehyde, linalool, 4-terpinenol, α-terpineol and 2-phenylethanol. The alcohols are responsible for the fruity and floral odour of the product (Chairote *et al.*, 1981; Guichard and Souty, 1988).

According to Guichard and Souty (1988), two unidentified sesquiterpene

TABLE II.II

Fractions of apricot volatiles collected after silica gel chromatography and pooled according to their odour (Chairote *et al.*, 1981)

Fraction No.	Odour	Main compounds identified by GLC
1	Plum, dried fruit	Benzaldehyde, non-3-en-2-one, *trans*-rose oxide, nerol oxide, *p*-hydroxybenzaldehyde, *p*-methoxybenzaldehyde, 1-phenylheptan-4-one (or 2-methyl-6-phenylhexan-3-one), damascenone
2	Floral, fruity	Linalool, 4-terpinenol, α-terpineol, geraniol, nerol
3	Floral	
4	Lactone, tea	γ-Caprolactone, γ-decalactone, γ-octalactone, γ-dodecalactone, dihydroactinidiolide
5	Rose	Epoxydihydrolinalool I and II (linalool oxides)
6	Cooked-apricot	2-Phenylethanol

ketones of apricot with mass spectra similar to that of pinocamphone, possessed floral odour as estimated from GC sniffing.

According to Chairote *et al.* (1981), γ- and δ-lactones reported in different cultivars of apricot (Molina *et al.,* 1974; Rodriguez *et al.,* 1980; Tang and Jennings, 1968; Guichard and Souty, 1988), as well as in different fruits: peach (Jennings and Sevenants, 1964; Sevenants and Jennings, 1966); pineapple (Creveling *et al.,* 1968); plum (Forrey and Flath, 1974); raspberry (Winter and Enggist, 1971); and strawberry (Tressl *et al.,* 1969), are responsible for the background aroma of apricot as in clingstone peach (Spencer *et al.,* 1978).

A biosynthetic sequence starting from acetate has been given for γ- and δ-lactones by Tang and Jennings (1968) (Fig. II.2). However, as indicated by the authors, the mechanism involved in the formation of the double bond in the $\gamma - \delta$ position is not elucidated.

2.2 Variation of Apricot Volatile Constituents

2.2.1 Influence of cultivar
In agreement with results obtained for other fruits, important differences are observed for different cultivars of apricot with respect to the quantities of aroma compounds present in extracts obtained by vacuum distillation of fresh fruit (Guichard and Souty, 1988, Table II.III). The Précoce de Tyrinthe cultivar contained less volatile components than the other cultivars studied, and is known to be less aromatic. Palsteyn is characterized by high concentrations of terpene alcohols (except for 4-terpinenol), benzaldehyde and phenylacetaldehyde.

The highest quantities of aliphatic aldehydes and alcohols, and more especially C_6 compounds responsible for an herbaceous odour, were present in cultivar Polonais. Ketones were found in high quantities in the Moniqui cultivar, which is characterized by a pleasant and floral aroma, as well as a peach- or apricot-like odour induced by the presence of lactones. β-Ionone, with its characteristic violet aroma, was detected only in this cultivar and Rouge du Roussillon.

Lactones, especially γ-decalactone, are characteristic of Rouge du Roussillon

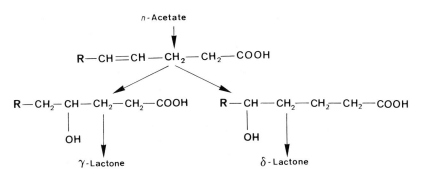

Fig. II.2. Biosynthetic formation of γ- and δ-lactones from acetate (Tang and Jennings, 1968).

TABLE II.III

Content and relative percent of major volatile components of several apricot cultivars (from Guichard and Souty, 1988)

Chemical class	Précoce de Tyrinthe		Palsteyn		Moniqui		Rouge du Roussillon		Polonais		Bergeron	
	mg/kg	Relative percent	mg/kg	Relative percent	mg/kg	Relative percent	mg/kg	Relative percent	mg/kg	Relative percent	mg/kg	Relative percent
Alcohols	0.3	4	11.0	14	22.2	16	39.4	27	51.9	32	6.2	5.5
Terpene alcohols	0.1	1	5.6	7	3.9	2	4.4	3	2.7	1.5	0.9	1
Aldehydes	2.3	32	21.0	27	48.0	34	25.9	18	95.2	59	25.6	23
Ketones	0.1	1	7.2	9	5.3	3	4.3	3	4.8	3	1.6	1.5
Esters	0.04	1	15.2	19	5.7	4	24.0	17	3.0	2	25.0	22
Lactones	4.3	60	18.3	23	55.4	39	45.0	31	2.8	22	52.5	46
Total	7.14		78.3		140.5		133.0		160.4		118.8	

49

and Bergeron, both known as aromatic varieties. However, terpene alcohols, responsible for a pleasant aroma, were only present in significant quantities in Rouge du Roussillon; the Bergeron cultivar contained only trace quantities of these compounds. These two cultivars are particularly rich in esters.

These varietal differences are also shown in the chromatograms obtained after gas stripping of aroma compounds from apricot purée, trapping on charcoal, and desorption using microwave radiation (Rolle and Crouzet, 1988) (Figs. II.3a,b).

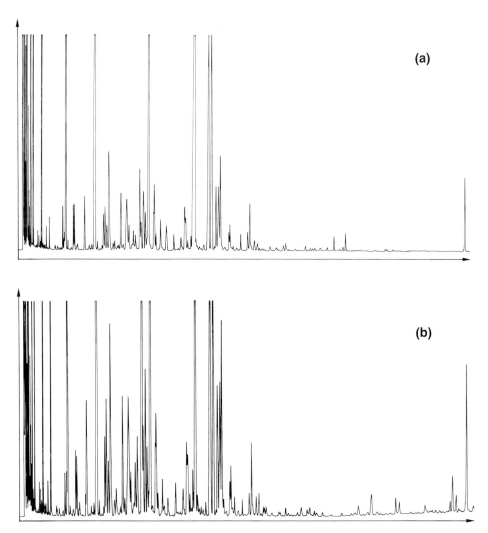

Fig. II.3. Headspace gas chromatograms of the volatiles of two apricot cultivars: (a) Moniqui, (b) Bergeron (Rolle and Crouzet, 1988).

50

2.2.2 Influence of heat treatment and processing

Differences between volatile components isolated from apricot purées obtained after heat treatment (10 min at 90°C) and crushing, and volatile components isolated from fresh fruit were found by Guichard *et al.* (1986).

In purées, 2-methylpropyl propanoate, pentyl acetate, hexa-2,4-dienal, benzaldehyde, hexane-2,5-dione, fenchone, pinocamphone, isopinocamphone, verbenone and *p*-cymen-8-ol were lacking, having been lost or transformed during heating.

On the other hand, thermally induced components were identified in apricot purées and more generally in heated apricot products (Table II.IV). 2-Furaldehyde (0 – 210 ppb) and 5-methyl-2-furaldehyde (84 – 1000 ppb) were found in apricot purées (Guichard *et al.*, 1986), whereas these compounds were not detected by Crouzet *et al.* (1983) after continuous heat treatment for 5 min at temperatures from 75 to 98°C of purées obtained from Rouge du Roussillon cultivar. Under these temperature conditions, modifications were found for terpene hydrocarbons: myrcene, limonene and γ-terpinene (Crouzet *et al.*, 1983; Guichard *et al.*, 1986). As the temperature was increased from 75 to 98°C, a decrease in the amount of myrcene, limonene and γ-terpinene was noticed,

TABLE II.IV

Concentrations of main volatile components in apricot (cultivar Rouge du Roussillon) before and after heat treatment (Crouzet *et al.,* 1983)

Compound	Non-treated purée	Heated purée		
		75°C, 5 min	90°C, 5 min	98°C, 5 min
Camphene	2.0	2.2	3.0	3.0
β-Pinene	2.3	2.8	4.1	3.5
Myrcene	6.3	15.0	9.0	12.8
Limonene	4.7	3.2	2.2	2.0
γ-Terpinene	11.9	13.4	11.1	8.1
p-Cymene	9.0	9.6	11.5	8.1
Epoxydihydrolinalool I	25.7	39.7	65.0	86.5
Epoxydihydrolinalool II	18.7	28.2	40.0	35.9
Benzaldehyde	2	2.2	7.9	8.2
Linalool	258	152.7	295.4	240.4
Terpin-1-en-4-ol	7.9	9.8	13.9	20.7
γ-Butyrolactone	8.9	20.7	43.0	35.4
α-Terpineol	221	225.0	346.0	257.6
Nerol	24.4	29.7	50.0	34.4
Geraniol	70.7	103.2	187.2	112.6
2-Phenylethanol	1.5	1.5	6.1	2.3
γ-Decalactone	17.5	24.8	46.7	34.7
Ethyl pentanoate	2.4	0.7	2.5	0.9
Farnesol	5.0	19.1	25.2	18.0

Concentration (μg/l)

whereas an increase of *p*-cymene occurs. These variations can be related to re-arrangement reactions of myrcene and limonene to γ-terpinene, followed by dehydrogenation of this compound to *p*-cymene.

Oxidation of monoterpenes during heat treatment is probably involved in the formation of α-terpineol, linalool, 4-terpinenol, nerol, geraniol, myrcenol and carveol (Crouzet *et al.,* 1983; Guichard *et al.,* 1986; Guichard and Souty, 1988). Some of these compounds can also be produced through acid hydrolysis of glycosidically bound terpenes (Salles *et al.,* 1988).

Epoxydihydrolinalool I and II may be produced during heat treatment

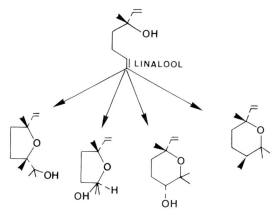

Fig. II.4. Formation of linalool oxides from linalool (Crouzet *et al.,* 1983).

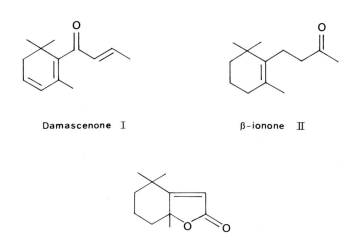

Damascenone I   β-ionone II

Dihydroactinidiolide   III

Fig. II.5. Degradation products of carotenoids found in apricot purée.

through oxidation of linalool (Fig. II.4), or may originate from degradation of hydroxylated linalool derivatives found in fruits such as grape (Williams *et al.*, 1980) or passion fruit (Engel and Tressl, 1983).

Damascenone, β-ionone and dihydroactinidiolide (Fig. II.5), found in heated apricot purées, are associated with heat treatment of several products, such as tea (Bricout *et al.*, 1967), tobacco (Fujimori *et al.*, 1976; Enzell, 1981), dried prune (Moutonnet, 1978), grapes (Schreier *et al.*, 1976) and mango (Sakho *et al.*, 1985). They are considered as degradative products of β-carotene (Demole and Berthet, 1972; Murray *et al.*, 1972). The different assumptions formulated concerning the presence of these compounds in vegetable products are founded on experimental results obtained from β-carotene degradation studies in model systems (Day and Erdman, 1963; Mulik and Erdmann, 1963; Mader, 1964; La Roe and Shipley, 1970; Schreier *et al.*, 1979; Kawamaki, 1982; Kanasawud and Crouzet, 1989).

Indirect enzymic reactions, such as those found during tea fermentation (Sanderson *et al.*, 1971), or photochemical degradation (Isoe *et al.*, 1969), may be involved in the formation of volatile compounds. However, according to Guichard and Souty (1988), these mechanisms could not be responsible for the presence of β-ionone and dihydroactinidiolide in fresh fruits of some apricot cultivars, as no heating or light irradiation were involved during the extraction procedure used. Moreover, the authors conclude the existence of different biogenetic pathways involved in the formation of these products, due to the fact that they are not detected simultaneously in the same cultivar, except for Rouge du Roussillon.

An increase in the amount of γ-butyrolactone during heat treatment of apricot purée was observed by Crouzet *et al.* (1983). This compound has been previously found to be formed during ascorbic acid degradation (Tatum *et al.*, 1969) or during fructose heating in basic medium (Shaw *et al.*, 1968). The first pathway is the more probable during apricot purée heating; moreover, a decrease in content (from 72 to 55%) was observed, according to the severity of heat treatment (75 – 98°C for 5 min).

2.3 *Glycosidically Bound Volatile Components*

The increase in concentration of linalool furan oxides and α-terpineol found during preparation of apricot purée (Chairote *et al.*, 1981) or during heat treatment of apricot purées, can be explained by hydrolysis of bound terpenic compounds. The occurrence of these compounds was first reported by Francis and Allcock (1969) in rose petals, and then they were found in several plants and, more particularly, in fruits such as grapes (Cordonnier and Bayonove, 1974; Williams *et al.*, 1981, 1982a,b; Gunata *et al.*, 1985), passion fruit (Engel and Tressl, 1983), papaya (Heidlas *et al.*, 1984), and mango (Salles *et al.*, 1987).

In apricot, evidence for the presence of glycosidically bound terpenes was given by Chalier (personal communication, 1987) using the rapid analytical technique described by Dimitriadis and Williams (1984) (see Table II.V).

The results obtained after acid hydrolysis of the glycoside fraction isolated from Rouge du Roussillon cultivar show the presence of glycosidically bound terpenes: epoxydihydrolinalool I and II, linalool, α-terpineol, nerol, citronellol, geraniol and glycosidically bound aromatic alcohols: benzyl alcohol and 2-phenylethanol (Salles *et al.*, 1988). These results were confirmed by enzymic hydrolysis of the monoglucoside and diglycoside fraction isolated from apricot (Rouge du Roussillon) using almond β-glucosidase and pectinol VR, respectively (Table II.VI).

In contrast with the results found for grapes (Williams *et al.*, 1982a), the monoglucosides are present in apricot in more significant quantities than the diglycosides. This fact was confirmed by HPLC studies on Cyclobond I and by D/CI positive mode mass spectrometry of the glycosides fraction purified extract of apricot (Salles *et al.*, 1988) (Figs II.6 and II.7).

TABLE II.V

Free and glycosidically bound terpenes in apricot cultivars, determined using vanillin-sulphuric acid reagent

Cultivar	Free volatile terpenes (mg/kg)[a] FVT	Potentially volatile terpenes (mg/kg)[a] PVT	PVT ——— FVT
Bergeron	1.98	3.20	1.6
Canino	2.80	3.58	1.3
Rouge du Roussillon	1.48	7.61	5.2
Précoce de Tyrinthe	0.58	3.15	5.3

[a] Expressed as linalool.

TABLE II.VI

Volatile compounds identified after enzymic hydrolysis of the monoglucoside fraction by almond β-glucosidase and of the glycoside fraction by pectinol VR: cultivar Rouge du Roussillon (from Salles *et al.*, 1988)

Volatile compound	Monoglucoside fraction[a]	Diglycoside fraction[a]
Linalool oxides	420	
Linalool	1039	13.5
α-Terpineol	331	22.5
Citronellol	66	5
Nerol	515	20
Geraniol	298	62.5
Benzyl alcohol	552	58
2-Phenylethanol	460	25

[a] Results expressed in μg/litre; octanol was used as internal standard.

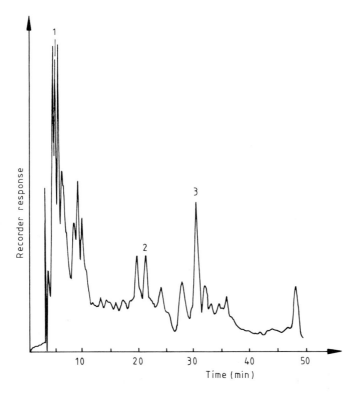

Fig. II.6. Separation of glycosidically bound volatile compounds of apricot cv. Rouge du Roussillon (whole purified fraction) on Cyclobond I column (25 × 0.46 cm) using an acetonitrile – water gradient (10/90 to 22.5/77.5, v/v) at 1 ml/min. Detection at 210 nm. (1) Benzyl β-D-glucoside, (2) Neryl β-D-glucoside, (3) Linalyl β-D-glucoside (Salles *et al.*, 1988).

3. PLUM

The first chemical investigation on plum aroma was published by Forrey and Flath in 1974, and concerned a particular species named *Prunus salicina*. Subsequent studies were made on a different species, *Prunus domestica*. They were mainly accomplished at the Long Ashton Research Station (University of Bristol, U.K.) (Ismail, 1977; Ismail *et al.*, 1977, 1980a,b,c, 1981a,b; Williams and Ismail, 1981) and at the Institut National de la Recherche Agronomique (INRA) in Toulouse, Dijon and Colmar (France) (Moutounet *et al.*, 1975; Moutounet, 1978; Etiévant *et al.*, 1986; Dirninger-Rigo, 1987; Le-Quéré *et al.*, 1987). Finally, there are two remaining publications, by Kereselidze and Mikeladze (1977) in the Soviet Union, and by Vernin *et al.* (1985) from the University of Marseille (France). As seen from Table II.VII, only a few of the numerous *Prunus* species and subspecies have been investigated.

Comparisons between these different results are almost impossible, as the dif-

TABLE II.VII

Relationship between the different plum cultivars studied for their aroma and reported in the literature

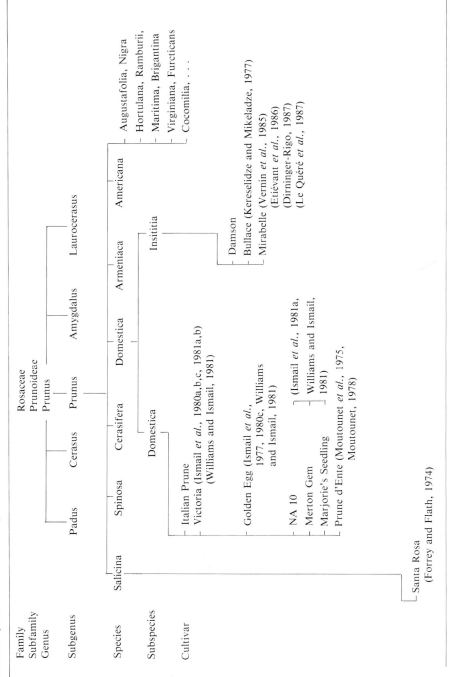

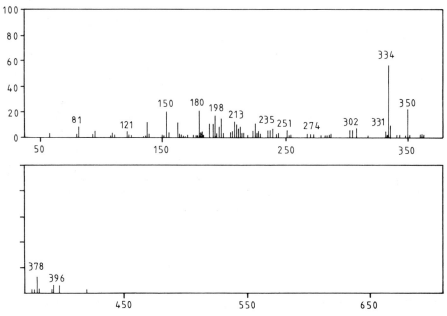

Fig. II.7. D/CI positive mode mass spectrum of heterosidic purified extract of apricot (Salles *et al.*, 1988).

TABLE II.VIII

Comparison of the different extraction methods used in plum aroma investigations

Publication	Substrate used for extraction	Extraction method
Forrey and Flath, 1974	Sliced fruits	Vacuum distillation + solvent extraction (ether)
Moutounet and Jouret, 1975, and Moutounet, 1978	Deep-frozen grated fruits	Vacuum distillation + solvent extraction (F11)
Ismail *et al.*, 1977	Whole fruits	Solvent extraction (CH_2Cl_2)
Dirninger-Rigo, 1987	Methanolic oxygen-free slurry	Vacuum distillation + solvent extraction (F11)
Ismail *et al.*, 1980b	Fermented plum juice	Climbing film evaporation + solvent extraction (F11)
Ismail *et al.*, 1980c	Canned fruits	Distillation + solvent extraction or headspace collection (Porapak Q)
Ismail *et al.*, 1981b	Enzymed juice	Climbing film evaporation + ether extraction
Ismail *et al.*, 1980a, 1981a	Whole fruits	Headspace (Porapak Q)
Vernin *et al.*, 1985	Pressed fruits	Atmospheric distillation + Likens – Nickerson extraction (ether)
Etiévant *et al.*, 1986	Whole fruits	Headspace collection (Porapak Q)
Le Quéré *et al.*, 1987	Methanolic oxygen-free slurry	Solvent extraction (F11)

ferent authors used different extraction techniques, which are itemized in Table II.VIII. However, most of these authors pointed out that enzymic oxidation of the fruits took place quickly, as soon as the fruits were milled or destoned, with a consequent browning of the tissues and a deterioration of the aroma, which should be avoided (Ismail *et al.*, 1981b; Etiévant *et al.*, 1986; Dirninger-Rigo, 1987). These authors consequently tried to avoid oxidation during aromatic extract isolation by using ascorbic acid as an oxygen trap (Ismail *et al.*, 1980b; Dirninger-Rigo, 1987), sulphur dioxide to combine with the phenolic substrates (Dirninger-Rigo, 1987), or methanol to deactivate the proteins responsible for the reaction (Etiévant *et al.*, 1986; Dirninger-Rigo, 1987). Etiévant *et al.* observed that milling the plums in methanol previously purged of oxygen and waiting a further 20 min under CO_2 atmosphere allowed isolation of a stable slurry, as judged by its colour and UV spectrum. This technique has also been chosen by Dirninger-Rigo (1987) after comparing the different methods cited above.

3.1   The Volatile Constituents of Plum

3.1.1   Qualitative and quantitative data
The different plum constituents identified in aromatic extracts obtained using different techniques are itemized in Table II.IX. Esters are qualitatively more important than any other class of compounds. Depending on the cultivar, or more probably on the technique of extraction used, alcohols or esters have been claimed to be quantitatively the major components of aromatic extracts (Table II.X). As none of these alcohols and esters by themselves exhibit a flavour similar to that of the plum, most authors have tried to identify other substances as potential contributors to plum aroma.

3.1.2   Flavour impact information
A first approach consisted of selecting components producing an olfactory stimulus, when sniffed at the exit of a gas-chromatographic column, and in describing their odours (Forrey and Flath, 1974; Ismail, 1977; Dirninger-Rigo, 1987; Ismail *et al.*, 1980b,c, 1981a,b; Williams and Ismail, 1981; Etiévant *et al.*, 1986).

Williams and Ismail (1981) were the only authors who detected with this method some odours reminiscent of that of fresh or cooked plums. A mass spectrometric investigation of three interesting regions found systematically in their chromatograms indicated that they were respectively associated with mixtures of linalool plus benzaldehyde and ethyl nonanoate (fresh plum odour), methyl cinnamate plus γ-decalactone (fresh plum aroma) and finally 2-phenylethanol plus γ-octalactone and damascenone (cooked plum aroma). Selective removal of some of these components was then performed in order to check if the odours were due to their fortuitous co-elution, or to other undetected components eluted at the same time. Since the characteristic odours disappeared after alkali treatment of the extract (lactone saponification) or after percolation of the ex-

TABLE II.IX

Volatile constituents identified in plum extracts

Volatile constituent	Reference[a]
Aliphatic hydrocarbons	
2-Methylbuta-1,3-diene	9, 15
Octane	7
Nonane	7
Decane	6
Undecane	1, 11
Dodecane	11
2,2,6,6-Tetramethyl-4-ethylheptane	7
2,2,9-Trimethyldecane	7
3,3,4-Trimethyldecane	7
4,4-Dimethylundecane	7
4,5-Dimethylundecane	7
Tridecane	1, 7, 11
Tetradecane	1, 6, 7, 11
Pentadecane	1, 7, 11
6-Propyltridecane	9, 15
Hexadecane	1, 6, 7, 11
Heptadecane	7, 11
Heptadecene	7
Octadecane	6
Alkanes (C_{20} – C_{31})	2,6
Aromatic hydrocarbons	
Vinylbenzene	1, 2
Ethylbenzene	1
Xylenes (o-, m- and p-)	1, 2
Propylbenzene	1
Naphthalene	1, 2, 3, 9, 15
1-Methylnaphthalene	5
2-Methylnaphthalene	7
Biphenyl	5

Volatile constituent	Reference[a]
Terpene hydrocarbons	
α-Pinene	7
β-Pinene	7
Limonene	7
p-Cymene	1, 6
Aliphatic alcohols	
Methanol	5
Ethanol	3, 4, 5, 6, 9, 10, 15
Propan-1-ol	3, 5, 9, 15
2-Methylpropan-1-ol	1, 3, 4, 5, 9, 10, 15
Butan-1-ol	1, 3, 4, 5, 6, 7, 9, 10, 11, 14, 15
But-2-en-1-ol	3, 9, 15
Butane-2,3-diol	9, 14, 15
2-Methylbutan-1-ol	5, 7
3-Methylbutan-1-ol	3, 4, 5, 7, 9, 10, 15
2-Methylbut-3-en-2-ol	5
Pentan-1-ol	1, 3, 5, 9, 14, 15
Pent-1-en-3-ol	5
Pent-3-en-2-ol (cis)	3, 9, 15
Hexan-1-ol	1, 3, 4, 5, 6, 7, 9, 10, 15
Hex-3-en-1-ol (cis)	1, 3, 5, 6, 7, 9, 15
Hex-2-en-1-ol (trans)	1, 5, 6, 7
Hex-3-en-1-ol (trans)	5
2-Butoxyethanol	1
Heptan-1-ol	3, 5, 6, 9, 15
2-Ethylhexan-1-ol	1, 9, 15
3-Ethylhexan-1-ol	3
Octan-1-ol	1, 3, 5, 7, 9, 15
Oct-1-en-3-ol	3, 9, 15
Nonan-1-ol	1, 3, 7, 9, 15

(continued)

| 1-(2-Methylpropoxy)-1-(3-methyl-butoxy)ethane | 3, 9, 15 |
| 1,1-Di(3-methylbutoxy)ethane | 9, 15 |

Ketones

Hydroxypropanone	7
Butanone	7
Butanedione	7
3-Hydroxybutanone	1, 3, 4, 5, 6, 7, 9, 10, 14, 15
Pentan-3-one	7
Pentane-2,3-dione	7, 9, 14, 15
Cyclopent-2-ene-1,4-dione	7
Hexan-2-one	9, 15
Hexane-2,5-dione	2
Furyl methyl ketone	11, 12, 14, 15
Acetophenone	9, 14, 15
Octan-3-one	7
Nonan-2-one	9, 15
Carvone	6, 9, 14, 15
Verbenone	3, 9, 15
Camphor	6
1-(2,6,6-Trimethylcyclohexa-1,3-dienyl)but-2-en-1-one (damascenone)	9, 12, 15
4-(2,6,6-Trimethylcyclohex-1-enyl)-but-3-en-2-one (α-ionone)	13
4-(2,6,6-Trimethylcyclohex-2-enyl)-but-3-en-2-one (β-ionone)	13
Noneicosan-10-one	2

Aliphatic aldehydes

Acetaldehyde	3, 5, 9, 15
Methylpropanal	7
2-Methylbutanal	7
3-Methylbutanal	7
2-Methylpent-2-enal	5
Hexanal	1, 3, 7, 9, 14, 15

Non-2-en-1-ol	3, 9, 15
Primary alcohols (C$_{18}$ – C$_{30}$)	2
Secondary alcohols (C$_{27}$ – C$_{31}$)	2

Aromatic alcohols

Benzyl alcohol	1, 3, 4, 5, 6, 9, 10, 11, 12, 14, 15
2-Phenylethanol	1, 3, 4, 5, 6, 7, 9, 10, 11, 12, 15
p-Cymen-8-ol	6

Terpene alcohols

Borneol	6
Geraniol	1, 3, 6, 9, 15
Linalool	1, 3, 4, 5, 6, 9, 10, 12, 15
Nerol	6
Terpin-1-en-4-ol	1, 3, 6, 7, 9, 15
Terpin-3-en-1-ol	1
α-Terpineol	1, 3, 5, 6, 9, 11, 12, 15
γ-Terpineol	1
β-Terpineol (*cis*)	1
β-Terpineol (*trans*)	1
Cadinol	6

Ethers and oxides

3,4-Dimethoxy-1-(prop-2-enyl)-benzene	1, 3, 9, 12, 15
2-Phenethyl ether	3, 9, 15
Cineol	11, 12
Linalool oxide (*cis*)	9, 15
Linalool oxide (*trans*)	9, 15

Acetals

1,1-Diethoxyethane	5, 9, 15
1-Ethoxy-1(3-methylbutoxy)ethane	9, 15
1,1-Di(2-methylpropoxy)ethane	9, 15
1-Ethoxy-1-hexoxyethane	3, 9, 15
1-Ethoxy-1-*cis*-hex-3-enoxyethane	3, 9, 15

TABLE II.IX (*continued*)

Volatile constituent	Reference[a]	Volatile constituent	Reference[a]
Hex-2-enal (*trans*)	1, 3, 7, 9, 15	Ethyl butanoate	2, 3, 5, 7, 9, 10, 15
Hex-3-enal (*cis*)	1, 7	Ethyl but-2-enoate	3, 7, 10, 15
Heptanal	9, 14, 15	C$_7$	
Octanal	1, 6, 9, 14, 15	2-Methylbutyl acetate	5
Nonanal	1, 2, 3, 4, 9, 10, 14, 15	3-Methylbutyl acetate	3, 4, 5, 9, 10, 15
Decanal	1	Pentyl acetate	5, 7
Deca-2,4-dienal (*trans, trans*)	9, 15	Butyl propanoate	1, 7
Deca-2,4-dienal (*trans, cis*)	15	Propyl butanoate	7
3,7-Dimethylocta-2,6-dienal	7	Ethyl 2-methylbutanoate	7
Neral	6	Ethyl 3-methylbutanoate	7
		Methyl hexanoate	1, 4, 7, 9, 10
Aromatic and cyclic aldehydes		C$_8$	
2-Furaldehyde	3, 7, 9, 11, 12, 14, 15	Benzyl acetate	3, 5, 9, 11, 15
5-Methyl-2-furaldehyde	9, 12, 14	Hexyl acetate	3, 4, 5, 6, 7, 9, 10, 15
Benzaldehyde	1, 3, 4, 5, 7, 9, 10, 11, 12, 14, 15	Hex-2-enyl acetate (*trans*)	7, 9
4-Methoxybenzaldehyde	11, 12	Hex-3-enyl acetate (*cis*)	3, 5, 7, 9, 15
Cinnamaldehyde (*trans*)	7	Pentyl propanoate	7
Cinnamaldehyde (*cis*)	7	2-Methylpropyl methylpropanoate	9, 15
Phenylacetaldehyde	1, 7, 11, 12	2-Methylpropyl butanoate	7
		Butyl butanoate	1, 5, 7, 9, 15
Aliphatic esters		Ethyl succinate	11
C$_3$		Ethyl hexanoate	1, 3, 4, 5, 7, 9, 10, 15
Ethyl formate	3, 9, 15		
C$_4$		C$_9$	
Ethyl acetate	3, 5, 6, 7, 9, 15	Heptyl acetate	5
C$_5$		Hexyl propanoate	1, 7
Diethyl carbonate	5	Hex-2-enyl propanoate (*cis*)	7
Propyl acetate	3, 5, 9, 15	Hex-3-enyl propanoate (*cis*)	7
Methyl butanoate	7	2-Methylbutyl methylpropanoate	4, 9, 10, 15
C$_6$		3-Methylbutyl methylpropanoate	7, 15
2-Methylpropyl acetate	3, 5, 15	Pentyl methylpropanoate	7, 15
Butyl acetate	5, 7, 9, 15	3-Methylbutyl butanoate	1, 4, 9, 10, 15
Ethyl methylpropanoate	3, 9, 15	Pentyl butanoate	7
		2-Methylpropyl 3-methylbutanoate	7

Compound	
Butyl 3-methylbutanoate	7
Butyl pentanoate	7, 9, 15
Propyl hexanoate	4, 9, 10, 15
Ethyl heptanoate	7
Methyl octanoate	1, 7
C_{10}	
2-Phenethyl acetate	3, 4, 9, 10, 15
Octyl acetate	9, 14, 15
Hexyl methylpropanoate	3, 7, 9, 15
Hexenyl methylpropanoate	7
Hexyl butanoate	1, 4, 7, 9, 10, 15
Hexyl but-2-enoate (*trans*)	7
Hex-2-enyl butanoate (*trans*)	7
Hex-3-enyl butanoate (*cis*)	3, 9, 15
3-Methylbutyl 3-methylbutanoate	7, 10, 25
2-Methylpropyl hexanoate	1, 3, 4, 9, 15
Butyl hexanoate	1, 3, 4, 5, 7, 9, 10, 15
Ethyl octanoate	7
Ethyl octenoate	
C_{11}	
Heptyl butanoate	7
Hexyl 2-methylbutanoate	1, 7, 8
Hex-3-enyl 2-methylbutanoate (*cis*)	7
Hexyl pentanoate	4, 9, 10, 15
Pentyl 4-methylpentanoate	3, 4, 7, 9, 10, 15
3-Methylbutyl hexanoate	4, 9, 10, 15
Pentyl hexanoate	9, 15
Butyl heptanoate	9, 15
Propyl octanoate	4, 7, 9, 15
Ethyl nonanoate	1, 7
Methyl decanoate	1, 7
C_{12}	
Octyl methylpropanoate	7
Octyl butanoate	7
Hexyl hexanoate	1, 3, 4, 7, 9, 10, 15
Hex-3-enyl hexanoate (*cis*)	9, 15
2-Methylpropyl octanoate	9, 15
Butyl octanoate	4, 7, 9, 10, 15
Ethyl decanoate	1, 3, 4, 5, 7, 9, 10, 15
Ethyl dec-4-enoate	4, 7, 9, 10, 15
C_{13}	
3-Methylbutyl octanoate	9, 15
Butyl nonanoate	9, 15
Propyl decanoate	9, 15
C_{14}	
3,7-Dimethylocta-2,6-dien-1-yl methylpropanoate	7
Octyl hexanoate	9, 15
Hexyl octanoate	7, 9, 15
3-Methylbutyl nonanoate	9, 15
2-Methylpropyl decanoate	15
Butyl decanoate	9, 15
2-Methylpropyl dec-4-enoate	7
Butyl dec-4-enoate	1, 7
Ethyl dodecanoate	9, 15
C_{15}	
Hexyl nonanoate	9, 15
3-Methylbutyl decanoate	9, 15
C_{16}	
Hexyl decanoate	3, 9, 15
2-Methylpropyl dodecanoate	15
Butyl dodecanoate	9, 15
Ethyl tetradecanoate	3, 9, 15
C_{17}	
3-Methylbutyl dodecanoate	9, 15
Ethyl pentadecanoate	3, 9, 15
Methyl hexadecanoate	1
Esters ($C_{34} - C_{50}$)	2
Miscellaneous esters	
Cyclic and aromatic	
Ethyl furoate	9, 15
Ethyl benzoate	1, 3, 9, 15

(continued)

62

TABLE II.IX (*continued*)

Volatile constituent	Reference[a]	Volatile constituent	Reference[a]
Ethyl anisate	5	5-Hydroxydodecanoic acid, lactone	7
Methyl salicylate	1, 3, 9, 14, 15	Dihydroactinidiolide	11, 12
Ethyl phenylacetate	5		
Methyl cinnamate	3, 4, 9, 10, 11	Acids	
Ethyl cinnamate	11, 12, 15	Formic acid	7
Methyl nicotinate (methyl pyridine-3-carboxylate)	1	Acetic acid	3, 4, 6, 7, 9, 10, 15
		2-Methylpropanoic acid	3, 6, 9, 15
Diethyl phthalate	7	Butanoic acid	6
		2-Methylbutanoic acid	6
		Furoic acid	6
Hydroxy-esters		Benzoic acid	6
Ethyl 2-hydroxypropanoate	3, 9, 15	Hexanoic acid	1, 3, 6, 9, 15
Ethyl 3-hydroxybutanoate	7	Nonanoic acid	1
Butyl 3-hydroxybutanoate	1	Oleanolic acid	2
		Fatty acids ($C_{16} - C_{36}$)	2
Terpene esters			
Bornyl acetate	6	Phenols	
Geranyl acetate	6	4-Methoxyphenol	9, 14, 15
Myrcenyl acetate	6	4-Vinylphenol	12
3,7-Dimethylocta-2,6-dien-1-yl ester	7	2-Methoxy-4-vinylphenol	12
		2-Methoxy-4-(prop-1-enyl)phenol	3, 6, 7, 9, 12, 15
		2-Methyl-5-(methylethyl)phenol	6
Lactones		5-Methyl-2-(methylethyl)phenol	6
4-Hydroxypentanoic acid, lactone	1	2,6-Di *t*-butyl-4-methylphenol	1, 5, 6
4-Hydroxyhexanoic acid, lactone	1, 5, 7, 9, 15		
5-Hydroxyhexanoic acid, lactone	7	Miscellaneous	
4-Hydroxyoctanoic acid, lactone	1, 3, 5, 7, 9, 15	Benzothiazole	1, 3, 7, 9, 11, 15
5-Hydroxyoctanoic acid, lactone	1	2-Hydroxymethylfuran	7, 12
4-Hydroxynonanoic acid, lactone	1, 3, 4, 5, 9, 10, 15	Chlorobenzene	7
4-Hydroxydecanoic acid, lactone	1, 3, 4, 5, 6, 7, 9, 10, 11, 15	1,2-Dichlorobenzene	7
5-Hydroxydecanoic acid, lactone	1, 7	Diphenylamine	11
4-Hydroxyundecanoic acid, lactone	7	Acetylpyrrole	12
4-Hydroxydodecanoic acid, lactone	6, 7		

a (1) Dirninger-Rigo, 1987; (2) Ismail *et al.*, 1977; (3) Ismail *et al.*, 1981b; (4) Ismail *et al.*, 1981a; (5) Forrey and Flath, 1974; (6) Vernin *et al.*, 1985; (7) Etiévant *et al.*, 1986; (8) Le Quéré *et al.*, 1987; (9) Ismail, 1977; (10) Ismail *et al.*, 1980a; (11) Moutounet *et al.*, 1975; (12) Moutounet, 1978; (13) Kereselidze

tract through an alumina column (linalool removal), more clues were thus obtained in favour of the first hypothesis.

The importance of these components, and of some others suspected to contribute to the overall aroma of the fruits, was estimated by the same authors, who calculated their threshold odour unit values in a synthetic acidic sweet solution (Table II.XI). This experiment showed that nonanal, *cis*-hex-3-en-1-ol, linalool, γ-decalactone, benzaldehyde, γ-octalactone and hexan-1-ol could be considered as contributors to the aroma of the plums studied. However, it must

TABLE II.X

Relative percentages of major volatile aroma constituents of some types of plum

Chemical class	Marjorie's Seedling (1)	Merton Gem (2)	NA 10 (3)	Victoria (4)	Mirabelle (5)	Mirabelle (6)	Mirabelle (7)
Hydrocarbons	–	–	–	–	1	4	–
Alcohols	62	48	48	48	8	6	> 3
Terpene alcohols	4	9	< 1	10	–	2	6
Aldehydes	21	13	21	17	6	13	87
Methyl esters	< 1	< 1	1	1	< 1	1	–
Ethyl esters	< 1	7	1	2	16	< 1	–
Butyl esters	< 1	< 1	< 1	< 1	32	43	–
3-Methylbutyl esters	3	4	2	5	< 1	< 1	–
Hexyl and hexenyl esters	4	7	10	7	43	< 1	–
Esters (total)	8	18	12	16	92	45	< 1
Lactones	1	1	< 1	1	< 1	4	–
Acids	< 1	5	< 1	1	–	< 1	–

(1) – (4) Ismail *et al.*, 1981a; (5) Etiévant *et al.*, 1986; (6) Dirninger-Rigo, 1987; (7) Vernin *et al.*, 1985.

TABLE II.XI

Threshold odour unit values of selected plum components (from Williams and Ismail, 1981)

Compound	Concentration in juice (ppm)	Thresholds value in sugar-acid base (ppm)	Odour unit value
Benzaldehyde	0.6	0.19	3.1
Linalool	0.7	0.10	7.0
γ-Octalactone	0.4	0.16	2.5
2-Phenylethanol	1.5	1.40	1.1
γ-Decalactone	0.9	0.17	5.2
Hexan-1-ol	5.0	4.00	1.25
cis-Hex-3-en-1-ol	2.3	0.22	10.4
Nonanal	0.7	0.056	12.5

be noted that only some of the substances selected by the sniffing method were investigated.

A further study was made by the same authors by asking a panel to score acidic sweet synthetic solutions for their similarity to the concept of fresh or cooked plums. These solutions contained some of the components cited above, separately or as mixtures. Different concentrations were tested above the calculated detection thresholds, and the relative concentrations of compounds tested in mixtures were determined from chromatographic analyses of juices. Three mixtures, i.e. methyl cinnamate plus γ-decalactone; benzaldehyde plus linalool; and ethyl nonanoate plus linalool, were scored once at a level approaching or exceeding the score given to a thawed plum juice introduced as standard. A mixture of these five substances was also scored higher than the thawed juice, but lower than a stewed plum juice. A sensory description obtained with this solution and with thawed and stewed juices showed that the mixture lacked greenness and sharpness. Addition of hexan-1-ol and *cis*-hex-3-en-1-ol, or acetic and hexanoic acids, to the solution marginally improved the similarity to fresh plums, but had a marked effect in the case of similarity to cooked plums.

Besides this interesting work on *Prunus domestica domestica* subspecies, Kereselidze and Mikeladze (1977) mentioned that esters, and α- plus β-ionone were probably responsible for the aroma of Bullace plums (*Prunus domestica insititia*). Other data about flavour impact of volatile constituents of plums relate to processed plums or comparisons between fresh and processed plums. They are therefore given in the next section.

3.2 Variation of Plum Volatile Constituents

3.2.1 Influence of cultivar

Using the same headspace technique, Ismail (1977) compared four different cultivars (Merton Gem, Victoria, Marjorie's Seedling, NA10) of *Prunus domestica domestica* L. He concluded that the volatile composition of these different plums only varies quantitatively and not qualitatively. An assessment of the odour of the different plums was made by a trained panel, and the cultivars were thus classified in the order indicated above from the scores given by the panel. When compared with the results of headspace quantification, and more particularly with components believed to have a sensory impact (see Section 3.2.2), the author observed that the scores given by the panel were positively correlated with the concentrations of benzaldehyde, methyl cinnamate, γ-decalactone and 2-phenylethanol, and negatively correlated with the amount of hexanal. From these results, increase in hexanal concentration was concluded to overpower the positive characteristics of the former components in plum aroma (Williams and Ismail, 1981).

Victoria and Golden Egg canned plums were also compared by Ismail *et al.* (1980c) from odour assessments given by 15 panelists. A significant difference between these varieties was thus established ($p < 0.01$), which was shown to be mainly related to two descriptors: almond-like and woody, respectively, scored higher for Victoria and for Golden Egg plums. These olfactory differences were believed to be due to higher relative concentrations of benzaldehyde in Victoria plums compared to Golden Egg plums (54.2 versus 45.1%), and to higher relative amounts of nonanal in Golden Egg plums when compared to Victoria plums (24 versus 14.5%).

Concerning *Prunus domestica insititia* plums, three publications give information about the characteristics of their aromas. As stated above (see also, Section 3.1.2), Kereselidze and Mikeladze (1977) indicated that esters and α- plus β-ionone could be responsible for the particular aroma of Bullace plums. As these components were not identified in other plum cultivars, they could consequently be specific for these plums.

Working on Mirabelle plums and using different extraction techniques, Etiévant *et al.* (1986) found that very few terpene alcohols could be identified when compared with results obtained using other plum cultivars (see Tables II.IX and II.X). This apparent qualitative, and quantitative, deficiency was also noticed by Dirninger-Rigo (1987). This author even suggested that their occurrence in the Mirabelle extracts she obtained could be partly due to chemical hydrolysis of the glucoside precursors during vacuum distillation, or to an enzymic cleavage of the same precursors during milling in ethanol. Both hypotheses can help to explain why Vernin *et al.* (1985) found high amounts of terpene alcohols (6%) in extracts obtained by atmospheric steam distillation of Mirabelle plums merely milled in water. These three studies thus indicate that Mirabelle plums contain, as other *domestica domestica* subspecies, terpene glycosides. Other precursors appeared nevertheless to be very little hydrolyzed in fruits, thus explaining the poor concentration of volatile terpenols in the headspace. Since Kereselidze and Mikeladze (1977) could not identify this type of volatile component in Bullace plums (see Table II.IX), glycosidase activity would be interesting to quantify in different *Prunus domestica* subspecies in order to try to characterize them.

3.2.2 Influence of maturity

An extensive study was performed by Dirninger-Rigo (1987), who quantified 56 constituents in half-ripe, ripe and overripe Mirabelle plums. During extraction, great care was taken in order to avoid any oxidation of the fruits or the juice, which could lead to subsequent artefact formation. To facilitate discussion, volatile components can be grouped into several chemical classes. Hydrocarbons became less abundant as the maturity of the fruits increased. Among them, naphthalene was said to be responsible for the decrease of the fresh aromatic character of half-ripe plums, since the intensity of its fresh odour estimated by sniffing the chromatographic effluent decreased with maturity. Except for nonanal, a more drastic decrease of aldehyde concentration was furthermore observed with maturity. As most of these compounds (C_6 saturated and unsaturated aldehydes) exhibit green and pungent odours at too high levels, their decrease can favour an increase in the quality of ripe plum aroma.

The observed development of the amount of esters with maturity was not so simple. It was shown to be positive (butyl butanoate and dec-4-enoate, hexyl hexanoate, 2-methylbutanoate and propanoate, and methyl nicotinate), negative (methyl hexanoate, ethyl octanoate, butyl propanoate and cis-hex-3-enyl butanoate) or positive until normal ripeness and then null (methyl octanoate and dec-4-enoate, butyl 3-hydroxybutanoate and 3-methylbutyl butanoate). Because of its low detection threshold, hexyl hexanoate was suggested by the author as an important contributor to the more fruity aroma of ripe and overripe plums.

Parallel to cis-hex-3-enal, the concentration of the corresponding alcohol was demonstrated to decrease with increasing maturity. Conversely, most of the terpene alcohols increased. The author noticed, however, that the amount of linalool first increased until normal maturity, and then decreased quickly after normal maturity. Nevertheless, an estimation of the intensity of the odour of this compound detected during the chromatographic analyses gave no significant difference between the extracts. The author concluded from this observation that linalool was probably more interesting in order to evaluate the degree of maturity of Mirabelle plums than for its sensory impact. Some other terpene alcohols, such as geraniol and α-terpineol, were more abundant than linalool in the different extracts, but were considered even less important in their aroma participation because of their lower detection thresholds.

The most drastic modification with maturity was observed for lactones, the concentration of which was 77 times higher in overripe than in half ripe fruits. Three of them, γ-octa-, nona- and decalactones, were easily detected by the sniffing technique, and the intensity of their odours was noticed to increase significantly with ripeness of samples. Because of its lower odour threshold, γ-nonalactone could be more important than the other lactones, and contribute to the more fruity aroma of overripe fruits. Finally, methyleugenol was demonstrated to be, as linalool, a good indicator of normal ripeness of Mirabelle plums. This compound was detected from effluent sniffing and its odour was described as peach-plum, reminiscent.

3.2.3 Influence of processing

Deep-freezing and thawing. A comparison of headspace composition of Mirabelle plums before and after deep-freezing was made recently by Etiévant et al. (1986). These authors, taking into account 65 gas-chromatographic peaks as variables and performing six replications of both experiments, observed a clear difference between the two types of samples using principal component analysis. Sixty-five percent of the variance was shown to be associated with this technological variation along the first principal component. The interpretation of the difference was made from vector loadings, and results were confirmed from raw data. They showed first that C_6 aldehydes and alcohols were much more abundant after thawing, thus indicating a favoured enzymic oxidation of linoleic and linolenic acids in the tissues damaged by ice crystals. Some of the esters arising from the alcohols (*trans*-hex-2-enyl acetate, propanoate and butanoate, *cis*-hex-3-enyl propanoate) were also shown to be more abundant after deep-freezing. Conversely, most of the other esters were less abundant in the thawed fruits, thus confirming previous observations made on peaches (Souty and Reich, 1978), strawberries (Ueda and Iwata, 1982) and Arctic brambles (Kallio, 1976). As esters produced from fatty acids and C_6 saturated and unsaturated alcohols were more abundant in thawed plums, whereas other identified esters were less abundant, it is possible that two different mechanisms were involved in their production. The first could be purely chemical and favoured by significant amounts of C_6 alcohols produced in the disrupted tissues, whilst the second could be enzymic, and either favour the hydrolysis of existing esters, or inhibit the production of these esters (esterase activities). As well as these major sources of variation, a decrease of aliphatic hydrocarbons and an increase of terpene hydrocarbons were also observed after deep-freezing and thawing.

A sensory evaluation of the fresh and of the thawed fruits was made using several descriptors of their odour quality and intensity. The intensity of Mirabelle plums was thus demonstrated to be unchanged by deep-freezing. However, the panelists scored differently ($p < 0.05$) the 'fresh Mirabelle', the 'oxidized' and the 'cooked' descriptors. When compared with the interpretation of the Principal Components Analysis (PCA) and the results of effluent sniffings, some possible explanations of these alterations could be given. First, the increase of oxidized character after deep-freezing could be a consequence of the important increase of hexan-1-ol, *trans*-hex-2-enal and hexanal. Secondly, the decrease of perception of fresh Mirabelle odour after thawing could be due to the decrease of ethyl, butyl and pentyl acetates, ethyl butanoate, hexanoate and octanoate, as well as of butyl octanoate and hexyl 2-methylbutanoate.

Preserving as prunes. Some varieties of plums (prunes d'Ente) are traditionally preserved as prunes (pruneau d'Agen) after dehydration down to $18-28\%$ water in a heated and ventilated tunnel. As seen from Table II.IX (Moutounet *et al.,* 1975; Moutounet, 1978), these fruits contain far less ketones, aliphatic aldehydes and alcohols, hydrocarbons and esters than unheated plums of other varieties. On one hand, this loss of the most volatile components could be expected, as it is the normal consequence of heating associated with high ventilation. On the other hand, high temperatures are known to activate chemical degradation of sugars, alone or by combination with amino acids. Some clues of such reactions in plums are given, since furfural (2-furaldehyde), methylfurfural, acetylfuran and furfuryl alcohol were identified in prunes (see Table II.IX). In order to confirm the occurrence of Maillard reactions, Moutounet and Jouret (1975) demonstrated a decrease of $75-80\%$ of the amino acid content after heating of the fruits. Some other compounds suspected to arise from thermal degradation of other precursors, the carotenoids, were also found, i.e. damascenone and dihydroactinidiolide. This degradation was confirmed by Moutounet (1976) when quantifying these pigments at different times of dehydration. This cleavage of carotenoids could be coupled to the enzymic oxidation of polyphenolics, as already observed for ethanol in wine (Wildenradt and Singleton, 1974), since addition of chlorogenic acid stimulated this degradation. Some phenols were also detected after heating. As far as we know, the main origin of these very aromatic substances is thermal degradation of the corresponding phenolic acids, ferulic (4-hydroxy-3-methoxycinnamic) and *p*-coumaric (4-hydroxycinnamic) acids, as demonstrated by Steinke and Paulson (1964) in grain. Another difference observed before and after heating was that the concentration of benzaldehyde increased significantly, probably by degradation of its well-known glycoside precursor, amygdalin. Finally, some terpene alcohols were identified in the prunes, but could not be detected in the plums by the same authors. Their origin in prunes can be explained by enzymic cleavage of terpene glycosides at the beginning of the heating, or by simple chemical hydrolysis of different precursors at the pH of the fruits, as studied by Williams *et al.* (1980, 1982a) on grapes.

Among all these volatile components of prunes, the importance of ketones in the cooked plum odour of the product was proved by the authors, since addition of hydroxylamine to the prune distillate made it disappear. The authors concluded that none of the identified ketones (anisaldehyde, benzaldehyde, furfural, furylmethyl ketone) could be directly responsible for this aroma, and that some trace components probably remained unidentified. It is, however, possible that these compounds did contribute to the aroma, with other compounds such as damascenone, later identified by Moutounet (1978) in prunes d'Agen, and demonstrated to exhibit a cooked plum aroma when sniffed from the chromatographic effluent with coeluted 2-phenylethanol and γ-octalactone (Williams and Ismail, 1981). We also noticed that a mixture of damascenone, ethyl cinnamate and γ-lactones, which are natural components of dried plums (pruneau d'Agen), produces an intense and pleasant odour, very much reminiscent of the aroma of cooked plums (Etiévant, unpublished results).

Preserving in cans. The importance of carbonyl compounds in heated plums was confirmed by Ismail *et al.* (1980c) when comparing canned and fresh plums (*Prunus domestica domestica* subspecies). After processing, the major constituents of the headspace were shown to be benzaldehyde, 2-furaldehyde, 2-furylmethyl ketone and nonanal, with pentane-2,3-dione, 3-hydroxybutanone, carvone and acetophenone as minor constituents. A sensory evaluation of Victoria and Golden Egg canned plums showed high scores for almond-like and woody descriptors. Comparing this result with results of GC effluent sniffing, the authors concluded that these particular aromas could be due to high levels of benzaldehyde and nonanal, respectively. The difference in aroma before and after processing could be a consequence of the absence in the heated plums of linalool, methyl cinnamate, γ-octalactone and γ-decalactone.

Preserving as jam. Ismail (1977) studied the aroma of jams made from Victoria plums. As for canned plums, benzaldehyde and nonanal were determined as major constituents of atmospheric distillation or headspace collection extracts. However, unlike similar extracts obtained from Victoria canned plums, and as for extracts obtained from fresh plums, esters formed numerically the largest class of volatile compounds. 2-Furaldehyde was also easily identified in the extracts obtained from headspace collection, and was one of the major components. These three substances probably arose from thermal treatment of the fruits, either by activation of enzymic degradation of precursors at the first stage of heating, or by chemical degradation of sugars and amino acids. These compounds could be important contributors to the overall aroma of the jam, as well as damascenone, the possible origin of which has been described in a previous section (3.3.2). The authors mentioned a difference of aroma between canned plums and jam, which could be due to the difference of ripeness of fruits used (unripe for canning and ripe or overripe for cooking in sugar), or to the difference of time or intensity of heating applied to the fruits in the two processes. Since heating is more intense when canning, inhibition of enzymes responsible for production of substances such as esters could have been favoured.

Fermenting. A few studies have concerned fermented plum brandies (fermented juice distillates), and these have been reviewed by Ismail *et al.* (1980b). As most of the volatile compounds identified in the distillates were fermentation by-products, they have been omitted from Table II.IX.

As expected, fusel alcohols, fatty acids and ethyl esters of fatty acids were found in the distillates. Of these, methanol was the major constituent, with concentrations varying from 2.2 to 3.2 g/l (Filajdic and Djukovic, 1973). Esters are qualitatively and quantitatively abundant in the distillates, as they were found to be in fresh plums. However, esters in distillates are mainly composed of ethyl esters of fatty acids, acetates or propanoates of fusel alcohols, and ethyl or methyl esters of organic acids. These particular esters have sometimes been identified in fresh plum juice or fresh plum headspace, but only as minor constituents. Ismail (1977) and Ismail *et al.* (1980b) noticed that the major esters identified in fermented plum distillates varied very much from one publication to another. They concluded from this observation that highly volatile esters were reported as major when a fractional distillation process was applied. Otherwise, less volatile esters derived from propan-1-ol, 2-methylpropan-1-ol or 3-methylbutan-1-ol with hexanoic, octanoic, decanoic or dodecanoic acids could be more abundant than the former, as clearly demonstrated by the analyses of fractions obtained during distillation of spirits (Postel and Adam, 1985). Beside these esters, ethyl lactate, arising from natural esterification of lactic acid in ethanolic solutions, was found by all authors in plum brandies, but at different levels.

Ismail (1977) noticed that the distillates he studied exhibited a strong and characteristic aroma of plums. He concluded that this odour was due to a mixture of linalool, benzaldehyde, methyl cinnamate and γ-decalactone, as in fresh plums (Section 3.1.1). Compared with the results of GC sniffing made with fresh plums, he noticed, however, that the odour of fresh plum detected during the elution of 2-phenylethanol could not be detected for distillate extracts, and was due to the absence of γ-octalactone.

Some other interesting conclusions concerning the flavour of plum distillates can be drawn from the studies of Crowell and Guymon (1973) and Ismail *et al.* (1980b).

First, methyleugenol (3,4-dimethoxy-1-(prop-2-enyl)benzene) identified as a fresh plum constituent could not be detected in distillates. Secondly, eugenol itself (2-methoxy-4-(prop-2-enyl)phenol) was found in larger quantities in fermented plum distillates than in the unfermented plums. These two components are known to arise from the enzymic degradation of coniferyl alcohol (4-(3-hydroxyprop-1-enyl)-2-methoxyphenol) (Schütte, 1976), but no evidence has so far been given for demethylation of methyleugenol into eugenol by microorganisms or enzymes. Moreover, the amounts of eugenol produced in the distillates were more than the average amounts of methyleugenol found in fresh plum juice. Eugenol is therefore believed to arise from the corresponding phenolic alcohol, or from ferulic acid (4-hydroxy-3-methoxycinnamic acid), during the fermentation (Ismail, 1977).

Thirdly, furfural (2-furaldehyde) was found in large quantities in plum brandies. As for jam production or plum canning, this aromatic substance arises from the heating process involved in distillation of the fermented plums.

3.3 Conclusions

The different investigations made on plum aroma have shown that the aromatic characteristics of different plum cultivars are related to qualitative and quantitative differences in volatile composition. Esters form the most numerous chemical substances identified in plum extracts. Depending on the cultivar, esters, alcohols or aldehydes can be the major volatile constituents of fresh plums. The relative percentage of terpene alcohols has also been demonstrated to vary with the cultivar, and this variation seemed to be associated with different enzymic activities.

The different technological processes applied to fresh plums have been shown to induce modifications of their aroma, but without masking it. Thermal processes (canning, jam making, dehydrating and distilling) induce degradation of precursors. These precursors are mainly glycosides (amygdalin, terpene glucosides), sugars and amino acids or carotenoids. Other processes (milling, deep-freezing and thawing) favour formation of large amounts of volatiles arising from enzymic activities (peroxidation, hydrolysis).

As is often the case, only differences induced by technological processes can be explained by chemical analysis, since few characteristics of the overall aroma are consequently altered. When trying to explain the quality of the overall aroma of plums, more studies should be performed correlating sensory and chemical analyses. Following the example given by Williams and Ismail (1981), more efforts should be directed towards determining the sensory importance of esters and some ketones, which have been systematically detected as potential aromatic contributors by the GC sniffing technique by most authors, but the contribution of which still remains unclear.

4. PEACH

All the specialists agree with a Chinese origin for peaches, and all types of known peaches grew wild or in a semi-cultivated state in central China (Hugard and Monet, 1985). The peach tree was planted in Italy during the Roman era, and it was introduced into France at nearly the same time. Subsequently, during several centuries, according to the domestic character of its cultivation, selection was operated by local growers or amateurs, so that a considerable number of cultivars was produced, and more than 1000 have been listed (Hugard and Monet, 1985). In the United States the peach tree appeared later, during the 19th Century; the variety Chinese Cling cultivated near Shanghai was introduced by an English botanist, and all the cultivars obtained later originate from this (Hugard and Monet, 1985).

According to its characteristics, different varietal groups of peach can be distinguished: peaches themselves, pavies, nectarines and brugnons:
— peaches possess a downy skin, a white or yellow soft pulp, and a free stone (freestone peach),
— cling peaches possess a downy skin, a white or yellow firm pulp, and a clingstone,
— nectarines possess a smooth skin, a soft pulp, and the stone is free,
— brugnons have a smooth skin, a firm pulp, and a cling stone.

However, it must be appreciated that the term peach is used to specify all the fruits belonging to these groups.

Production of peaches extends to several countries, but remains concentrated in Mediterranean-like areas. It reaches about 7 Mt per year, two-thirds of which is produced in Europe and North America. The main producers are Italy, the United States, Greece, Spain and France (Anonymous, 1987; Vidaud et al., 1987).

Because of market requirements, the creation and selection of peach cultivars is very important. The orchards of short life provide an important varietal renewal at each replantation, so that the range is rapidly transformed, and cultivars available at one time disappear from the markets to be replaced by others (Hugard and Monet, 1985; Vidaud et al., 1987). This situation has probably played an important rôle in the limited interest given to aroma differences between cultivars. In these circumstances, the selection criteria used have often been oriented according to commercial requirements rather than according to flavour quality (Hugard and Monet, 1985).

It appears that initially the aromatic fraction of peach was studied in order to improve natural formulations of peach aroma (Broderick, 1975). Interest was also focused on the behaviour of aroma compounds during fruit processing (Paillard, 1976; Mori, 1977; Souty and Reich, 1978) or during post-harvest

TABLE II.XII

Volatile constituents identified in peach extracts

Volatile constituent	Reference[a]	Volatile constituent	Reference[a]
Hydrocarbons		trans-Hex-2-en-1-ol	9, 10, 11, 17, 20, 21, 25, 26, 31
Ethylene (ethene)	4, 10, 11, 26	cis-Hex-3-en-1-ol	21, 25, 26
Non-1-ene	30	Octan-1-ol	22, 26
Dec-1-ene	30	Octan-2-ol	22
Hexadecane	27	Nonan-1-ol	22
Heptadecane	27	Decan-1-ol	22
Octadecane	27	Benzyl alcohol	5, 9, 10, 11, 12, 13, 14, 17, 21, 22, 26, 27, 29
Nonadecane	27	2-Phenylethanol	8, 10, 21, 22, 26
Heneicosane	27	Linalool	8, 10, 11, 20, 21, 22, 23, 25, 26, 27, 30, 31
Tricosane	30		
Limonene	12, 13, 26, 30	Ho-trienol	23, 25, 31
Cadinene	1, 10	4-Terpineol	21, 26
(E, Z)-Undeca-1,3,5-triene	32	α-Terpineol	8, 10, 21, 22, 23, 25, 26, 30, 31
(E, Z, Z)-Undeca-1,3,5,8-tetraene	32	Citronellol	22, 23, 25
Vinylbenzene	21, 26	Nerol	22, 23, 25
1,1,6-Trimethyl-1,2-dihydro-naphthalene	15, 26	Geraniol	21, 22, 23, 25, 26, 30
1,1,6-Trimethyl-1,2,3,4-tetra-hydronaphthalene	15, 26	Linalool oxides	31
		Carbonyl compounds	
Alcohols		Acetaldehyde	1, 3, 4, 8, 9, 10, 11, 14, 17, 21, 26
Methanol	1, 10, 11, 26	Butanal	21, 26
Ethanol	4, 9, 10, 11, 14, 17, 22, 24, 26, 27	3-Methylbutanal	12, 13, 21, 26
Propan-1-ol	22	Furfural (2-furaldehyde)	1, 8, 10, 11, 21, 26, 27, 30
Butan-1-ol	7, 22, 24, 26	Hexanal	21, 24, 25, 26
2-Methylpropan-1-ol	22	trans-Hex-2-enal	24, 25, 26
3-Methylbutan-1-ol	8, 10, 21, 22, 26, 27	Heptanal	21, 26
Pentan-1-ol	7, 21, 22, 26	Nonanal	26, 30
Hexan-1-ol	8, 9, 10, 11, 14, 17, 21, 22, 23, 25, 26, 27, 31	Benzaldehyde	5, 9, 10, 11, 12, 13, 14, 17, 21, 23, 25, 26, 27, 30, 31

73

Compound	References
2-Phenylacetaldehyde	21, 26
Cinnamaldehyde	27
Carvomenthenal	30
Heptan-2-one	8, 10, 21, 26
Non-3-en-2-one	30
Undecan-2-one	21, 26
α-Ionone	21, 26
β-Ionone	21, 26, 27

Acids

Compound	References
Formic	1, 10, 11, 21, 26
Acetic	1, 3, 8, 9, 10, 11, 17, 21, 26, 27
Butanoic	8, 10, 21, 26
3-Methylbutanoic	8, 9, 10, 17, 21, 26
Pentanoic	1, 11, 26
Hexanoic	8, 9, 10, 11, 14, 17, 21, 26
Hexenoic	21, 26
Octanoic	1, 8, 11, 21, 26
Decanoic	8, 21, 26

Esters

Compound	References
Methyl formate	22
Methyl acetate	9, 10, 11, 14, 17, 24, 26
Methyl 3-methylbutanoate	12, 13, 22, 26
Methyl salicylate	8, 10, 12, 13, 21, 26
Ethyl formate	22
Ethyl acetate	4, 7, 9, 10, 11, 14, 17, 20, 22, 24, 26, 27, 30
Ethyl propanoate	22
Ethyl butanoate	8, 10, 21, 22, 24, 26
Ethyl pentanoate	22
Ethyl 3-methylbutanoate	8, 10, 21, 22, 26
Ethyl hexanoate	8, 10, 21, 22, 26
Ethyl heptanoate	22
Ethyl octanoate	21, 22, 26
Ethyl nonanoate	22
Ethyl decanoate	22
Ethyl dodecanoate	22
Ethyl benzoate	9, 10, 11, 12, 13, 14, 17, 21, 22, 26
Ethyl phenylacetate	22
Ethyl cinnamate	21, 26
Propyl acetate	22
Butyl acetate	21, 22, 24, 26
Pentyl acetate	10, 11, 22, 26, 30
3-Methylbutyl acetate	3, 8, 10, 12, 13, 21, 22, 26, 27, 30
Hexyl formate	9, 10, 11, 14, 17, 22, 26, 27
Hexyl acetate	4, 8, 9, 10, 11, 12, 13, 14, 17, 20, 21, 22, 26, 27, 30
trans-Hex-2-enyl acetate	9, 10, 11, 14, 17, 21, 26, 30
cis-Hex-3-enyl acetate	30
Hexyl benzoate	9, 10, 11, 12, 13, 14, 17, 26
Octyl acetate	22
Benzyl acetate	9, 10, 11, 12, 13, 14, 17, 21, 26
Linalyl formate	2, 10, 21, 26
Linalyl acetate	2, 10, 21, 26
Linalyl pentanoate	2, 10, 21, 26
Linalyl octanoate	2, 10, 21, 26

Lactones

Compound	References
γ-Pentalactone	12, 13, 17, 18, 20, 23, 25, 26, 28
γ-Hexalactone	5, 9, 10, 11, 14, 17, 19, 20, 21, 26, 27, 28, 30, 31, 33
γ-Heptalactone	9, 10, 11, 14, 17, 19, 21, 26, 27, 28, 30, 31
γ-Octalactone	5, 9, 10, 11, 12, 13, 14, 17, 18, 19, 20, 21, 25, 26, 27, 28, 30, 31, 33
δ-Octalactone	28, 31
γ-Nonalactone	9, 10, 11, 14, 19, 20, 22, 23, 25, 26, 27, 28, 30, 31
γ-Decalactone	5, 8, 9, 10, 11, 12, 13, 14, 17, 18, 19, 20, 21, 22, 23, 25, 26, 27, 28, 29, 31, 33

(continued)

TABLE II.XII (*continued*)

Volatile constituent	Reference[a]	Volatile constituent	Reference[a]
δ-Decalactone	5, 8, 9, 10, 11, 12, 13, 14, 17, 20, 21, 23, 25, 26, 27, 28, 31	δ-Dodecalactone	10, 12, 13, 17, 26, 27, 28
		α-Pyrone	9, 10, 11, 14, 17, 20, 23, 25, 26
γ-Undecalactone	22, 26, 28	6-Pentyl-α-pyrone	16, 17, 26
δ-Undecalactone	31	Coumarin	22
γ-Dodecalactone	10, 12, 13, 17, 19, 21, 26, 27, 28, 30, 31, 33		

[a] (1) and (2) Power and Chesnut, 1921, 1922; (3) Daghetta *et al.*, 1956; (4) Lim, 1963; (5) Jennings and Sevenants, 1964; (6) Lim and Romani, 1964; (7) Spanyar *et al.*, 1964, 1965; (8) Broderick, 1966; (9) Sevenants and Jennings, 1966; (10) Antoine Chiris, 1967 (review); (11) Nursten and Williams, 1967 (review); (12) Do, 1968; (13) Do *et al.*, 1969; (14) Bolin and Salunkhe, 1971; (15) Kemp *et al.*, 1971; (16) Sevanants and Jennings, 1971; (17) Jennings, 1972 (review); (18) Bayonove, 1973; (19) Molina *et al.*, 1973; (20) Bayonove, 1974; (21) Broderick, 1975b (review); (22) Soboleva and Yazvinskaya, 1975; (23) Bayonove, 1976; (24) Paillard, 1976; (25) Bayonove, 1977; (26) Maarse, 1977, 1984; (27) Mori, 1977; (28) Fresneda *et al.*, 1978; (29) Soboleva and Epifanov, 1978; (30) Spencer *et al.*, 1978; (31) Souty and Reich, 1978; (32) Berger *et al.*, 1985; (33) Mori, 1986.

storage and ripening (Daghetta *et al.,* 1956; Lim and Romani, 1964; Raulston, 1969; Bayonove, 1973, 1974, 1976, 1977; Watada *et al.,* 1979; Mori, 1986).

4.1 *The Volatile Constituents of Peach (Table II.XII)*

About a dozen volatile compounds, including linalyl esters, were identified using classical methods of chemical analysis from an aromatic extract obtained by distillation of peaches of the cultivar Georgia Belle (Power and Chesnut, 1921). This extract also contained a saturated hydrocarbon and cadinene, or a similar compound. According to the authors, the peach essential oil obtained was very unstable and was quickly transformed into a black viscid mass, completely losing its original fragrance by oxidation.

Later, most of the interest was focused on substances produced by the fruit during post-harvest ripening (Daghetta *et al.,* 1956; Lim, 1963; Lim and Romani, 1964). However, few compounds, and only the most volatile, were identified. The same was true for the chromatographic comparisons of several fruit families including peach (Spanyar *et al.,* 1964).

The study of peach volatile compounds was then undertaken on an extract obtained by distillation of seven tons of fruit of the cultivar Red Globe (Jennings and Sevenants, 1964; Sevenants and Jennings, 1966). Using gas-liquid chromatography and infrared spectroscopy, the authors identified 24 compounds. Some of them are of little interest, being present in several other fruits; on the other hand, lactones, including α-pyrone and 6-pentyl-α-pyrone, appear characteristic of peach (Sevenants and Jennings, 1971).

Attempts at reproduction of peach aromatic essence using previously identified compounds have failed (Power and Chesnut, 1921; Sevenants and Jennings, 1966). In the more favourable cases, the aroma remained discrete, analogous to that obtained with clingstone peaches, but it never reached the level of intensity found in freestone peaches. Peach aroma is not attributable to one or several compounds, but it is considered as the integrated response of the olfactive organ to a series of contributing flavour compounds (Sevenants and Jennings, 1966).

Advances were made with the work of Broderick (1966), which, after identification of the same compounds, gave particular importance to lactones, and more particularly to γ-decalactone, for the formation of peach aroma. This conclusion was based on results of a fortuitous discovery which led to the production of an essential oil with a very pronounced peach odour from castor oil containing crude γ-undecalactone. This substance, called 'peach aldehyde', is currently used in peach formulations. On the other hand, this compound is rarely cited among peach natural compounds (Soboleva-Yazvinskaya, 1975; Maarse, 1977; Souty and Reich, 1978). In the fruit its presence is uncertain and it is always in trace quantities. A combination of lactones identified in peaches has been used to replace γ-undecalactone in peach formulations (Broderick, 1975b).

In peach, lactones are well represented, with all the γ-lactones from C_5 to C_{12}, some δ-lactones and unsaturated lactones (Do *et al.,* 1969; Bayonove,

1974). The only quantitative work (Do *et al.*, 1969) indicated that lactones represent 26.5% of the estimated constituents, and that γ-decalactone (95 ppm) was the second most abundant component after benzaldehyde (115 ppm). From chromatographic results (Jennings and Sevenants, 1964; Broderick, 1966; Sevenants and Jennings, 1966; Do *et al.*, 1969; Bayonove, 1973; Mori, 1977, 1986; Souty and Reich, 1978) it can be seen that, for several cultivars, γ-decalactone is, or is among, the major peaks, with benzaldehyde, δ-decalactone, α-pyrone and sometimes γ-hexalactone. When chromatographic results include substances with low boiling point, this lactone may be supplanted by other compounds (Antoine Chiris, 1967). This is the case, for example, for the cultivar Halford (Spencer *et al.*, 1978), where the major peak is linalool, followed by *cis*- and *trans*-hex-3-enyl acetate, an unknown terpene, nonan-1-ol and benzaldehyde. In some peach extracts (Broderick, 1966), esters, such as ethyl hexanoate and 3-methylbutanoate, are more important, followed by γ-decalactone, benzaldehyde and benzyl alcohol.

Lactones certainly play an important rôle in peach aroma, but these compounds act in association with other compounds. Studying the odour of components separated by GLC, some volatile fractions corresponding to peach aroma and containing benzyl alcohol and γ-decalactone were found (Bayonove, 1973; Soboleva and Yazvinskava, 1975). Statistical relations between sensory characteristics and concentrations of volatile compounds established for 10 cultivars of clingstone peaches show that four peaks, which varied considerably among cultivars, contributed to peach aroma. These four peaks were those of linalool, an unknown monoterpene, α-terpineol and γ-decalactone. Differences were due more to the relative concentrations of esters and monoterpenes than to relative concentrations of γ-lactone (Spencer *et al.*, 1978). The regression equations obtained between organoleptic appreciations and γ-decalactone content for different peach cultivars showed a narrow parallelism between flavour scores and γ-decalactone content (Soboleva and Epifanov, 1978). So, lactones have been recognized by several authors as key components of peach aroma, and they are used as reference compounds for aroma studies of fresh fruit, canned fruit or ripening fruit (Do *et al.*, 1969; Molina *et al.*, 1973; Bayonove, 1973, 1974; Mori, 1977, 1986).

The lactone odorous notes are generally described as being of 'tropical fruit', in some cases 'coconut', 'liquor-like' or 'peach-like'. Intense and characteristic odours are attributed to these compounds. Differences between γ- and δ-lactones have been observed, so γ-decalactone develops a peach odour, whereas the δ-lactone has rather a coconut odour. The thresholds of these substances are low (Maga, 1976). In Gleason Early Elberta peach, γ-decalactone was found in concentration equal to about 1000 times its threshold; this value is 700 for γ-octalactone, 250 for δ-dodecalactone, and 43 for δ-decalactone (Do *et al.*, 1969; Maga, 1976). 6-Pentyl-α-pyrone has a strong odour typical of coconut (Nobuhara, 1969; Sevenants and Jennings, 1971; Bayonove, 1974). This compound, identified as the major volatile component present in cultures of *Trichoderma viride* developing the same odour (Collins and Halim, 1972;

Latrasse *et al.*, 1985), is considered as a contributor to the whole peach aroma.

The biogenesis of peach lactones has not been particularly studied, but the hypothesis formulated by Tang and Jennings (1968) concerning apricot lactones could be relevant.

4.2 Variation of Peach Volatile Constituents

4.2.1 Influence of cultivar

The aroma content of only 20 cultivars has been studied, out of the approximately 360 cultivars used in horticulture (Delassale, 1985).

Most of the investigations have been performed using very different analytical conditions, and few quantitative data are available. Under these circumstances, comparison of results between cultivars is very difficult. However, a relative constancy in the content of lactones and α-pyrone for different cultivars of canned peaches has been found, and this fact is consistent with the good retention of aroma found in peach products (Souty and Reich, 1978). Differences in the intensity of perception are probably attributable to quantitative differences, as observed in artificially ripened peaches, which are less aromatic than tree ripe fruits (Do *et al.*, 1969).

More important variations between cultivars have been observed concerning other compounds such as esters or terpenols, and in some cases several compounds are lacking. This is the case with linalool, which is missing in the Cardinal cultivar. Similarly, several such compounds are present in freestone peaches but absent in clingstone peaches, the latter recognized as less aromatic than freestone peaches (Bayonove, 1974; Soboleva and Yavinskaya, 1975; Spencer *et al.*, 1978).

Varietal characteristics, if present, must be looked for among aroma components which are the most variable between different cultivars. This is the case for some components such as benzaldehyde, some terpenols, and other minor compounds.

Benzaldehyde is interesting due to its peach stone odour. It is relatively abundant in fresh fruit (Bayonove, 1974, 1976, 1977) as well as in canned fruit (Souty and Reich, 1978). The origin of this compound in the fruit is not clearly known, but it can originate from enzymic hydrolysis of amygdalin present in the stone or from phenylalanine (Souty and Reich, 1978), or even from oxidation of benzyl alcohol (Blaise, 1986).

Linalool, sometimes found in abundant quantities with other terpene alcohols, probably plays an important part in olfactory quality (Bayonove, 1974, 1976, 1977; Souty and Reich, 1978; Spencer *et al.*, 1978). In some cultivars, such as Fayette and Spring Crest, the linalool content is low (1% of volatiles), or even absent as in Cardinal; in other cultivars, the content is more important, such as in Red Top (3%) or Early Spring (8%). α-Terpineol, nerol, geraniol and ho-trienol are generally present in trace quantities (one-tenth that of linalool). However, the relative importance of these compounds, and more particularly of linalool, is dependent on the growth phase reached when the fruit is gathered. Linalool represents 60% of the volatiles present in Early Sungrand 10 days before full ripening (Bayonove, 1976). These compounds have been found in fruits of the same botanical family such as apricot (Tang and Jennings, 1967; Rodriguez *et al.,* 1980; Chairote *et al.,* 1981) or in very different fruits such as muscat grapes (Bayonove and Cordonnier, 1971; Williams et al., 1981; Marais, 1983; Gunata *et al.,* 1985).

Some carotenoid derivatives, such as 1,2-dihydro-1,1,6-trimethylnaphthalene and 1,2,3,4-tetrahydro-1,1,6-trimethylnaphthalene, can also contribute to olfactory nuances. The second compound, also found in peach leaves, contributes to strawberry (Kemp *et al.,* 1971) and aged wine (Simpson, 1978) aroma. β-Ionone, known as possessing a very low aroma threshold (Etiévant *et al.,* 1983) has also been identified in some freestone and clingstone peach cultivars. In some cases its concentration is as important as that of linalool (Mori, 1977). Since peach is relatively rich in carotenoids, and more particularly in β-carotene, the presence of such compounds is not surprising (Goodwin and Goad, 1970).

Two linear undecenes recently identified by Berger *et al.* (1985) in peach and in several fruits and vegetables must be mentioned, namely (*E, Z*)-undeca-1,3,5-triene and (*E, Z, Z*)-undeca-1,3,5,8-tetraene, which show, respectively, a balsamic, pleasant aroma with strong fruit undernotes and an exotic fruity odour. Due to their very low odour detection thresholds, these compounds may contribute to the aroma of the fruit at very low concentrations near to the analytical threshold.

4.2.2 Influence of maturity

Peaches are generally gathered before fully ripe, and consumed or used only after a post-harvest ripening period. This procedure, imposed by commercial considerations, generally gives fruit with a very low level of taste and aroma. Several studies are related to this mode of ripening, and some of them take into account the volatile emission during ripening (Daghetta *et al.,* 1956; Lim and Romani, 1964; Raulston, 1969; Watada *et al.,* 1979). In others, numerous compounds are isolated after extraction (Do, 1968; Do *et al.,* 1969) or only some compounds, such as lactones, terpenols and benzaldehyde are studied (Bayonove, 1973, 1974, 1976, 1977; Mori 1986). The organoleptic quality of peaches is dependent on two essential factors: the maturity stage at which the fruits are picked, and the post-harvest storage and the artificial ripening conditions used. Harvest maturity and temperature management have been identified as the most important factors (Shewfelt *et al.,* 1987). The best flavour and the best volatile reducing substances index were obtained for a post-harvest ripening temperature of 25°C (Chung and Luh, 1971). Beyond this value and for lower temperatures, less good results are obtained, and at 0°C wrong taste can develop; on the other hand, the peach flesh can become 'cottony' and lose its juiciness. According to Pech and Fallot (1972), an important rôle would be played by pectic enzymes in the appearance of this structure.

During post-harvest ripening, volatile substances are emitted by fruits, and the first volatile produced is ethylene (Lim and Romani, 1964), then CO_2, acetaldehyde and ethyl acetate appear (Daghetta *et al.*, 1956). In general, this emission is less at 20°C than at 3°C, though the ripening rate of the picked fruit is three times faster at 20°C (Daghetta *et al.*, 1956). The production rate and the number of volatiles detectable are dependent on the maturity at harvest, and probably on the cultivar. The concentration of volatile substances emitted and aromatic quality of the fruit are in good correlation (Lim and Romani, 1964; Raulston, 1969).

When the ripening occurs in controlled atmosphere, peaches are more fruity (Watada *et al.*, 1979). Compared to tree-ripened fruits, fruits picked before ripening are quantitatively and qualitatively lacking in volatile compounds. For the peach Gleason Early Elberta, the lactone content and the benzaldehyde content are respectively 40- and 52-fold less at the hard mature stage than at the tree ripe stage. Moreover, γ-decalactone is only present in very small amounts, and γ- and δ-dodecalactone are lacking in artificially ripened peaches (Do *et al.*, 1969). It is well established for several cultivars that post-harvest-ripened fruits produce volatiles (particularly γ-decalactone and α-pyrone) in relatively important quantities only when picked at an advanced growth phase. The production rate and the quantity of γ-decalactone produced are very important when harvest occurs near the tree ripening stage (Bayonove, 1974, 1976, 1977) (see Fig. II.8).

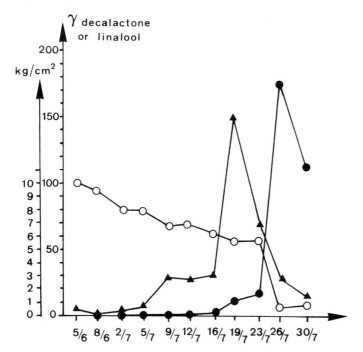

Fig. II.8. Variation in γ-decalactone (●———●) and linalool (▲———▲) content (expressed as peak area) and firmness (○———○) (expressed in kg/cm² for Early Sungrand nectarine, gathered during tree ripening.

It appears that an induction phase takes place, and that this is necessary before picking, in order to allow the further production of γ-decalactone. According to the cultivar, a decrease in fruit firmness is associated with an important increase in γ-decalactone (Bayonove, 1974, 1976, 1977) (see Fig. II.9). A different situation may be found for other compounds, such as linalool or benzaldehyde. For fruits harvested at different times before tree ripening, differences in volatiles measured immediately after picking and after 10 days of artificial ripening were found. The differences are important, and positive for γ-decalactone and α-pyrone; for other compounds, such as linalool and benzaldehyde, this difference may be important and negative (see Fig. II.10). It appears that, during ripening of the picked fruit, the disappearance of some compounds reused in metabolic pathways was observed. Losses of linalool were found in several cultivars, and these losses were more important when harvest occurred later. They are particularly high for Early Sungrand nectarine. However, for the Merril Sundance cultivar, a slight increase in linalool, as well as lactones, was observed under the same conditions. On the other hand,

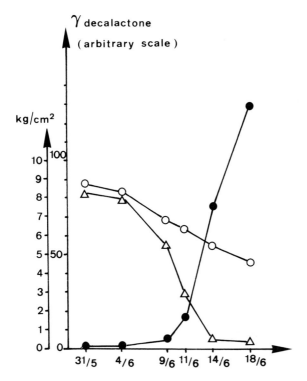

Fig. II.9. Evidence of the induction phase for γ-decalactone production in peach. Variation in γ-decalactone content ($\bullet$———$\bullet$) (arbitrary scale) after 10 days of artificial ripening and of firmness (kg/cm^2) at gathering ($\circ$———$\circ$) and after 10 days of artificial ripening ($\triangle$———$\triangle$) for different gathering times.

volatile components determined at different times for tree-ripened fruit showed a constant increase in linalool and, to a lesser extent of other terpenols, which then become (as lactones) more and more abundant, near the ripening. However, benzaldehyde decreases strongly at the same time, and this compound disappears almost completely in fully ripened fruit. The benzaldehyde content of some cultivars, such as Spring Crest, is very low. In the tree-ripened fruit, a great increase in the γ-decalactone production rate was observed when fruit firmness decreased. If the fruit stays on the tree after full ripening, a decrease of γ-decalactone and linalool content was noticed (Bayonove, 1976, 1977). A similar phenomenon was observed during over-ripening of muscat grapes (Bayonove and Cordonnier, 1970; Terrier, 1972).

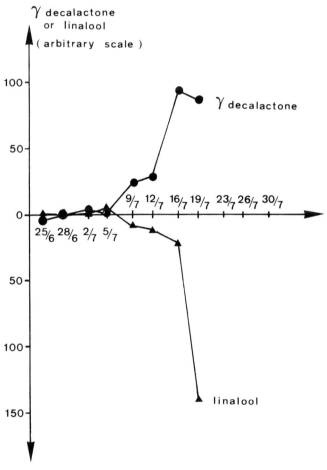

Fig. II.10. Variation in γ-decalactone (●——●) and linalool (▲——▲) content (arbitrary scale) after 10 days of artificial ripening for different gathering times.

82

TABLE II.XIII

Volatile constituents identified in cherry extracts

Volatile constituent	Reference[a]	Volatile constituent	Reference[a]
Hydrocarbons		Propanal	6, 8
Methylbuta-1,3-diene	7	Methylpropanal	6, 8
Myrcene	7, 8	Hexanal	9, 10
		trans-Hex-2-enal	9, 10
Alcohols		cis-Hept-4-enal	10
Methanol	1, 6, 8	Benzaldehyde	1, 3, 4, 5, 7, 8, 9, 10
Ethanol	1, 3, 5, 6, 8	Octanal	9
Propan-1-ol	5, 7	Phenylacetaldehyde	9, 10
2-Methylpropan-1-ol	5, 7, 8	trans-Oct-2-enal	9, 10
2-Methylpropane-1,2-diol	5	Nonanal	9, 10
Propan-1,2,3-triol	5	Nona-trans-2,cis-6-dienal	10
Butan-1-ol	3, 5, 7, 8	trans-Non-2-enal	10
3-Methylbutan-1-ol	7, 8	Decanal	10
Pentan-1-ol	5, 7, 8		
Hexan-1-ol	7	Ketones	
cis-Hex-3-en-1-ol	9	Propanone	6, 8
trans-Hex-2-en-1-ol	9, 10	Heptan-2-one	9
Heptan-2-ol	9	Octan-2-one	9
Octan-1-ol	3, 5, 8	Carvone	9
Geraniol	1, 10	4-Methylacetophenone	10
Linalool	4, 9, 10	α-Ionone	9
Benzyl alcohol	7, 8, 9	β-Ionone	10
2-Phenylethanol	9, 10		
cis-Ocimenol	10	Esters	
trans-Ocimenol	10	Ethyl formate	5, 8
α-Terpineol	9, 10	Methyl acetate	6, 8
Terpin-1-en-4-ol	9	Ethyl acetate	3, 5, 6, 8, 9, 10
p-Menth-1-en-9-ol	10	Ethyl acrylate (ethyl propenoate)	9
		Ethyl butanoate	5, 8, 9
Aldehydes		Butyl acetate	5, 8
Acetaldehyde	5, 6, 8, 9	Ethyl 3-methylbutanoate	5

Ethyl 2-methylbutanoate	10
Pentyl acetate	5, 8
Ethyl hexanoate	8, 9
Ethyl octanoate	7
Ethyl decanoate	5, 7, 8
Methyl benzoate	7, 8, 9
Ethyl benzoate	5, 7, 8, 9
Benzyl acetate	9
Propyl benzoate	7, 8
2-Methylpropyl benzoate	7, 8
3-Methylbutyl benzoate	7, 8
Methyl salicylate	9
Acids	
Formic	2, 5, 8
Acetic	2, 4, 5, 8, 9
Butanoic	5, 8
3-Methylbutanoic	4
Hexanoic	5, 8
Octanoic	4, 5, 8
Decanoic	5, 8
Benzoic	5, 8
2-Hydroxybenzoic	2, 5, 8
Hydrogen cyanide	5, 8
Lactones	
γ-Octalactone	10
γ-Decalactone	10
Phenols	
p-Allylphenol	10
Guaiacol	9, 10
Eugenol	9, 10
Miscellaneous	
1,1-Diethoxyethane	5, 8
Furfural (2-furaldehyde)	5, 7, 8

[a] (1) McGlumphey, 1951; (2) Mehlitz and Matzik, 1956; (3) Spanyar et al., 1964; (4) Jorysch and Broderick, 1967; (5) Gierschner and Bauman, 1967; (6) Stinson et al., 1969a; (7) Stinson et al., 1969b; (8) Bauman and Gierschner, 1974; (9) Broderick, 1975a; (10) Schmid and Grosch, 1986a.

No explanation can be given concerning these observations, but there is a narrow parallelism between the appearance of γ-decalactone and the transformation of protopectins into soluble pectins (Bayonove, 1974). An increase in the production rate of this lactone was also found when the softening of the fruit takes place. According to several authors, an increase in ethylene emission was observed during ripening (Lim and Romani, 1964). This production marks, in the climacteric, the change from growth to senescence. During this period, biochemical changes are initiated and probably accelerated by the selective breakdown of some cellular structures and the reorganization of new ones (Rhodes, 1970; Romani and Jennings, 1971). Under these conditions, action of enzymes on substrates, precursors of volatile compounds, as well as oxygen transfer, are greatly increased.

Be that as it may, these observations show that the aroma content of the fruit varies according to the stage of maturity at harvest, the storage conditions, and the post-harvest ripening. In particular, the aromatic equilibrium reached by the post-harvest-ripened fruit will not be the same at eating or utilization time, depending on the gathering date. Furthermore, there are additional effects induced by possible technological processes used before consumption (Paillard, 1976; Souty and Reich, 1978).

5. CHERRY

The sweet cherry, *Prunus avium* L., first grew wild in nothern Persia and in the Russian provinces south of the Caucasus. It was spread rapidly from these areas by birds, which have given their name to the fruits: bird cherries.

The sour cherry, *Prunus cerasus* L., also originated in Asia, but it was known in Europe at the beginning of the Greek civilization, and is now widespread throughout the world (Romani and Jennings, 1971). These fruits are also known as tart cherries.

The more popular cultivars are Morello and Montmorency. The Morello cherries have a much deeper colour and are more sour, more aromatic, and richer in benzaldehyde, than the fruits of the cultivar Montmorency.

5.1 The Volatile Constituents of Cherries

5.1.1 Qualitative data

To date, 82 volatile compounds have been described, including 2 hydrocarbons, 23 aldehydes, 7 ketones, 19 esters, 10 acids, 2 lactones, 3 phenols and 2 miscellaneous compounds (MacGlumphy, 1951; Mehlitz and Matzik, 1956; Spanyar *et al.*, 1964; Jorysch and Broderick, 1967; Gierschner and Bauman, 1967; Stinson *et al.*, 1969a,b; Bauman and Gierschner, 1974; Broderick, 1975; Schmid and Grosch, 1986a,b) (see Table II.XIII).

5.1.2 Flavour impact information

Benzaldehyde is considered as a character-impact compound for cherry aroma (Schmid and Grosch, 1986b), and is the key for flavour compositions (Broderick, 1975). According to this author, this compound is produced by acid or enzymic hydrolysis of the glucoside amygdalin present in cherry pips and cherry bark. Hydrogen cyanide present in the volatile fraction (Spanyar *et al.,* 1964; Bauman and Gierschner, 1974) is also released during this reaction.

The importance of this compound in the aroma of cherry is emphasized by a recent study of Schmid and Grosch (1986a). Sniffing stepwise diluted extracts obtained by simultaneous distillation extraction (Likens and Nickerson apparatus) (extract I) and by vacuum distillation followed by extraction using dichloromethane (extract II) revealed seven compounds with highest aroma values (see Table II.XIV): benzaldehyde, linalool, hexanal, *trans*-hex-2-enal, phenylacetaldehyde, nona-*trans*-2,*cis*-6-dienal and eugenol. Extract I also contained an unknown fruity flavour compound of high aroma value.

5.2 Variation of Cherry Volatile Constituents

5.2.1 Influence of cultivar

The quantitative study of some important volatile components present in the juices of five cultivars of sour and of five cultivars of sweet cherries shows differences in concentrations (Schmid and Grosch, 1986b) (see Table II.XV).

TABLE II.XIV

Stepwise dilution of extracts I (simultaneous distillation-extraction) and II (vacuum distillation followed by extraction) (Schmid and Grosch, 1986a)

Compound	Aroma at sniffing port n dilution ratio (in volume)					
	$n = 4$		$n = 10$	$n = 20$	$n = 200$	$n = 1000$
	I	II	II	I	I	I
Hexan-1-ol	+	+		+		
trans-Hex-2-enal	+	+		+		
Benzaldehyde	+	+	+	+	+	
Octanal	+					
Phenylacetaldehyde	+	+		+		
Compound X	+			+	+	
Linalool	+	+		+	+	+
Nona-*trans*-2,*cis*-6-dienal	+	+	+			
trans-Non-2-enal	+					
Geraniol	+					
p-Menth-1-en-9-ol	+					
Eugenol	+	+		+		
β-Ionone	+					

86

TABLE II.XV

Concentration (μg/l) of volatile components of high aroma values isolated in five cultivars of sweet cherries after vacuum distillation and solvent extraction

Compound	Cultivar				
	Tragana	Duroni Marca	Neckaperle	Bütner	Hedelfinger Knorpelbirs-che
Benzaldehyde	115	18.0	388	317	165
Linalool	nd	nd	nd	nd	0.5
Hexanal	10.4	13.0	54.7	9.0	8.8
trans-Hex-2-enal	22.9	17.6	220	10.0	8.9
Nona-trans-2,cis-6-dienal	2.4	0.6	0.1	0.6	0.1
Phenylacetaldehyde	nd	2.9	5.2	5.6	3.2
Eugenol	1.2	1.0	1.5	22.2	15.3

5.2.2 Influence of heat treatment

Furfural (2-furaldehyde), isolated by several authors (Gierschner and Bauman, 1967; Stinson et al., 1969b; Bauman and Gierschner, 1974), must be considered as an artifact produced by heat during extraction procedures. On the other hand, Schmid and Grosch (1986b) have found an increase in benzaldehyde and in linalool during jam preparation or during simultaneous distillation – extraction of juices. According to these authors, the observed increase is attributed to hydrolysis of the β-glucoside and β-gentiobioside of mandelonitrile (hydroxyphenylacetonitrile) present in the flesh of sweet cherries for the first compound, and to hydrolysis of linalool glycoside for the second compound (Williams et al., 1982b; Engel and Tressl, 1983; Heidlas et al., 1984; Salles et al., 1988).

REFERENCES

Anonymous, 1987. Monde. Production et commercialisation de quelques fruits d'eté. Marché International des Fruits et Légumes Frais (MIFL), No. 2: 18 – 19.
Antoine Chiris, 1967. La pêche et son arôme. Ind. Aliment. Agric., T2: 1358 – 1663.
Baldrati, G. and Gianone, L., 1967. Ricerca dei componenti carbonicili volatili negli alimenti vegetali mediante cromatografia su strato sottile. Ind. Conserve (Parma), 42: 252 – 254.
Bauman, C. and Gierchner, K., 1974. Technology of the juice of black currents and morello cherries by using a new apparatus for obtaining fruit product aromas for gas chromatographic analysis. Riechst., Aromen, Körperpflegem., 24: 62 – 68, 92 – 99, 128 – 136.
Bayonove, C., 1973. Recherches sur l'arôme de la pêche. Evaluation des constituants volatils au cours de la maturation de la variété. Cardinal. Ann. Technol. Agric., 22: 35 – 44.
Bayonove, C., 1974. Evolution des composés volatils de la pêche pendant la maturation, après récolte. Coloques Internationaux CNRS, No. 238, Facteurs et régulation de la maturation des fruits, pp. 327 – 333.

Bayonove, C., 1976. Evolution des constituants volatils de la pêche: variétés Spring Crest, Red Top, Early Sungrand. Commission Qualité Pêche. Association Française des Comités Economiques Agricoles de Fruits et Légumes (AFCOFEL).

Bayonove, C., 1977. Evolution des constituants volatils de la pêche: variétés Fayette, Merrils. Commission Qualité Pêche. Association Française des Comités Economiques Agricoles de Fruits et Légumes (AFCOFEL).

Bayonove, C. and Cordonnier, R., 1970. Recherches sur l'arôme du Muscat. I. Evolution des constituants volatils au cours de la maturation du 'Muscat d'Alexandrie'. Ann. Technol. Agric., 19: 79 – 93.

Bayonove, C. and Cordonnier, R., 1971. Recherches sur l'arôme du Muscat. Etude de la fraction terpénique. Ann. Technol. Agric., 20: 347 – 355.

Berger, R. G. Drawert, F., Kollmannsberger, H. and Nitz, S., 1985. Natural occurrence of undecaenes in some fruits and vegetables. J. Food Sci., 50: 1655 – 1656.

Blaise, A., 1986. Le goût d'amande amère dans les vins. Journées de rencontres oenologiques, Fac. Pharma. Montpellier (France).

Bolin, H.R. and Salunkhe, D.K., 1971. Physicochemical and volatile flavour changes occurring in fruit juices during concentration and foam-mat drying. J. Food Sci., 36: 665 – 668.

Bricout, J., Viani, R., Marion, J.P., Muggler-Chavan, F., Raymond, D. and Egli R.H., 1967. Sur la composition de l'arôme de thé noir. 2. Helv. Chim. Acta, 50: 1517 – 1522.

Broderick, J.J., 1966. What is important in peach flavor? Am. Perfum. Cosmet., 81: 43 – 45.

Broderick, J.J., 1975a. Cherry common denominators. Int. Flav. Food Add., 2 April: 103 – 109.

Broderick, J.J., 1975b. Peach dynamic adaptations. Int. Flav. Food Add., 6 : 243 – 247.

Chairote, G., Rodriguez, F. and Crouzet, J., 1981. Characterization of additional volatile flavor components of apricot. J. Food Sci., 46: 1898 – 1901 and 1906.

Chung, J.I. and Luh, B.S., 1971. Effect of ripening temperature on chemical composition and colour of canned freestone peaches. Confructa, 16: 275 – 280.

Collins, R.P. and Halim, A.F., 1972. Characterization of the major aroma constituent of the fungus Trichoderma Viride (Pers.). J. Agric. Food. Chem., 20: 437 – 438.

Cordonnier, R. and Bayonove, C., 1974. Mise en évidence dans la baie de raisin, variété Muscat d'Alexandre de Monoterpenes liés révélables par une ou plusieurs enzymes du fruit. C.R. Acad. Sci. 278, Sér. D: 3387 – 3390.

Creveling, R.K., Silverstein, R.M. and Jennings, W.G., 1968. Volatile components of pineapple. J. Food Sci., 33: 284 – 287.

Crouzet, J., Chairote, G., Rodriguez, F. and Seck, S., 1983. Volatile components modifications during heat treatment of fruit juices. In: G. Charalambous and G. Inglett (Editors), Instrumental Analysis of Foods, Vol. 2. Academic Press, New York, pp. 119 – 135.

Crowell, E.A. and Guymon, J.F., 1973. Aroma constituents of plum brandy. Am. J. Enol. Vitic., 24: 159 – 165.

Daghetta, A., Forti, G. and Monzini, A., 1956. Il problema dei prodotti di respirazione della frutta conservate in celle frigorifere. Ann. Sper. Agrar. (Roma), 10: 321 – 327.

Day, W.C. and Erdman, J.G., 1963. Ionene: Thermal degradation product of β-carotene. Science, 141: 808.

Delasalle, B., 1985. Le point de vue d'un responsable de la gestion des marchés sur la situation variétale actuelle. In A.C.F.E.V./B.R.G. Les variétés locales d'espèces fruitières. Edit. Lavoisier, Paris.

Demole, E. and Berthet, D., 1972. Identification de la damascone et de la β-damascenone dans le tabac Burley. Helv. Chim. Acta, 54: 681 – 682.

Dimitriadis, E. and Williams, P.J., 1984. The development and use of rapid analytical technique for estimation of free and potentially volatile monoterpene flavorants of grapes. Am. J. Enol. Vitic., 35: 66 – 71.

Dirninger-Rigo, N., 1987. Contribution à l'étude de l'évolution des constituants aromatiques de la mirabelle de Lorraine au cours de la maturation. Thesis for the degree 'diplome de recherche spécialisée', University L. Pasteur of Strasbourg, France.

Do, J.Y., 1968. Isolation, identification and comparison of the volatiles of peach (Prunus persica, Cultivar. Gleason Early Elberta) fruit as related to harvest maturity and artificial ripening. Thesis (DABSAQ).

Do, J.Y., Salunke, D. and Olson, L., 1969. Isolation, identification and comparison of the volatiles of peach fruits as related to harvest maturity and artificial ripening. *J. Food Sci.,* 34: 618 – 621.

Engel, K.H. and Tressl, R., 1983. Formation of aroma components from non-volatile precursors in passion fruit. *J. Agric. Food Chem.,* 31: 958 – 1002.

Enzell, C.R., 1981. Influence of curing on the formation of tobacco flavour. In: P. Schreier (Editor), Flavour '81. W. Gruyter, Berlin, pp. 449 – 477.

Etiévant, P., Issanchou, S.N. and Bayonove, C.L., 1983. The flavour of muscat wine: the sensory contribution of some volatile compounds. *J. Sci. Food Agric.,* 34: 497 – 504.

Etiévant, P.X., Guichard, E.A. and Issanchou, S.N., 1986. The flavour components of Mirabelles plums. Examination of the aroma constituents of fresh fruits: variation of head-space composition induced by deep-freezing and thawing. *Sci. Aliment.,* 6: 417 – 432.

Filajdic, M. and Djukovic, J., 1973. Gas-chromatographic determination of volatile constituents in Yugoslav plum brandies. *J. Sci. Food Agric.,* 24: 835 – 842.

Forrey, R.R. and Flath, R.A., 1974. Volatile components of *Prunus salicina,* var. Santa Rosa. *J. Agric. Food Chem.,* 22: 496 – 498.

Francis, M.J.O. and Allcock, C., 1969. Geraniol β-D-glucoside occurrence and synthesis in rose flowers. *Phytochemistry,* 8: 1339 – 1347.

Fresneda, P., Molina, P. and Soler, A., 1978. Identification de Gamma y Delta-lactonas and el aroma de melocoton (*Prunus persica*) for cromatographia gas-liquido. *An. Bromatol.,* 30: 235 – 240.

Fujimori, T., Kasuga, R., Matsushita, H., Kneko, H. and Noguchi, M., 1976. Neutral aroma constituents in Burley tobacco. *Agric. Biol. Chem.,* 40: 303 – 315.

Gierschner, K. and Bauman, G., 1967. Aroma constitue 's of fruit. In: J. John (Editor), Aroma. Geschmacksstoffe Lebensm. Forsbildungskurs. Forster Verlage A.G., Zurich, pp. 49 – 89.

Goodwin, T.W. and Goad, L.J., 1970. Carotenoids and triterpenoids. In: A.C. Hulme (Editor), The Biochemistry of Fruits and their Products, Vol. 1. Academic Press, London, New York, pp. 305 – 358.

Guichard, E. and Souty, M., 1988. Comparison of the quantities of aroma compounds found in fresh apricot (*Prunus armeniaca*) from six different varieties. *Z. Lebensm.-Unters.-Forsch.,* 186: 301 – 307.

Guichard, E., Symonds, P., Richard, H., Crouzet, J. and Souty, M., 1986. Etude quantitative des composés volatils de l'arôme d'abricot. Compte rendu A.I.P., INRA.

Gunata, Y.Z., Bayonove, C., Baumes, R. and Cordonnier, R., 1985. The aroma of grapes. Extraction and determination of free and glucosidically bound fractions of some grape aroma components. *J. Chromatogr.,* 311: 83 – 90.

Heidlas, J., Lehr, M., Idstein, H. and Schreier, P., 1984. Free and bound terpene compounds in papaya (*Carica papaya* L.) fruit pulp. *J. Agric. Food Chem.,* 32: 1020 – 1021.

Hugard, J. and Monet, R., 1985. L'évolution variétale du pêcher. In: A.C.F.E.V./B.R.G. Les vairétés locales et espèces fruitières. Edit. Lavoisier, Paris.

Ismail, H.M.M., 1977. A study of the volatile constituents of plums (*Prunus domestica* L.). Thesis, University of Bristol, (U.K.).

Ismail, H.M.M., Brown, G.A., Tucknott, O.G., Holloway, P.J. and Williams, A.A., 1977. Nonanal in epicuticular wax of golden egg plums (*Prunus domestica*). *Phytochemistry,* 16: 769 – 770.

Ismail, H.H., Tucknott, O.G. and Williams, A.A., 1980a. The collection and concentration of aroma components of soft fruits using Porapak Q. *J. Sci. Food Agric.,* 31: 262 – 266.

Ismail, H.M., Williams, A.A. and Tucknott, O.G., 1980b. The flavour components of plum: An examination of the aroma components present in a distillate obtained from fermented plum juice. *Z. Lebensm.-Unters.-Forsch.,* 171: 24 – 27.

Ismail, H.M., Williams, A.A. and Tucknott, O.G., 1980c. The flavour components of plum: An investigation into the volatile components of canned plums. *Z. Lebensm.-Unters.-Forsch.,* 171: 265 – 268.

Ismail, H.M., Williams, A.A. and Tucknott, O.G., 1981a. The flavour of plums (*Prunus domestica* L.): An examination of the aroma components present in the headspace above four cultivars of intact plums, Marjorie's Seedling, Merton Gem, NA 10 and Victoria. *J. Sci. Food Agric.,* 32: 498 – 502.

Ismail, H.M., Williams, A.A. and Tucknott, O.G., 1981b. The flavour of plums (*Prunus domestica* L.): An examination of the aroma components of plums from the cultivar Victoria. *J. Sci. Food Agric.,* 32: 613 – 619.

Isoe, S., Hyen, S.B. and Sakan, T., 1969. Photooxygenation of carotenoids. 1. The formation of dihydroactinidiolide and β-carotene. *Tetrahedron Lett.,* 4: 279 – 281.

Jennings, W.G., 1972. Constituants volatils des abricots, des pêches et des poires Bartlett. *Ind. Aliment. Agric.,* 89: 121 – 126.

Jennings, W.G. and Sevenants, M., 1964. Volatile components of peach. *J. Food Sci.,* 29: 796 – 801.

Jorysch, D. and Broderick, J.J., 1967. Methodology of flaun development, before and after introduction of dermatography. *Cereal Sci. Today,* 12: 292 – 294, 312 – 313.

Kallio, H., 1976. Development of volatile aroma compounds in Artic bramble (*Prunus articus* L.). *J. Food Sci.,* 41: 563 – 566.

Kanasawud, P. and Crouzet, J., 1989. Mechanism of formation of volatile compounds by thermal degradation of carotenoids in aqueous medium. I. β-Carotene degradation. *J. Agric. Food Chem.,* to be published.

Kawamaki, M., 1982. Ionone series compounds from β-carotene by thermal degradation in aqueous medium. *Nippon Nogeikagaku Kaishi,* 56: 917 – 921 (in Japanese).

Kemp, T.R., Stoltz, L.P. and Packett, L.V., 1971. Aromatic hydrocarbons: examination of fruit and foliage volatiles. *Phytochemistry,* 10: 478 – 479.

Kereselidze, Ts.G. and Mikeladze, G.G., 1977. Flavor substances of Bullace plums of various degrees of ripeness. *Konservn. Ovoshchesush. Promst.,* 2: 17 – 19, (in Russian) via *Chem. Abstr.,* 1986, 21: 154193c.

Kovacs, A.S. and Wolf, H.O., 1964. The effect of processing on the aroma of fruit and vegetable. *Ind. Obst-Gemueseverwert.,* 49: 53 – 57.

La Roe, E.G. and Shipley, P.A., 1970. Whiskey composition: formation of α- and β-ionone by thermal decomposition of β-carotene. *J. Agric. Food Chem.,* 18: 174 – 175.

Latrasse, A., Degorce-Dumas, J.R. and Leveau, J.Y., 1985. Production d'arôme par les micro-organismes. *Sci. Aliment.,* 5: 1 – 26.

Le Quéré. J.-L., Sémon, E., Latrasse, A., A. and Etiévant, P., 1987. Gas chromatography – Fourier transform infrared spectrometry: applications in flavour analysis. *Sci. Aliment.,* 7; 93 – 109.

Lim, L.S., 1963. Studies on the relationship between the production of volatiles and the maturity of peaches and pears. Thesis, University of California, Davis.

Lim, L.S. and Romani, R.J., 1964. Volatiles and the harvest maturity of peaches and nectarines. *J. Food Sci.,* 89: 246 – 253.

Maarse, H., 1977. Volatile Compounds in Food. 4th edition 1977. Division for Nutrition and Food Research (T.N.O.), The Netherlands.

Maarse, H., 1984. Volatile Compounds in Food. Quantitative Data, Vol. 3. Division for Nutrition and Food Research (T.N.O.), The Netherlands.

MacGlumphy, J.M., 1951. Fruit flavors. *Food Technol.,* 5: 353 – 355.

Mader, I., 1964. Thermal degradation of β-carotene. *Science,* 14: 533 – 534.

Maga, J.A., 1976. Lactones in foods. *Crit. Rev. Food Sci. Nutr.,* 8: 1 – 56.

Marais, J., 1983. Terpenes in the aroma of grapes and wine. A review *S. Afr. J. Enol. Vitic.,* 4: 49 – 47.

Mehlitz, A. and Matzik, 1956. Volatile acids in fruit juices. *Ind. Obst-Gemueseverwert.,* 41: 227 – 229, 308 – 314.

Molina, P., Soler, A. and Cambromero, J., 1973. Volatile aroma components of peach (*Prunus persica*): determination of gamma lactones by GLC. *An. Bromatol.,* 25: 403 – 410.

Mori Mitsukuni, 1977. Studies on quality index of canned agricultural foods. Part V. Volatile components of canned peach. *Nippon Shokuhin Kogyo Gakkaishi,* 24: 215 – 220 (in Japanese).

Mori Mitsukuni, 1986. Changes in the lactone content of peach fruits after picking. *Kanzume Jiho,* 65: 163 – 170 (in Japanese).

Moutounet, M., 1976. Les carotenoïdes de la prune d'Ente et le pruneau d'Agen. *Ann. Technol. Agric.,* 25: 73 – 84.

Moutonnet, M., 1978. Formation des substances volatiles au cours de l'élaboration du pruneau de l'élaboration du pruneau in International Federation of fruit juice producers. In: Flavours of Fruits and Fruit Juices. Juris Druck and Verlag, Zurich, pp. 363 – 372.

Moutounet, M. and Jouret, C., 1975. Les acides aminés de la prune d'Ente et du pruneau d'Agen. *Fruits,* 30: 345 – 348.

Moutounet, M., Dubois, P. and Jouret, C., 1975. Les composés volatils majeurs du pruneau d'Agen. *C.R. Acad. Agric. France,* May: 581 – 585.

90

Mulik, J.D. and Erdman, J.G., 1963. Genesis of hydrocarbons of molecular weight in organic rich aquatic systems. *Science,* 141: 806 – 807.

Murray, K.E., Shipton, J. and Whitfield, F.B., 1972. The chemistry of food flavour. I. Volatile constituents of passion fruit, *Passiflora edulis Aust. J. Chem.,* 25: 1921 – 1933.

Nobuhara, A., 1969. Synthesis of unsaturated lactones, Part III. Flavorous nature of some lactones having the double bonds at various sites. *Agric. Biol. Chem.,* 33: 1264.

Nursten, H.E. and Williams, A.A., 1967. Fruit aromas: a survey of components identified. *Chem. Ind.,* 3: 486 – 497.

Paillard, N., 1976. Effets de la congélation et de différents traitements annexes sur les constituants volatils des pêches congelées. *Rev. Gén. Froid,* 12: 845 – 854.

Pech, J.C. and Fallot, J., 1972. Les enzymes et la qualité des pêches. *Ann. Technol. Agric.,* 21: 81 – 93.

Postel, W. and Adam, L., 1985. Quantitative determination of volatiles in distilled alcoholic beverage. In: R.G. Berger, S. Nitz and P. Schreier (Editors), Topics in Flavour Research. H. Heichorn, Marzling-Hangenham, pp. 79 – 107.

Power, F.B. and Chesnut, V.K., 1921. The odorous constituents of peaches. *J. Am. Chem. Soc.,* 43: 1725 – 1739.

Power, F.B. and Chesnut, V.K., 1922. Confirmation of the occurrence of linalyl esters in peaches. *J. Am. Chem. Soc.,* 44: 2966 – 2967.

Raulston, J., 1969. Direct chromatographic measurement of peaches headspace volatiles as a rapid, objective measurement for correlation with maturity and/or quality. Thesis, University of Maryland.

Rhoades, J.W. and Millar, J.D., 1965. Gas chromatographic method for comparative analysis of fruit flavours. *J. Agric. Food Chem.,* 13: 5 – 9.

Rhodes, M.J.C., 1970. The climacteric and ripening of fruits. In: A.C. Hulm (Editor), The Biochemistry of Fruits and their Products. Vol. 1. Academic Press, London, New York, pp. 521 – 533.

Rodriguez, F., Seck, S. and Crouzet, J., 1980. Constituants volatils de l'abricot, variété du Rouge du Roussillon. *Lebensm.-Wiss. Technol.,* 13: 152 – 155.

Rolle, S. and Crouzet, J., 1988. Etude des composés volatils de l'abricot. Influence de la variété sur la composition de l'émace de tête. *Fruits,* to be published.

Romani, R.J. and Jennings, W.G., 1971. Stone fruits. In: A.C. Hulme (Editor), The Biochemistry of Fruits and their Products. Academic Press, London, Vol. 2, pp. 411 – 436.

Sakho, M., Crouzet, J. and Seck, S., 1985. Evolution des composés volatils de la mangue au cours du chauffage. *Lebensm.-Wiss. Technol.,* 18: 89 – 93.

Salles, C., Essaied, H., Chalier, P., Jallageas, J.C. and Crouzet, J., 1988. Evidence and characterization of glycosidically bound volatile components in fruits. In: P. Schreier (Editor), Bioflavour '87. W. de Gruyter, Berlin, pp. 145 – 160.

Sanderson, G.W., Co, H. and Gonzalez, J.G., 1971. Biochemistry of tea formation: the role of carotenes in black tea aroma formation. *J. Food Sci.,* 36: 231 – 236.

Schmid, W. and Grosch, W., 1986a. Identification of highly aromatic volatile flavour compounds from cherries (*Prunus cerasus* L.). *Z Lebensm.-Unters.-Forsch.,* 182: 407 – 412.

Schmid, W. and Grosch, W., 1986b. Quantitative analysis of the volatile flavour compounds having high aroma values from sour (*Prunus cerasus* L.) and sweet (*Prunus avium* L.) cherry juices and jams. *Z. Lebensm.-Unters.-Forsch.,* 183: 39 – 44.

Schreier, P., Drawert, F. and Junker, A., 1976. Identification of volatile constituents from grapes. *J. Agric. Food Chem.,* 24: 331 – 336.

Schreier, P., Drawert, F. and Bhiwapurkar, S., 1979. Volatile compounds formed by thermal degradation of β-carotene. *Chem. Mikrobiol. Technol. Lebensm.,* 6: 90 – 91.

Schütte, H.-R., 1976. Secondary plant products: Special topics of the phenylpropanoid metabolism. In: Progress in Botany, Vol. 40. Springer Verlag, Berlin.

Sevenants, M. and Jennings, W.G., 1966. Volatile components of peach. II. *J. Food Sci.,* 31: 81 – 86.

Sevenants, M.R. and Jennings W.G., 1971. A research note: occurrence of 6-pentyl-alphapyrone in peach essence. *J. Food Sci.,* 36: 536.

Shaw, P.E., Tatum, J.H. and Berry, R.E., 1968. Base-catalyzed fructose degradation and its relation to nonenzymic browning. *J. Agric. Food Chem.,* 16: 979 – 982.

Shewfelt, R.L., Meyers, S.C., Prussia, S.E. and Jordan, J.L., 1987. Quality of fresh-market peaches within the postharvest handling system. *J. Food Sci.,* 52: 361 – 364.

Simpson, R.F., 1978. 1,1,6-Trimethyl-1,2-dihydronaphthalene: an important contributor to the bottle aged bouquet of wine. *Chem. Ind.,* 7: 37.

Soboleva, I.M. and Epifanov P.V., 1978. Correlation between organoleptic and gas-chromatographic evaluations of the peach aroma. *Sadovod. Vinograd. Vinodel. Mold.,* 33: 53 – 55 (in Russian).

Soboleva, I.M. and Yazvinskaya, T.M., 1975. Aroma substances of peaches. *Sadovod. Vinograd. Vinodel. Mold.,* 30: 54 – 57 (in Russian).

Souty, M. and Reich, M., 1978. Effets de traitements technologiques – congélation et appertisation – sur certains constituants de l'arôme des pêches. *Ann. Technol. Agric.,* 27: 837 – 848.

Spanyar, P., Kevei, E. and Blazovich, M., 1964. Recherches effectuées par chromatographie en phase gazeuse sur la composition des arômes de fruits et de produits de fruits. *Ind. Aliment. Agric.,* 81: 1063 – 1071.

Spanyer, P., Kevei, E. and Blazovich, M., 1965. Recherches effectuées par chromatographie en phase gazeuse sur la composition des arômes de fruits et de produits de fruits. *Ind. Aliment. Agric.,* 82: 213 – 214.

Spencer, M.D., Pangborn, R.M. and Jennings, W.G., 1978. Gas chromatographic and sensory analysis of volatiles from cling peaches. *J. Agric. Food Chem.,* 26: 725 – 732.

Steinke, R.D. and Paulson, M.C., 1964. The production of steam-volatile phenols during the cooking and alcoholic fermentation of grain. *J. Agric. Food Chem.,* 12: 381 – 387.

Stinson, E.E., Dooley, C.J., Filipic, V.J. and Hills, C.H., 1969a. Composition of Montmorency cherry essence. 1. Low-boiling components. *J. Food Sci.,* 34: 246 – 248.

Stinson, E.E., Dooley, C.J., Filipic, V.J. and Hills, C.H., 1969b. Composition of Montmorency cherry essence. 2. High-boiling components. *J. Food Sci.,* 34: 544 – 546.

Tang, C.S. and Jennings, W.G., 1967. Volatile components of apricot. *J. Agric. Food Chem.,* 15: 24 – 28.

Tang, C.S. and Jennings, W.G., 1968. Lactonic compounds of apricot. *J. Agric. Food Chem.,* 16: 252 – 254.

Tatum, J.H., Shaw, P.E. and Berry, R.E., 1969. Degradation products from ascorbic acid. *J. Agric. Food Chem.,* 17: 38 – 40.

Terrier, A., 1972. Les composés terpéniques dans l'arôme des raisins et des vins de certaines variétés de *Vitis vinifera.* Thesis, Université de Bordeaux I.

Tressl, R., Drawert, F. and Heimann, W., 1969. Gas chromatographic and mass spectrometric composition of strawberry aroma. *Z. Naturforsch.,* 24: 1201 – 1202.

Ueda, Y. and Iwata, T., 1982. Off odor of strawberry by freezing. *Engei Kakkai Aasshi.,* 51: 219 – 223.

Vernin, G., Metzger, J., Suon, K.-N. and Fraysse, D., 1985. Arôme de mirabelle: Etude par CG-SM-Banque Specma. *Parfum., Cosmét., Arômes,* 62: 69 – 74 (in French).

Vidaud, J., Jacoutet, I. and Thivend, J., 1987. Le pêcher: références techniques. Edit. Centre Interprofessionnel des Fruits et Légumes, 22 Rue Bergère, 75009 Paris.

Watada, A., Anderson, R. and Aulenbach, B., 1979. Caractéristiques sensorielles, composition et composés volatils des pêches stockées en atmosphère contrôlée. *J. Soc. Hortic. Sci.,* 104: 626 – 629.

Wildenradt, H.L. and Singleton, V.L., 1974. The production of aldehydes as a result of oxidation of polyphenolic compounds and its relation to wine aging. *Am. J. Enol. Vitic.,* 25: 119 – 126.

Williams, A.A. and Ismail, H.M.M., 1981. The volatile flavour components of plums and their sensory evaluation. In: J. Solms and R.L. Hall (Editors), Criteria of Food Acceptance. Foster Verlag, Zürich, pp. 333 – 354.

Williams, P.J., Strauss, C.R. and Wilson, B., 1980. Hydroxylated linalool derivatives of volatile monoterpenes of muscat grapes. *J. Agric. Food Chem.,* 28: 766 – 771.

Williams, P.J., Strauss, C.R. and Wilson, B., 1981. Classification of monoterpenoid composition of muscat grapes. *Am. J. Enol. Vitic.,* 32: 230 – 235.

Williams, P.J., Strauss, C.R., Wilson, B. and Massy-Westropp, R.A., 1982a. Novel monoterpene disaccharide glycosides of *Vitis vinifera* grapes and wines. *Phytochemistry,* 21: 2019 – 2020.

Williams, P.J., Strauss, C.R., Wilson, B. and Massy-Westropp, R.A., 1982b. Studies of the hydrolysis of *Vitis vinifera* monoterpene precursor compounds and model monoterpene β-D-glucosides rationalizing the monoterpene composition of grapes. *J. Agric. Food Chem.,* 30: 1219 – 1223.

Winter, M. and Enggist, P., 1971. Recherches sur les arômes, sur l'arôme de framboise IV. *Helv. Chim. Acta,* 54: 1881 – 1898.

Chapter III

FACTORS AFFECTING THE FLAVOUR OF CITRUS FRUIT

STEVEN NAGY and PHILIP E. SHAW

Florida Department of Citrus, 700 Experiment Station Road, Lake Alfred, FL (U.S.A.), U.S. Citrus & Subtropical Products Laboratory, Post Office Box 1909, Winter Haven, FL (U.S.A.)

1. INTRODUCTION

Nearly all species of citrus fruits originated in southeast Asia, primarily in India, China and the Malay Archipelago. The ancient dynasties of China regarded citrus fruits as prized tributes, with the earliest mention of citrus occurring in the Chinese literature around 2205 – 2197 BC (during the reign of Ta Yu). It is evident from ancient writings that sour oranges, citrons, kumquats, mandarins, pummelos, limes, lemons and oranges were cultured and esteemed. The only citrus fruits which appear to have originated in China were the trifoliate orange and the kumquat. Citron, lemon, lime and some mandarin types spread from the Himalayas southward into India and eastward into the Malayan region. The sweet orange developed in southeastern China and is undoubtedly a unique, apomictically perpetuated biotype of hybrid origin (Barrett and Rhodes, 1976). Cooper and Chapot (1977) suggest that the sweet orange might have developed by natural hybridization between the *kan* mandarin and a pummelo cultivar in the Canton region. The Chinese never accorded much significance to this new cultivar, considering it simply as another selection of *kan* with a tight skin. Their Sekkan and Yinkan are classified as sweet oranges by westerners, but the *kan* suffix belies mandarin origin (Cooper and Chapot, 1977).

After a period of geographical isolation, citrus fruits found their way to Africa and Europe by way of caravan traders, conquering armies and early sea explorers. It is difficult to trace the exact geographical path and time sequence by which citrus reached North Africa and Europe, but the citron, prized for its fragrant rind, appears to be the first citrus introduced into Europe (Theophrastus, around 310 BC). The next citrus fruits introduced into Europe were the sour orange, lemon, lime and sweet orange. The exact dates of fruit introduction are not known, but probably occurred centuries apart. Apparently, the sweet orange reached India from China before the Christian era because, during the 1st century, AD, Romans were aware of the sweet orange and, in fact, called it Indian fruit. From Europe, citrus fruits spread to North, Central and South America by Spanish and Portugese seafarers during the late 15th to

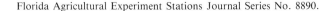

Florida Agricultural Experiment Stations Journal Series No. 8890.

mid-16th centuries. The grapefruit is a natural hybrid of probably pummelo and sweet orange ancestry (Albach and Redman, 1969), and is the only citrus fruit to have originated in the Western Hemisphere (Hughes, 1750).

2. CITRUS FRUIT CULTIVARS

A citrus fruit is botanically a hesperidium, a particular kind of berry with a leathery rind and divided internally into segments. This unique fruit belongs to six genera: *Citrus, Fortunella, Poncitrus, Microcitrus, Eremocitrus* and *Clymenia*. However, only the *Fortunella* and *Citrus* genera have fresh fruit cultivars of commercial importance. Citrus fruits range from less than 2.5 cm for calamondin or kumquats (*Fortunella* spp.) to more than $12-18$ cm in diameter for grapefruit (*C. paradisi*) or pummelo (*C. grandis*). Shape is variable: oblate in grapefruit, mandarins and mandarin hybrids; globose in sweet orange and sour orange; prolate in lemon, lime and citron; distinctly pyriform in pummelo cultivars, Ponderosa lemon, and obovoid in some limes. Citrus fruits typically have $8-15$ segments; occasionally 17 or 18 in grapefruit or pummelo; $3-5$ in kumquats and $6-8$ in trifoliate orange.

It is beyond the scope of this chapter to describe the many edible citrus cultivars and their unique flavour qualities; however, information on this subject can be obtained by reading the comprehensive works of Hodgson (1967), Cooper and Chapot (1977) and Young (1986). An exellent compilation of general flavour qualities of some commercially important citrus cultivars has been reported by Fellers (1985) (Table III.I). Diversity of citrus flavour ranges from the acidic, zesty and sensitive aroma and taste of Key lime to the rich, unique, sweet taste, aromatic aroma and negligible body of the Temple orange. Even within a specific citrus group, for example sweet oranges, the aroma and taste will vary from moderate, as exemplified by Hamlin orange, to a full, rich, fruity aroma and taste characteristic of Valencia orange.

3. AGRICULTURAL PRACTICES AFFECTING FLAVOUR

Throughout the ages, man learned which crops grew best in his environment and, with time, developed an appreciation of the interplay of complex variables associated with plant growth. Soils, physiography and climate essentially control where citrus may be grown; if one or more of these factors is limiting, the chances for high productivity are diminished (Black, 1968). Citrus trees grow on a wide variety of soil types but thrive best on well-drained, sandy loam soils which permit deep root penetration, and in soils of slight acidity (pH $6-7$).

A number of agricultural parameters has been identified which impact on the flavour quality of citrus fruit. Harding (1964) listed variety, rootstock, maturity, number of fruit per tree, soil moisture and type, fertilization, cultivation, climate and spray materials. In this chapter, we will discuss only the more im-

TABLE III.I

General flavour comments on some major citrus cultivars (Fellers, 1985)

Cultivar	General flavour comments
Sweet oranges	
Hamlin	Moderate orange flavour with no top notes
	Attained fair juice quality late in their commercial shipping season
	Lacking in acid; flavour sweet; eating quality is generally somewhat disappointing
	Acidity and sweetness well blended; flavour excellent
	Poor to medium quality
Marrs (Marrs Early)	Flesh . . . juice, lacking in acid; flavour sweet
Parson Brown	Attained high flavour about the first of November
	Pulp melting; acidity and sweetness not well blended unless picked quite early
	Well flavoured
	Low to moderate quality
Pera	Flavour rich
	Pera Rio and Pera Coroa of Brazil – produce high-quality juices of good colour and flavour
	Pera Natal of Brazil – the juice has even better colour and flavour than the common Pera
Pineapple	It is noted for . . . the rich flavour of the juice
	Moderate to full orange flavour with a fruity top note
	Flavour rich though sweet
	Pulp melting; acidity and sweetness well blended; flavour excellent; quality excellent; flavour rich
	Of excellent flavour
Shamouti (Jaffa)	The fruit has an excellent flavour
	Flesh firm, tender, juicy; fragrant and pleasantly sweet-flavoured
	Pulp melting; acidity and sweetness normal and well blended; flavour rich; quality excellent
Valencia	The juice is of excellent quality
	Juice made from freshly squeezed Valencia oranges is so good that it becomes the standard against which all other juices must be compared
	Full orange flavour with good body
	The juice . . . had a desirable blending solids and acids by Feb. 15 to Mar. 15, when it was rated as pleasantly tart
	Flavour good but commonly somewhat acid; excellent for processing
	Acidity and sweetness well combined; pulp melting; flavour rich, sprightly and vinous; quality excellent
	The best common orange, both for fresh and processed markets; Valencia's excellence is uncontested
	Flesh is firm with good sweet juice. It is primarily a superior-quality juice orange
	(In Brazil) The juice has even better colour and flavour than the common Pera
Navel (Washington)	Under favourable conditions, it is of excellent quality; the fruit is particularly good for eating from the hand
	. . . Distinctive characteristics . . . include crispness of flesh texture . . . and richness of flavour, which under favourable climatic conditions combine to make Navel oranges among the finest of dessert fruits

(continued)

TABLE III.I (*continued*)

Cultivar	General flavour comments
	Pulp melting; acidity and sweetness well blended; flavour rich, vinous; quality excellent
	Their flesh texture, flavour and ease of peeling make Navels one of the premier fresh fruits of the world
	Juicy, delicious seedless fruits that are wonderful for salads or eating out of hand

Mandarins, tangerines, tangelos, and allied cultivars

Cultivar	General flavour comments
Dancy tangerine	Has a lively, sweet-tart flavour
	The pulp is very tender and the flavour rich, with a definite aroma, somewhat spicy
	Flesh . . . tender and melting; moderately juicy; flavour rich and sprightly (acidity moderately high)
	Pulp melting; flavour rich and sprightly; acidity and sweetness well blended; quality excellent
King (mandarin)	The flavour is good, more like an orange than a tangerine
	Flesh . . . tender; moderately juicy; flavour rich
	Pulp melting; flavour agreeable, sprightly; acidity and sweetness well blended; quality very good
	The tree does not bear heavily, but the flavour and quality of its fruit tend to compensate for this less desirable feature
Lee tangerine	Flesh . . . tender and melting . . . flavour rich and sweet
Murcott	It is exceedingly sweet, rich
	Flesh . . . tender, very juicy; flavour very rich and sprightly
	Its . . . flesh is firm and possesses ample rich, sweet juice
	Juice extracted by hand-squeezing has an excellent flavour
Minneola tangelo	Pulp is tender and fine textured; flavour fair to good with grapefruit-like tartness
	Flesh . . . tender, juicy, aromatic; flavour rich and tart
	Has excellent flavour
Nova tangelo	Juicy; flavour pleasant
	Sweeter . . . than Orlando Tangelo
Orlando tangelo	Flesh . . . tender, very juicy; flavour mildly sweet
	Flavour good, with acid and sweet combined
Osceola tangelo	Flesh . . . juicy, flavour rich and distinctive
	Delicious fruit
Page	Flesh . . . tender and juicy; flavour rich and sweet
Ponkan	Flesh . . . tender and melting, juicy; flavour mild and pleasant, and aromatic
Robinson tangerine	Flesh . . . juicy; flavour rich and sweet
Satsuma mandarin	Flesh . . . tender and melting, flavour rich but subacid
(*Unshiu mikan*)	Pulp melting; flavour sprightly, agreeable; acidity and sweetness well balanced; quality excellent
Seminole tangelo	The flavour of the juice is tart and sprightly
	Flesh . . . tender, juicy; flavour sprightly and acid
Temple	The flavour is distinct, rich and spicy, sweet-tart
	The flavour is different from that of the ordinary sweet orange, and the pulp and juice have a pronounced aroma; it is the finest orange in Florida for eating from the hand

TABLE III.I (*continued*)

Cultivar	General flavour comments
	The fruit has a notably fine eating quality, is exceptionally juicy . . . it contains aromatic constituents that impart to flesh and juice a unique and unusually desirable bouquet, spicy and rich
	Flesh . . . tender, moderately juicy; flavour rich and spicy
	Flesh . . . melting, free from rag, very juicy . . .; flavour spicy, rich vinous and very characteristic; acidity and sweetness well blended
	The Temple is one of the most beautiful and highly flavoured citrus fruits
Grapefruit and K-Early	
Duncan	In prime eating condition . . . the flavour was classified as pleasantly tart and pleasantly tart to sweet
	Flesh . . . tender, very juicy; flavour pronounced and excellent
	Bitterness well masked; acidity and sweetness good
K-Early	Rather acid, of comparatively poor quality
Marsh	In prime eating condition . . . the flavour was classified as pleasantly tart and pleasantly tart to sweet
	Flesh . . . tender, very juicy; flavour good though not so pronounced as in some seedy varieties
	Bitter principle not strongly marked; acidity and sweetness medium
Ruby (Red Blush)	Fruit similar to Thompson (below) in all respects except for much deeper pigmentation in the flesh
Thompson (Pink Marsh)	Flesh texture tender and juicy; flavour good, similar to Marsh
Lemons	
Eureka	Flesh . . . tender, juicy; flavour highly acid
	Pulp melting; acid pure and strong; flavour excellent
Meyer	Flesh . . . tender, very juicy; lemon-flavoured and acid
	Flesh tender, juicy; juice abundant, acid
Limes	
Bearss	(Corresponds closely with Tahiti below)
Key (Mexican or West Indian)	The juice is very excellent in flavour and very acid
	Flesh . . . tender, juicy; highly acid with distinctive aroma
	Pulp melting; acid very strong; flavour distinctly of the lime
Tahiti (Persian)	The juice is very acid and of excellent flavour
	Flesh . . . tender, juicy; very acid and with true lime flavour
	Pulp melting; acid, pure, strong; flavour agreeable, distinctly lime

portant factors, namely fertilization, climate, rootstock, maturity and spray materials (abscission chemicals; growth regulators).

3.1 Fertilization

Nutrients required for citrus tree growth are customarily divided into three groups on the basis of normal requirements (Ochse *et al.,* 1961; Childers, 1975):

(a) major elements, nitrogen, phosphorus and potassium; (b) secondary elements, calcium, magnesium, sulphur and chlorine; and (c) minor or trace elements, iron, manganese, copper, zinc, boron and molybdenum. When a citrus tree is unable to obtain sufficient supplies of these essential nutrients, either because of insufficient quantity or availability, the tree and fruit manifest a number of deficiency symptoms. Replacement of soil nutrients by fertilization is principally directed to restoring the health of the plant's vegetative parts and to improving fruit yield; however, direct as well as indirect evidence show that fertilization also affects the nutrient content and flavour quality of the citrus fruit.

Flavour quality or palatability of citrus fruits is generally equated to the fruit's contents of total soluble solids and total acids (Harding, 1947). The soluble constituents of citrus, primarily composed of sugars and acids (about 85%), are of major importance in the taste sensation of the fruit. However, about 15% of the soluble solids is composed of inorganic compounds, amino acids, water-soluble vitamins (including ascorbic acid), essence oils, water-soluble pectins, glycosides, esters and other compounds which also contribute to the taste and aroma sensations of the fruit (Sinclair and Bartholomew, 1947). Most studies that relate an agricultural practice to fruit quality stress measurements of total soluble solids (primarily sugars; as measured by Brix), total acidity (primarily citric acid) and the Brix/acid ratio (or sugar/acid ratio). The sweet taste of fruit is due to the content of soluble sugars, whereas the sour taste (or tartness) is associated with the hydrogen ion concentration (total titratable acidity). Since the Brix/acid ratio correlates with the pH of citrus juice, either term is used to express the relative tartness of the fruit (Kilburn, 1958).

3.1.1 Nitrogen

No other element influences citrus fruit quality as much as does nitrogen. Although soils vary considerably in their content of organic matter, and hence nitrogen, even the richest soil soon becomes impoverished if not supplemented with this important element. The concentration of nitrogen in citrus fruit is usually increased by the application of nitrogen fertilizers (Sinclair, 1961). In addition, high nitrogen fertilization tends to increase total soluble solids (Koo, 1979a), total titratable acidity (Jones and Parker, 1949) and juice content. However, fruit size and weight vary inversely with nitrogen fertilization (Jones and Embleton, 1960). As a general rule, increasing the amount of nitrogen fertilization will cause negligible to small decreases in the Brix/acid ratio (sensory perception: none to slight increase in tartness of the flesh). Juice colour, as measured by citrus red and citrus yellow, was enhanced with nitrogen fertilization (Koo, 1979b). Colour of juice contributes significantly to taste preference. Consumers equate colour with ripeness and sweetness; the deeper the orange colour of the juice or flesh, the more preferred the fruit (Fellers, 1980).

3.1.2 Potassium

Potassium plays a dominant rôle in the mineral nutrition of citrus. Its requirements are not easily determined, because citrus can be grown within a wide range of potassium. However, the need for potassium fertilization may be more widespread than is generally believed, because more potassium is removed by citrus fruit than any other element; about 2.1 kg of potassium is removed from the soil for every ton of fruit harvested (Chapman, 1968). Potassium has only a minor influence on juice colour, but the soluble solids/acid ratio is consistently reduced by increased potassium fertilization, because of the positive relationship between juice acidity and potassium (Table III.II; Koo and Reese, 1977; Smith, 1966). Essentially, an increase in potassium fertilization will increase the tartness of the juice. Citric and ascorbic acid contents increase with higher potassium levels (Table III.II; Embleton et al., 1973). Potassium effects on other juice characteristics are minimal. Large fruit will usually have lower juice levels, soluble solids and acid content than small fruit. Consequently, potassium effects on these juice characteristics are indirect rather than direct (Koo, 1985).

3.1.3 Phosphorus

Phosphorus is essential for normal growth of citrus fruit. It plays a direct rôle as a carrier of energy, takes part in photosynthesis, and is a component of both storage and structural compounds, for example phospholipids, nucleic acids, phytin and phosphoproteins. Increasing the phosphorus content of soils from deficient to adequate levels markedly affects fruit quality, but increasing the phosphorus content above those adequate levels results in debatable benefits (Embleton et al., 1973). Both soluble solids and acid content show a decrease with increasing phosphorus fertilization, but the decrease in acid is much more apparent than the decrease in soluble solids. The result is a juice with a higher soluble solids/acid ratio and, thus, a sweeter tasting juice (Koo, 1979b). An increase in phosphorus fertilization causes an increase in the percentage of juice in a fruit and an increase in the soluble solids/acid ratio; however, peel thickness, soluble solids content, total acid and vitamin C all show decreases (Embleton et al., 1973).

TABLE III.II

Effects of potassium (K) rates due to fertilization on juice quality of Hamlin orange (Koo, 1985)

K rate (kg ha^{-1})	Juice content (%)	Soluble solids (%)	Titratable acid (%)	SS/A ratio[a]	Colour[b]	
					CY	CR
63	53.8	10.6	0.69	15.5	66.5	27.8
126	53.9	10.5	0.77	13.9	65.1	26.5
189	53.7	10.5	0.78	13.4	64.8	26.4
251	54.0	10.6	0.80	13.1	64.6	26.0

[a] SS/A ratio = soluble solids/acid ratio.
[b] CY = citrus yellow; CR = citrus red.

3.1.4 Secondary and trace elements

Flavour enhancement of fruit by soil or foliar application of secondary (Ca, Mg, S and Cl) or trace (Fe, Mn, Cu, Zn, B and Mo) elements is not as noticeable when compared with supplementation with the major elements (N, P and K). Over a five-season period, Sites (1944) applied nitrogen, potassium and phosphorus fertilizers to four plots, and the same materials plus magnesium, zinc, manganese and copper to four other plots of Marsh and Duncan grapefruit. The plots that received the supplemental elements yielded fruit with higher values for citric acid, soluble solids, Brix/acid ratios and ascorbic acid. These fruit also had a sweeter taste and a general improvement in flavour. Apart from a few minor reports, little information is available on the response of citrus to micronutrients in terms of fruit quality and flavour. It is assumed that once soil deficiencies of micronutrients are corrected, fruit quality is improved, but actual data to substantiate these assumptions are lacking (Koo, 1979b).

3.2 Rootstocks

Citrus trees are propagated vegetatively and are a combination of different scions budded to different rootstocks. The selection of a rootstock is based on many factors, such as resistance to specific diseases, compatibility with the scion, drought resistance, tolerance to soil conditions (e.g., salinity), and the rate at which nutrients are absorbed from the soil. Essentially, growers select rootstocks from among those giving adequate fruit quality, but make their final selection on the basis of tree survival and yield.

The chemical composition and flavour properties of citrus fruit are often influenced by the type of rootstock to which the scion is attached. Early studies by Hodgson and Eggers (1938) showed that trifoliate orange rootstock produced the highest concentration of sugars in the juice of grapefruit and five other commercial kinds of citrus, whereas rough lemon rootstock produced the lowest concentration of sugars. In Florida, Harding (1947) noted that Duncan and Marsh grapefruit grown on sour orange rootstock contained more total, reducing and nonreducing sugars than when grown on rough lemon. More recent experiments by Castle and Phillips (1980) confirm that grapefruit grown on rough lemon produce fruit of poor flavour quality, whereas fruit grown on Koethen sweet orange × Rubidoux trifoliate orange, and Rangpur lime × Troyer citrange rootstocks produce high quality juice.

Cook et al. (1952) reported a rootstock trial comparing Parson Brown orange on sour orange and with the same budwood on rough lemon. Trees on sour orange produced fruit with smaller size, thinner rind, more juice and higher total soluble solids. Hutcheson (1977) reported on the total soluble solids levels in juice of Valencia oranges on 26 different rootstocks over a 12-year period in Florida. The soluble solids levels averaged 10.2% on rough lemon and 12.6% on Rusk citrange. Because of a tree decline malady associated with Rusk, the Carrizo citrange with an average soluble solids content of 11.6% was recommended. As a general rule, scions budded on fast-growing rootstocks (e.g., rough lemon, Palestine sweet lime, Rangpur lime, citron, *Citrus macrophylla*) produce large fruit with thicker peel, rougher peel texture and lower concentrations of total soluble solids and acids. Cultivars budded on slower-growing rootstocks (trifoliate orange and some of its hybrids, such as citranges and citrumelos) produce smaller fruit but with higher total soluble solids and acids (Reitz and Embleton, 1986).

Limonin bitterness is a major factor contributing to quality loss of citrus fruit juice. Citrus contain limonin and other limonoid bitter principles primarily in the seeds, the albedo and the segment walls (Maier *et al.*, 1977). In general, no bitterness is apparent if the fruit is eaten out-of-hand; however, if juice is expressed from the fruit and allowed to stand, an intensely bitter sensation is elicited. Extensive research by many investigators (Marsh and Cameron, 1950; Marsh, 1953; Kefford and Chandler, 1961; Bowden, 1968) has clearly demonstrated the central rôle of rootstock on fruit bitterness. Marsh (1953) budded Washington Navel scion to six different rootstocks, namely sour orange, sweet orange, rough lemon, grapefruit, trifoliate orange and Washington Navel cutting, and evaluated the rate at which limonin disappeared from the fruit. At commercial maturity, fruit grown on the grapefruit rootstock was essentially free of bitterness; trifoliate orange and sweet orange exhibited some minor bitterness; sour orange and Washington Navel cutting showed intermediate bitterness; and bitterness (intense) never completely disappeared from fruit grown on rough lemon. Kefford and Chandler (1970) categorized rootstocks into three groups based on the time (after commercial maturity) for the fruit to yield a nonbitter or low limonin juice. Rapid growth or low-limonin stocks were trifoliate orange, tangelo, Cleopatra mandarin, Navel orange cuttings; the only intermediate stock was sweet orange; slow growth or high-limonin stocks were rough lemon, sweet lime, Kusaie lime and East India lime.

4. CLIMATE (SPECIFICALLY TEMPERATURE)

Temperature, moisture and light are three important factors affecting growth of citrus plants, and together with wind velocity and atmospheric pressure are collectively termed climate. Growth, maturation and fruit quality are influenced by the climate of the region in which citrus fruits are grown. Fruit produced in a specific climatic region (e.g., Mediterranean) share a common set of quality characteristics. Total available heat is probably the single most important factor in determining growth rate, time of ripening and fruit quality (Jones, 1961). Scora and Newman (1967) followed seasonal changes in the ratios of total soluble solids to titratable acidity for Valencia oranges in six major citrus-producing regions of the United States (Weslaco, Texas; Orlando, Florida; Tempe, Arizona; and Riverside, Indigo and Santa Paula, California). From November to March, the highest Brix/acid ratios were found in fruit from Weslaco (climate classified as warm, semiarid, subtropical, steppe) and the lowest ratios in fruit from Santa Paula (subtropical climate; cool, dry summers; limited rainfall occurring in late autumn, winter and early spring). In general, fruit develops more rapidly and is larger in size when grown in warmer climates. The juice of these fruit is generally sweeter, because warmer climates favour a more rapid drop in fruit acidity when contrasted to cool, Mediterranean-type climates (Reuther *et al.*, 1969).

4.1   Microclimate (Sunlight Exposure and Fruit Location)

The rôle of the microclimate within a tree in determining fruit quality has long been recognized. Sites and Reitz (1949, 1950) studied the effects of light exposure on the rates of chemical changes in Valencia oranges, and correlated various chemical constituents with the position of the fruit on the tree. Total soluble solids content was highest in outside fruit, intermediate in fruit located in the canopy of the tree, and lowest in fruit located on the inside. In general, sun-exposed upper sectors of the canopy yield more fruit of higher quality than shaded lower or inside canopy sectors (Reitz and Sites, 1948). Fruit grown under full exposure to the sun were smaller, weighed less, had less juice, higher total soluble solids, lower acidity and more granulated carpels than fruit grown in the shade of the tree (Ketchie and Ballard, 1968). Extensive research by Syvertsen and Albrigo (1980) with grapefruit confirmed the relationship of direction of exposure to light and composition of fruit (Table III.III). Fruit from more exposed upper and southern canopy sectors apparently manifested higher maturation and internal respiration rates and, thus, produced fruit with lower acidity and higher Brix/acid ratios (sweeter juice).

TABLE III.III

The effect of canopy position on the ranking of °Brix/acid ratios and corresponding °Brix, percent acid and percent juice values. Each value is from 40 fruit from all three trees which were harvested in April 1978. N, S, E, and W correspond to the four cardinal directions, and t, m, and b refer to top, middle, and bottom, respectively (Syvertsen and Albrigo, 1980)

Sectors	°Brix/acid	°Brix	Acid (%)	Juice (%)
Nt	9.06	9.2	1.01	52.2
St	8.85	9.3	1.05	51.5
Sm	8.54	9.4	1.10	52.4
Em	8.36	9.3	1.12	51.9
Nm	8.13	8.9	1.09	53.1
Wm	8.07	9.2	1.14	52.1
Eb	7.96	9.3	1.17	51.2
Wb	7.79	8.8	1.13	50.9
Sb	7.68	9.3	1.21	49.9
Nb	7.47	9.0	1.21	51.7

5. MATURITY

Biochemical changes occur throughout a citrus fruit's growth and development periods, with the result that its sensory flavour quality varies considerably, depending upon its degree of ripeness or date of harvest. Geographical influences, seasonal weather conditions, rootstocks, fertilization and tree age are superimposed on the maturation process and profoundly affect external and internal fruit qualities. Lack of a well-defined ripening mechanism results in a more or less prolonged period of flavour acceptability during which the fruit of a given cultivar may be marketed.

Essentially, as the fruit matures, the changes affecting flavour include an increase in total sugars, a decrease in total acidity, an increase in orange or yellow

TABLE III.IV

Typical flesh condition as taste profiles through the 1936–1937 season of the chief Florida sweet orange cultivars on mature trees on rough-lemon rootstock in norfolk fine sandy soil in Windemere, central Florida (Harding et al., 1940)

Sampling dates	Flesh condition[a]			Taste[a]		
	Hamlin	Pineapple	Valencia	Hamlin	Pineapple	Valencia
Aug 31 – Sept 1	Ricey	Ricey		Very acid	Very acid	
Sept 13 – 16	Ricey	Ricey		Acid	Very acid	
Sept 24 – 30	Ricey	–		Tart	–	
Oct 7 – 14	Ricey	Ricey		Tart	Acid	
Oct 21 – 28	Ricey	Coarse		Tart	Acid	
Nov 4 – 11	Coarse	Coarse		Tart	Tart	
Nov 13 – 19	–	Coarse		–	Tart	
Dec 3 – 10	–	Coarse	Ricey	–	Tart	Very acid
Dec 16 – 26	Coarse	Coarse	Ricey	Pleasantly tart	–	Acid
Dec 28 – Jan 6	Coarse	Normal	Coarse	Pleasantly tart	Pleasantly tart	Acid
Jan 13 – 19	Coarse	Normal	Coarse	Pleasantly tart	Pleasantly tart	Tart
Jan 24 – 29	–	–	Coarse	–	–	Tart
Feb 10	Coarse	Normal	Coarse	Pleasantly tart	Pleasantly tart	Tart
Mar 1 – 7	–	–	Coarse	–	–	Pleasantly tart
Mar 13 – 16	Slightly dry	Normal	Coarse	Sweet	Pleasantly tart	Pleasantly tart
Apr 6 – 14			Normal			Pleasantly tart
Apr 21 – 28			Normal			Pleasantly tart
May 3 – 12			–			Pleasantly tart
May 21 – 27			Normal			Pleasantly tart
June 24 – July 1			Normal			Normal
July 27 – Aug 6			Slightly dry			Insipid

[a] As judged by 'several people'.

colour, an increase in volatile flavour constituents, and a decrease in bitter and astringent compounds. The internal flesh of immature fruits is ricey and coarse textured, very acidic or tart, and often associated with a green or immature sensory flavour (Fellers, 1985). For oranges, as the fruit matures, the internal flesh becomes juicy and smooth; the acidity decreases to partially tart to sweet; and typical aromatic and fruity flavours develop to varying degrees depending upon cultivar type. Table III.IV correlates the typical maturation progression for three major Florida orange cultivars with changing flavour and flesh conditions (Harding *et al.*, 1940). Table III.IV indicates the need to allow oranges to reach full maturity for optimum flavour quality.

The date at which fruit is harvested also has an important bearing on the metabolism of limonin in specific citrus biotypes, notably grapefruit. As noted in Fig. III.1, the limonin content of three grapefruit cultivars harvested early in the season, i.e. November, contained about three times as much limonin as fruit harvested in April (Scott, 1970). On the average, Marsh grapefruit (a seedless cultivar) contained more limonin than either the Ruby or Duncan (seedy cultivars). It has been observed in oranges as well as grapefruit that seedy cultivars generally have lower juice limonin values than seedless cultivars, such as the Washington Navel orange. Unfortunately, market and economic considerations many times preclude allowing the fruit to remain on the tree until the limonin level is reduced to an acceptable level.

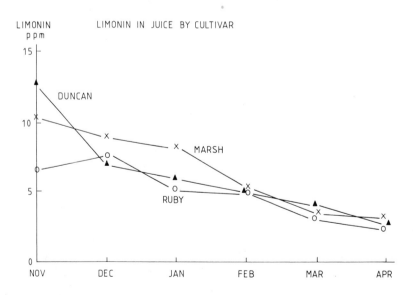

Fig. III.1. Effect of maturation on the limonin contents of Duncan, Marsh and Ruby grapefruit.

6. FRUIT COMPOSITION AND FLAVOUR QUALITY

6.1 Sugar – Acid Relationship

The total amount of sugar present and the ratio of total sugars to total acids present in the juice are so important to desirable flavour that these values are the primary criteria for determining legal maturity for oranges, grapefruit and tangerines in Florida (Citrus Fruit Laws, 1986). The most common measurement for total sugars in citrus juices is the degrees Brix value, which actually measures total soluble solids content in the juice. As the fruit matures, the increase in the total sugars parallels the increase in total soluble solids; the total sugars represents from 63 to 80% of the total soluble solids (Sinclair, 1961). Table III.V lists the approximate chemical composition of typical juices from Valencia orange, tangerine, and two grapefruit cultivars grown commercially in Florida, the Marsh seedless and a seedy variety (presumably Duncan). Of the 7 – 14% soluble solids, represented by degrees Brix in Table III.V, total sugars comprised about 70% of these soluble solids.

In citrus juices, the only sugars found in significant quantities are sucrose, glucose and fructose. In orange juice, sucrose, glucose and fructose occur in the approximate ratio of 2 : 1 : 1 (Curl and Veldhuis, 1948), although sucrose is the predominant sugar in the Valencia orange juice sample shown in Table III.V. Sucrose comprises more than 70% of the total sugars in the tangerine juice shown in Table III.V, while in grapefruit juice the reducing sugars predominate (Sinclair, 1972).

Florida maturity standards for oranges, tangerines and grapefruit involve total soluble solids (°Brix), ratio of Brix to acids, juice yield, and peel colour. Maturity standards for fresh oranges harvested early in the season (August 1 through October 31), when the fruit can have an immature flavour, require a

TABLE III.V

Chemical composition of some citrus juices (Roberts and Gaddum, 1937)

Analysis	Valencia orange	Tangerine	Grapefruit	
			Marsh seedless	Seedy
Water (%)	88.20	85.67	93.05	90.26
Degrees Brix	11.80	14.33	6.95	9.74
Citric acid (%)	0.98	1.26	0.89	1.35
Maturity ratio	12.04	11.37	7.81	7.22
pH	3.5	3.2	3.0	3.0
Sucrose (%)	5.02	7.89	1.34	2.24
Reducing sugar (%)	3.46	3.09	3.44	3.96
Total sugar (%)	8.48	10.98	4.78	6.20
Protein (N × 6.25) (%)	0.56	0.43	0.27	0.33
Ash (%)	0.412	0.401	0.218	0.414

minimum Brix of 9.0 and a minimum Brix/acid ratio of 10 to 1. For tangerines, the minimum Brix is 9.0 and the minimum ratio is 9 to 1. With fresh grapefruit, a distinction is made between seedless and seeded fruit. Seedless fruit must have a Brix of at least 7.5 and a ratio of 7 to 1, while for seeded grapefruit, the minimum Brix is 8.0 and the minimum ratio is 7 to 1.

The exact composition of the sugars has not been considered as important in citrus juice flavour as the total sugars and the sugar–acid ratio. Reducing sugars that are an equal mixture of glucose and fructose (invert sugar) would have a greater sweetness than sucrose. Only in the case of stored, canned grapefruit juice where hydrolysis has converted virtually all the sucrose present to invert sugar has the relative sweetness of the sugars present been considered (Nagy and Shaw, 1980).

Acidity in citrus juices is primarily due to the citric acid present, and total acids are the second most abundant class of soluble solids present (Table III.V). Malic acid is the only other organic acid present in juice in significant quantity. It is present at a rather constant level of about 0.2% in orange juice (Clements, 1964; Shaw et al., 1987), but the total acid value is usually reported as % citric acid as in Table III.V. Potassium and other salts of the acids are present in sufficient quantity to create a highly buffered solution in citrus juices. Thus, even though the acid content can vary from 0.7 to 2.6%, the pH of orange, tangerine and grapefruit juices generally is in the range of 3.0–3.5. Grapefruit juice is usually at the lower end of this range, while orange and tangerine juices are usually at the higher end (Table III.V). As fruit becomes overly mature, the total soluble solids decreases and the total acidity decreases disproportionately, creating the bland, watery, undesirable flavour character of overripe fruit (Fellers, 1985).

As citrus fruit mature, the total acidity drops and the total sugars increases gradually, and, at the same time, other desirable changes are taking place in the fruit. These changes include: increase in juice content, decrease in green colour as well as development of orange colour in oranges and tangerines and development of yellow colour in grapefruit, decrease in bitter and astringent juice components, development of desirable water-soluble and oil-soluble flavour components, and a change in texture of the flesh from coarse and ricey to smooth and tender. As fruit becomes overly mature, the texture can become ricey and dry, while the flavour becomes more bland (acidity decreases) and a stale, over-ripe off-flavour develops (Fellers, 1985).

An important factor having impact on the flavour of fresh citrus fruit is the variation in composition within a single fruit, as well as fruit-to-fruit variability (Sinclair, 1961). Generally, the stylar end of mature citrus fruit, especially Valencia orange, was found to have a higher concentration of soluble solids than the stem end. The stylar end also tended to have higher acids and higher sugar content than did the stem end. Variation among segments within a single fruit could be pronounced, with the Brix varying from 12.61 to 16.11% between individual segments. Size of fruit has been an important factor in soluble solids content as well. In Valencia oranges from a single grove, the Brix increased with a decrease in fruit size. These differences indicate that even while eating a single fruit, the consumer can experience flavour differences between bites.

6.2   Overripe and Freeze-damaged Fruit

As fruit are held on the tree past maturity, several undesirable changes occur. The total soluble solids and the juice yield decrease, and the fruit becomes dry and ricey in texture (Sinclair, 1961). The total acidity drops, resulting in a bland, watery flavour. In addition, the flavour becomes characteristically stale, old, or overripe, and in cases where seed sprouting occurs, a strong off-flavour develops (Fellers, 1985).

Similar changes can occur in freeze-damaged fruit, along with additional undesirable changes (Kefford and Chandler, 1970). The changes in orange and grapefruit juice composition resulting from freeze-damage (summarized by Bissett, 1958) include loss of juice, reduction in soluble solids and acid, an increase in the pH, physical breakdown of fruit tissue resulting in drying or granulation of juice sacs and formation of unsightly white hesperidin crystals. Other changes noted by Kefford and Chandler (1970) include a rapid increase in the Brix/acid ratio and increases in flavonoids, limonoid bitter principles, and pectinesterase activity. Soon after freeze-damage has occurred, the flesh becomes dry and translucent in appearance. The flavour is not affected initially, but the above changes intensify as the fruit remains on the tree.

6.3   Lemon and Lime Juices

With respect to the composition of the juice from fresh lemons and limes, the sugar/acid balance is the reverse of that found in other citrus fruit described above. As lemon and lime fruit mature, the total soluble solids and total acids increase, while the total sugars decreases (Sinclair, 1984; Swisher and Swisher, 1980). In these mature fruit, the acid, which is largely citric acid, comprises 60 – 75% of the total soluble solids, and the total sugars constitutes about 1% of the juice by weight (Table III.VI). Lemons are harvested while still green, and cured under controlled storage conditions (Swisher and Swisher, 1980). During storage, the fruit colour changes to whitish yellow, the juice sacs enlarge, the peel becomes thinner and it develops a thicker waxy layer that retards moisture evaporation and extends its shelf life. Limes are generally not stored, but utilized for the fresh fruit and juice markets shortly after harvest. Long-term and controlled atmosphere storage are not recommended for Florida limes (Grierson and Ben-Yehoshua, 1986); however, for California limes, Kader (1985) recommended a storage of no longer than 6 – 8 weeks at 10 – 12°C. Commercial Tahiti lime maturity in Florida is arbitrarily based on a minimum juice content of 42% by volume, and a minimum diameter of 45 mm; limes must be green to be sold as U.S. No. 1 grade (Campbell, 1979). Although lemon and lime juices are not consumed directly in undiluted form, they are widely used as flavourful and healthful acidulants in foods. Historically, they have been widely used in the diet to prevent scurvy; lemon juice has probably been more widely used as a flavour enhancer than any other food item, except sugar and salt (Swisher and Swisher, 1980).

7. PECTIN AND STRUCTURAL CARBOHYDRATES

The higher molecular weight carbohydrates that do not contribute to sweetness of the fruit or juice still play some rôle in citrus flavour. The water-soluble polysaccharides add body to the juice, and thus contribute to a desirable juice quality. Juice lacking such body is considered thin, watery and inferior, because it lacks a desirable expected mouth-feel (Baker, 1980). These water-soluble polysaccharides are perhaps more important to the flavour of freshly expressed juice than they are to the flavour of citrus flesh consumed directly. The insoluble carbohydrates are primary components of the structural material in the fruit. They include albedo, rag, core, segment membranes and juice sac walls. Expressed juice contains relatively uniform small pieces of all these structural materials (Scott *et al.,* 1965). As a consumer macerates a fruit segment, varying amounts and sizes of all the above structural carbohydrates become mixed with the juice and create mouth-feel as well as provide certain flavour components to the overall sensation. More insoluble fiber is consumed by a person eating a fresh fruit than in expressed juice from the same fruit.

TABLE III.VI

Chemical composition of lemon and lime juice

Constituent	Juice (g/100 g)	
	Lemon[a]	Lime[b]
Moisture	92.36	–
Soluble solids (°Brix)	8.30	10.0
Citric acid	5.98	5.97
pH	2.2	–
Sugar		
Total	1.17	0.86
Sucrose	0.09	0.14
Reducing	1.09	0.72
Ash	0.25	0.35

[a] Adapted from 'The Lemon', University of California Press, Oakland, CA, 1984, p. 81.
[b] Adapted from 'Citrus Science and Technology', Vol. 1, S. Nagy, P.E. Shaw and M.K. Veldhuis (Editors), AVI, Westport, CT, p. 172.

8. BITTERNESS AND ASTRINGENCY

Fresh grapefruit have a natural bitterness that is a desirable part of the flavour profile, provided it is not excessive. Bitterness in fresh grapefruit is mostly due to the presence of the flavanone glycoside, naringin. Smaller amounts of two other bitter glycosides, neohesperidin and poncirin, are also present (Table III.VII). Naringin and poncirin are about equal in bitterness, while neohesperidin was about one-tenth their bitterness (Horowitz and Gentili, 1977). These substances are at their highest levels in juice sacs of immature grapefruit, but the levels decrease until the fruit reaches maturity (late November in Florida), and then remain constant throughout the harvesting season. These changes are probably caused by dilution (increase in juice volume) as the fruit matures (Hagen et al., 1966).

Limonin (Fig. III.2) is the major bitter component of processed juices from navel and Shamouti oranges, the Natsudaidai (natural pummelo hybrid) in Japan, and it is the second major bitter component of processed grapefruit juice. Limonin is present only in trace quantities in fresh fruit or in freshly ex-pressed juice (Maier et al., 1977). A nonbitter precursor, limonoic acid A-ring lactone (Fig. III.2), is present in the fruit and is converted to the bitter principle, limonin, by acid-catalyzed and enzyme-catalyzed ring closure once the juice sacs are ruptured during juice extraction. These reactions are slow enough so that it takes about 30 min standing at room temperature after extraction for juice to develop a sufficient quantity of limonin to impart a bitter flavour. The flavour threshold of limonin in navel orange juice is in the 6 – 9 ppm range, but the threshold is highly dependent on the individual taster, as well as on sugar and acid content and pH of the juice (Guadagni et al., 1973). Many orange and mandarin cultivars contain low levels of the precursor of limonin, therefore the

TABLE III.VII

Monthly variation in concentration of bitter flavanone glycosides in juice sacs of Texas Ruby Red grapefruit (Hagen et al., 1966)

Harvest date	Concentration in juice sacs (μg/g wet weight)		
	Naringin	Neohesperidin	Poncirin
July 28	1630	79.5	59.9
Sept 28	1010	45.4	42.7
Oct 28	786	35.5	31.7
Nov 28	615	27.6	22.4
Dec 28	642	31.4	23.9
Jan 28	494	24.2	18.8
Feb 28	499	29.4	21.3
Mar 28	507	21.4	20.0
April 28	484	26.2	25.7

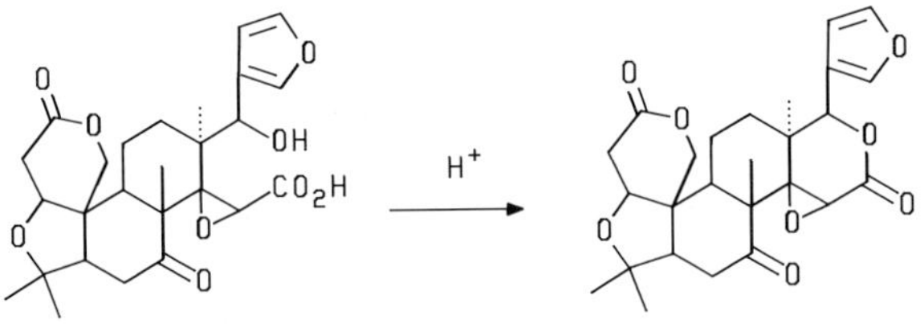

Fig. III.2. Conversion of limonoic acid A-ring lactone to limonin by acid catalysis.

level of limonin upon juice extraction rarely reaches the threshold level of 6 – 9 ppm. Guadagni *et al.* (1974) showed the bitterness of naringin and limonin to be additive in water solution. Since both naringin and the limonin precursor occur in grapefruit, the bitterness of grapefruit juice increases when freshly-expressed juice is allowed to stand under conditions favouring limonin formation.

Astringency is an unpleasant flavour encountered in barely mature citrus fruit that is usually associated with bitterness and with green, immature flavour. Astringency is actually a sensation of touch from shrinking of exposed tissues in the mouth. Coumarins and psoralins that are present in relatively high levels in immature grapefruit and sour orange have been implicated in immature flavour in grapefruit (Berry and Tatum, 1986). Coumarins and psoralins are not present in appreciable quantities in juice from immature orange and mandarin cultivars, but an astringent glycoside, coniferin, has been identified in pulp from Valencia orange, Duncan grapefruit and Murcott tangor (Tatum and Berry, 1987). These astringent compounds are in very low concentration in citrus juice sacs. Thus, a person consuming citrus sections or carefully hand-squeezed fresh juice would not experience the degree of astringency associated with processed early season fruit, where these astringent compounds have been introduced into the juice from peel juice and peel oil.

9. VOLATILE FLAVOUR COMPONENTS

The volatile flavour components in all citrus varieties can be separated into two broad and distinct categories, the oil-soluble constituents present in peel oil and in juice oil, and the water-soluble constituents present in the juice.

9.1 Oil-soluble Components

Citrus fruits have peel oil located in small, ductless glands present in the flavedo, or outer portion of the peel. Volatiles emitted from these glands are responsible for the characteristic, pleasant citrus aroma of the intact fruit (Kealey and Kinsella, 1978). Norman *et al.* (1967) used gas chromatography of air samples surrounding intact and injured Valencia oranges to show that volatile peel oil components were emitted from the fruit. The number of volatile components detected and the quantity of each component increased with increasing temperature, and increased dramatically (75-fold) when the fruit was injured. Attaway and Oberbacher (1968) studied two types of aromas emanating from intact Hamlin oranges. Oranges stored in a closed container emitted a fruity aroma characteristic of ethyl butanoate. Gas chromatographic analysis showed the major volatile components in the headspace gases to be ethyl acetate, ethyl butanoate, ethyl hexanoate, ethyl octanoate, ethanol and limonene. The cuticle (external surface) of the fruit was found to contain sesquiterpene hydrocarbons, mainly valencene, plus smaller amounts of elemene, caryophyllene, farnesene, humulene and cadinene. The fruit was dipped in dichloromethane to isolate this mixture of sesquiterpene hydrocarbons. These hydrocarbons were responsible for the persistent odour that remained on the hands of personnel handling the fruit.

The peel oil from each citrus variety provides much of the characteristic aroma and flavour in juice products made from that fruit. However, the peel oil has no influence on the flavour of the fruit sections containing the juice sacs, unless some peel oil is mechanically transferred to the fruit sections or juice as the fruit is either peeled or juiced. Careful hand-peeling of fruit would introduce little or no peel oil into the fruit sections. Hand-juicing of the fruit invariably introduces a small amount of peel oil into the juice (Huet, 1969). If one considers the flavour of the intact fruit, however, then the juice oil present in the juice sacs contributes the oil-soluble components to the flavour of the fruit.

Juice sacs in most citrus species contain an oil that is somewhat different in composition from that of peel oil. Juice oil droplets were first reported by Pennzig (1887) to be present in globular bodies within the juice sacs from several citrus species. Davis (1932) also observed oil droplets in juice sacs from orange and several other citrus fruit. Rice et al. (1952) compared the juice oil content of juice from: (1) oranges that had been carefully hand-peeled prior to hand extraction to exclude introduction of peel oil; and (2) unpeeled oranges similarly hand-extracted (Table III.VIII). Data in Table III.VIII show juice oil to be present in orange and in grapefruit juices at about 0.005%. Even hand-extraction of unpeeled Valencia oranges introduced peel oil into the juice, approximately doubling the total measurable oil present. The oil content of a commercially extracted juice cited in Table III.VIII was about equal to that in hand-extracted juice. However, commercial extractors can introduce too much oil into the juice. If the juice is to be concentrated, the oil is removed during the concentration process. However, if the juice is to be sold as a single-strength product, then it must be de-oiled, usually by flash distillation under vacuum (Sinclair, 1984). Florida regulations require that juice reconstituted from frozen concentrated orange juice has an oil content of 0.010 – 0.035%, while that from grapefruit juice has an oil content of 0.008 – 0.020%. Thus, processed citrus products depend on peel oil to provide some of the characteristic flavour, while this is not the case when fresh citrus is consumed.

TABLE III.VIII

Oil content of orange and grapefruit juices

Juice sample	Oil content	Reference	
Orange			
Hand-extracted			
Peeled Calif. Valencia	0.004 – 0.006	Rice et al.	(1952)
Unpeeled Calif. Valencia	0.008 – 0.012	Rice et al.	(1952)
Commercially extracted	0.008 – 0.014	Rice et al.	(1952)
Reconstituted from FCOJ[a]	0.010 – 0.035	Florida Department of Citrus	(1975)
Grapefruit			
Hand-extracted	0.005	Sinclair	(1961)
Reconstituted from FCGF[b]	0.010 – 0.035	Florida Department of Citrus	(1975)

[a] FCOJ = frozen concentrated orange juice.
[b] FCGF = frozen concentrated grapefruit juice.

In the few studies that have been carried out on juice oil, significant differences were found when composition of juice oil and peel oil were compared (Shaw, 1977a). Wolford et al. (1971) compared Valencia orange juice oil with cold-pressed Valencia orange oil, and found the ester content of juice oil to be 7 to 18 times greater than that of peel oil, while the aldehyde content was relatively low in juice oil. The high ester content probably explained the 'fruity note' present in the aroma of juice oils. Hunter and Brogden (1965) found differences in the sesquiterpene hydrocarbons of grapefruit juice and peel oils. The major difference was that the principle sesquiterpene in juice oil, valencene, was not detected in the peel oil.

One other consideration involving juice oil in freshly expressed juice, is the physical location of the oil. Scott et al. (1965) found that, immediately after juice extraction, oil was present in the clear serum of the juice. Upon standing, the oil became associated with the pulp with a resultant decrease in the stability of the oil components. A rapid loss in aroma of freshly extracted juice may be caused by this oil – pulp complex formation (Scott et al., 1965).

It is clear that flavour associated with the consumption of fresh citrus fruit differs from that in juice extracted from the fruit, in that little or no flavour from the peel oil is present when fresh fruit is consumed. Thus, the volatile oil-soluble flavour components important to fresh fruit flavour are those present in the juice sacs.

9.1.1 Effects of abscission chemicals

When citrus fruit is harvested mechanically (less than 0.5% of fruit in Florida), abscission chemicals are applied several days before harvest to loosen the fruit. Most of these abscission chemicals act by damaging the peel, causing the release of 'wound' ethylene which promotes abscission. These abscission agents have been found to cause flavour changes in the peel oil and juice from treated oranges (Moshonas *et al.*, 1976; Moshonas and Shaw, 1977). An objectionable 'overripe' flavour is noted in juice or peel oil from treated fruit; introduction of peel oil into the juice during extraction could have been the cause of the overripe flavour in juice (Moshonas and Shaw, 1980).

The abscission agents caused formation of six phenolic ethers in peel oil which had not previously been detected in citrus (Moshonas and Shaw, 1978). One of these, eugenol, was present at about its threshold level in orange juice (21 ppb). When added to orange juice, these phenolic ethers contribute an overripe flavour, similar to that noted in juice from oranges treated with abscission chemicals.

9.1.2   Effects of growth regulators

The major contribution of growth regulators is in enhancing external fruit qualities (rind colour, size, firmness, general fruit appearance) and shipping quality; they have limited impact on internal quality (arsenic, which reduces acidity, is a notable exception). Yokoyama *et al.* (1986) have used a growth regulator to change the composition of lemon oil during maturation of the fruit. Peel oil from lemons treated with 2-(3,4-dichlorophenoxy)triethylamine contained greater proportions of aldehydes, alcohols and esters than did peel oil from control fruit. The aldehydes (mostly citral) are the most important lemon oil flavour components.

9.2   Water-soluble Components

Most of the characteristic flavour of fresh citrus fruit comes from the water-soluble volatile components present in the juice sacs (Huet, 1969). Only recently have analytical techniques become sensitive enough through the development of capillary gas chromatography and headspace analysis techniques for the volatile flavour components present in citrus juices to be determined directly and quantitatively. A large amount of work has been carried out on extracts of juices or commercially prepared aqueous essences from juices (Shaw, 1977b). Using these techniques, the volatile flavour components have been concentrated to make their detection and identification easier.

9.2.1   Orange

The most thorough investigation of volatile components in fresh orange fruit was carried out by Schreier (1981) and Schreier *et al.* (1977). They blended carefully peeled fresh Sanguinello oranges with methanol to prevent enzymic degradation, extracted the methanol – water mixture, chromatographed the extract, used internal standards to compensate for losses during the experimental procedures, and obtained the results shown in Table III.IX. The contribution of most of these compounds to orange flavour has not been determined.

A study at the University of Florida was conducted using taste panels to evaluate the contribution of 15 of the components of orange juice and peel oil

114

TABLE III.IX

Concentration of components in peeled Sanguinello orange fruit (Schreier *et al.*, 1977)

Component	$\mu g\ l^{-1}$	Component	$\mu g\ l^{-1}$
α-Pinene	178	Butyl octanoate	36
β-Pinene	8	Linalyl acetate	12
Myrcene	690	Menthyl acetate	6
Limonene	70 000	Citronellyl acetate	12
γ-Terpinene	30	Neryl acetate	20
p-Cymene	17	Geranyl acetate	15
Terpinolene	10	Ethyl 3-hydroxybutanoate	650
allo-Ocimene	85	Methyl 3-hydroxyhexanoate	100
β-Caryophyllene	64	Ethyl 3-hydroxyhexanoate	700
Farnesene	48	Ethyl 3-hydroxyoctanoate	30
α-Humulene	30	Linalool	215
Valencene	5000	4-Terpineol	47
δ-Cadinene	31	α-Terpineol	55
Selinadiene	150	Nerolidol	10
Hexanal	20	Geraniol	11
Octanal	25	Citronellol	21
Nonanal	15	Nerol	20
Decanal	77	*trans-p*-Menthen-9-ol	22
Citral	58	*cis*-Linalool oxide	10
β-Ionone	25	*trans*-Linalool oxide	50
Perillaldehyde	30	3-Methylbutan-1-ol	569
Ethyl butanoate	900	Hexan-1-ol	30
Ethyl 2-methylbutanoate	22	*cis*-Hex-3-en-1-ol	54
Butyl butanoate	13	Octan-1-ol	17
Ethyl hexanoate	52	2-Phenylethanol	20
Ethyl octanoate	27		

TABLE III.X

Best mixtures evaluated for orange juice flavour (Ahmed, 1975; Shaw *et al.*, 1977; Ahmed *et al.*, 1978)

Mixture[a]
Three components
Acetaldehyde + ethyl butanoate + limonene
Citral + limonene + α-pinene
Ethyl butanoate + limonene + α-pinene
Four components
Acetaldehyde + citral + ethyl butanoate + limonene
Five components
Acetaldehyde + citral + ethyl butanoate + limonene + linalool
Acetaldehyde + citral + ethyl butanoate + limonene + α-pinene

[a] Each mixture tested in bland orange juice.

115

to orange flavour (Ahmed, 1975; Shaw *et al.,* 1977; Ahmed *et al.,* 1978). A high quality, commercial frozen concentrated orange juice that contained added aqueous essence, essence oil and peel oil flavour fractions, was used as the reference standard. The bland, concentrated orange juice containing no added flavour fractions that had been used as a base in blending the reference standard was used for evaluating mixtures of 2 – 15 known orange flavour components. Several of the three- to five-component mixtures shown in Table III.X received the highest flavour scores, which were 80 – 85% of the scores received by the reference juice. In these studies, a processed orange juice flavour, rather than a fresh orange juice flavour, was being evaluated.

TABLE III.XI

Volatile flavour components in fresh Marsh grapefruit juice (Nunez *et al.,* 1985)

Component	Component
Butan-1-ol	Umbellulone[a] (1-isopropyl-4-methylbicyclo-
3-Methyltetrahydropyran[a]	[0.1.3]hex-3-en-2-one)
Methoxypropoxymethane[a]	*p*-Menth-1-en-9-ol[a]
3-Methylbut-2-en-1-ol[a]	*p*-Menth-3-en-9-ol[a]
Furfural (2-furaldehyde)	*p*-Cymen-α-ol[a]
trans-Hex-2-enal	α-Terpineol
Ethyl 2-methylbutanoate	Decanal
α-Pinene	*p*-Mentha-4,8-dien-9-ol[a]
cis-Limonene oxide	*p*-Mentha-1,8-dien-9-ol[a]
2,6,6-Trimethyl-6-vinyltetrahydropyran[a]	*cis*-Carveol
Myrcene	*trans*-Carveol
α-Phellandrene[a]	Carvone
trans-Anhydrolinalool oxide[a]	Linalyl acetate
Phenylacetaldehyde[a]	Geraniol
α-Phellandrene[a]	2-Methoxy-4-vinylphenol[a]
Limonene	Mentha-1,8-dien-7-ol
cis-β-Ocimene	α-Copaene
trans-β-Ocimene	*cis*-Caryophyllene[a]
cis-*p*-Mentha-2,8-dien-1-ol	α-Cubebene
cis-Linalool oxide	β-Cubebene
1-Isopropenyl-4-methylbenzene[a]	*trans*-Caryophyllene[a]
trans-Linalool oxide	β-Caryophyllene
Linalool	α-Humulene
2,6,6-Trimethylcyclohex-2-enone[a]	β-Copaene
β-Terpineol	δ-Cadinene
4-Hydroxy-2,6,6-trimethyl-2-vinyltetra-hydropyran[a]	β-Caryophyllene alcohol[a]
	Elemol
5-Hydroxy-2,6,6-trimethyl-2-vinyltetra-hydropyran[a]	β-Nerolidol
	δ-Cadinol[a]
2-Isopropenyl-3-methylenecyclohex-4-enol[a]	Nootkatone

[a] Newly reported for grapefruit flavour.

9.2.2 Grapefruit

Volatile components in freshly expressed juice from Marsh grapefruit were recently reported by Nunez *et al.* (1985) as listed in Table III.XI. These workers did not comment on the flavour characteristics of individual components identified. The grapefruit juice components that have been shown to be important to juice flavour include aldehydes, esters and nootkatone (Shaw and Wilson, 1980). A sulphur-containing compound, *p*-menth-1-ene-8-thiol, isolated from grapefruit juice, was shown to possess the aroma of fresh grapefruit juice (Demole *et al.*, 1982). This compound has a taste threshold in water of 10^{-4} ppb, which is the lowest taste threshold recorded so far for any naturally-occurring flavour compound (Demole *et al.*, 1982).

9.2.3 Mandarin

The mandarin, or tangerine, group contains a number of species, and is a more diverse citrus group than are oranges or grapefruit (Hodgson, 1967). Mandarin peel oil has been studied extensively (Kugler and Kovats, 1963), and methyl *N*-methylanthranilate (methyl 2-(*N*-methylamino)benzoate), thymol, α-pinene and γ-terpinene have been shown to contribute to mandarin flavour in the oil (Wilson and Shaw, 1981). As with the other citrus flavours discussed above, the flavour of the fresh fruit would be little affected by peel oil.

Volatile components in Satsuma mandarin juice were determined using fruit peeled prior to extraction to minimize peel oil in the juice. Yajima *et al.* (1979) identified 68 volatile juice components, and listed their GC area percent values in a concentrated extract from juice. The juice volatiles contained a much higher percentage of oxygenated compounds than did peel oil (40 versus 2%). These oxygenated components were thought to contribute to the characteristic sweet aroma of the juice. The main juice volatiles were 3-methylbutan-1-ol, *trans*-hex-2-enal and hexanal. Juice components identified that were not in the peel oil included *trans*-pent-2-enal, *cis*-hex-3-enal, *trans*-hex-2-enal, *trans*-oct-2-enal and *trans*-non-2-enal. Some of the unsaturated aldehydes and alcohols identified by Yajima *et al.* (1979) contribute 'fruity' and 'green' flavour notes in low concentrations (Arctander, 1969). Among the identified components were thymol, α-pinene and γ-terpinene discussed above.

Headspace gases above juice from Satsuma mandarins extracted with a screw press were analyzed by Ohta *et al.* (1982), and 20 volatile components were identified (Table III.XII). The screw press used peeled fruit, and extracted juice contained less peel oil than juice from an FMC in-line extractor. The authors cited linalool, citronellal and aliphatic aldehydes as important to Satsuma mandarin flavour.

9.2.4 Lemon

Volatile components of lemon juice were studied extensively by Mussinan *et al.* (1981). They carefully hand-extracted juice to minimize introduction of peel oil, treated the juice with ethanol to prevent enzymic reactions, extracted the mixture with Freon 11, concentrated the extract, and used trained flavourists to

TABLE III.XII

Volatile components in headspace gases above fresh Satsuma mandarin juice (Ohta *et al.*, 1982)

Compound	Peak area (%)
Acetaldehyde	0.09
Acetone	1.32
Ethanol	67.23
α-Pinene	0.17
Camphene	0.08
β-Pinene + sabinene	0.08
Myrcene	0.66
α-Terpinene	0.13
d-Limonene	26.96
γ-Terpinene	1.14
p-Cymene	0.09
Terpinolene + octanal	0.07
A methylheptenone	0.10
Nonanal	Trace
Citronellal	0.07
Decanal	0.01
Linalool	0.13
Caryophyllene	0.09

evaluate the extract to ensure the presence of a fresh lemon juice aroma and flavour. Over 300 compounds were identified in this study, and the identities of those listed in Table III.XIII were reported. Quantitative data on the major components of the extract were also reported (Table III.XIII), as well as aromas of the ethers identified. One of the ethers, p-cymen-8-yl ethyl ether, possessed a lemon juice-like flavour. Only one sulphur-containing compound, dimethyl sulphide, was detected in this study through the use of a flame photometric detector especially sensitive to sulphur compounds. Shaw and Wilson (1982), using a similar detector in an all-glass gas chromatographic system, identified hydrogen sulphide as the main sulphur-containing component in fresh juice from all major citrus species. They also identified traces of dimethyl sulphide, methanethiol and dimethyl disulphide in a few samples.

The biphenyl identified in lemon juice was probably an artifact (Table III.XIII). Biphenyl is a fungicide commonly used with fresh citrus fruit, and its presence in aqueous citrus essences has been reported (Shaw, 1977b).

Most of the compounds listed in Table III.XIII have been identified in lemon peel oil or in aqueous lemon essence (Shaw, 1977a,b). The most important lemon flavour compound is citral, which is a mixture of neral and geranial. In lemon peel oil, the ratio is about 40% neral to 60% geranial (Shaw, 1977a). Quantitative data in Table III.XIII show citral in lemon juice to contain 95% geranial and only 5% neral.

TABLE III.XIII

Volatile components of lemon juice (Mussinan *et al.,* 1981)

Component	Component
Hydrocarbons	**Ethers**
Limonene (70.49)[a]	α-Terpinyl methyl ether
β-Pinene + myrcene (5.03)	α-Terpinyl ethyl ether
α-Pinene (0.56)	β-Terpinyl methyl ether
γ-Terpinene (7.52)	β-Terpinyl ethyl ether
	4-Terpinyl ethyl ether
Aldehydes	Bornyl methyl ether
Geranial (1.44)	Bornyl ethyl ether
Neral (0.07)	Isobornyl methyl ether
Octanal (0.12)	Isobornyl ethyl ether
Nonanal (0.10)	Myrcenyl methyl ether
	Fenchyl ethyl ether
Esters	*p*-Cymen-8-yl ethyl ether
Geranyl acetate (0.44)	Geranyl ethyl ether
Neryl acetate	Linalyl ethyl ether
Citronellyl acetate (0.04)	Sabinyl ethyl ether
α-Terpinyl acetate	*p*-Menth-1-en-3-yl ethyl ether
	Carvyl ethyl ether
Monoterpene alcohols	1,8-Dimethoxy-*p*-menthane
4-Terpineol (2.5)	
α-Terpineol (2.5)	**Acids**
trans-Carveol (0.8)	2-Methylbutanoic acid
Linalool (0.28)	Hexanoic acid
	Heptanoic acid
Sesquiterpene alcohols	Octanoic acid
α-Bisabolol	Decanoic acid
Campherenol	Geranic acid
Isocampherenol	Tetradecanoic acid
Caryophyllene alcohol	Hexadecanoic acid
Ledol	
Nerolidol	**Miscellaneous**
1,3-Dimethyl-3-(4-methylpent-3-enyl)norbornan-2-ol	Dimethyl sulphide
	Biphenyl[b]
2,3-Dimethyl-3-(4-methylpent-3-enyl)norbornan-2-ol	

[a] Quantities in parentheses are percentage of concentrated extract from juice as determined by GC.
[b] Possible artifact.

9.2.5 Lime

Limes are similar to lemons in structure and composition, and the two juices have about the same citric acid content (Swisher and Swisher, 1980). Although lime juice has many of the same uses as an acidulant as lemon juice, the flavour of lime beverages is distinctly different from that of lemon beverages. The components in lime peel oil and distilled lime oil have been studied extensively

(Shaw, 1977a; Chamblee *et al.*, 1985). However, no comprehensive study on lime juice volatiles has been carried out in which peel oil has been excluded. Aqueous lime essence has been studied, and the 30 components identified are listed in Table III.XIV (Moshonas and Shaw, 1972). Because of the method of collection, aqueous essence contains a mixture of peel oil and juice components (Shaw, 1977b).

The presence of citral is important to lime flavour, as well as to lemon flavour. Clark *et al.* (1987) isolated a sesquiterpene hydrocarbon, germacrene B (Fig. III.3), from cold pressed lime oil that was judged to be important to the 'fresh lime peel character'. The importance of the other components in Table III.XIV to fresh lime flavour is not well documented, although geranyl acetate is used in compounded lime flavours (Arctander, 1969). γ-Terpinene, which is

TABLE III.XIV

Compounds identified in lime essence

Alcohols	Esters
Borneol	Ethyl acetate
p-Cymen-8-ol	Neryl acetate
Ethyl alcohol	
Geraniol	Hydrocarbons
cis-Hex-3-en-1-ol	*p*-Cymene
3-Methylbutan-1-ol	Limonene
Isopropyl alcohol	γ-Terpinene
Linalool	
cis-*p*-Menth-2-en-1-ol	Ketones
Methyl alcohol	Acetone
2-Methylbutan-2-ol	Piperitenone
2-Methylbut-3-en-2-ol	
3-Methylbut-2-en-1-ol	Oxides
Nerol	1,8-Cineol
	cis-Linalool oxide
Aldehydes	*trans*-Linalool oxide
Acetaldehyde	
Geranial	
Hexanal	
Hex-2-enal	
Neral	
Perillaldehyde	

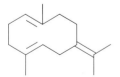

Fig. III.3. Germacrene B.

present in cold pressed lime oil in greater quantity than in any other citrus oil (Shaw, 1977a), has a citrus-like aroma and flavour (Arctander, 1969), and perhaps it modifies the basic lemon-like flavour contributed by citral.

In conclusion, the preponderance of the more volatile flavour components in fresh citrus juice sacs, as opposed to the heavier flavour notes present in peel oil, tend to give the fresh fruit a more fruity, estery flavour than is present in juice extracted under conditions where peel oil is present also. Citrus product formulation has relied on peel oil to provide basic flavour notes characteristic of the specific citrus fruit, and aqueous essence and essence oil to provide 'top-notes' necessary to make the juice taste more like that in juice sacs. Enzymic reactions and chemical reactions caused by acid, oxygen, and heat all act to change the flavour of the juice present in the intact fruit as soon as it is released during extraction. Better understanding of the composition of juice within the fresh fruit will help processors to produce products retaining more of the fresh fruit flavour.

REFERENCES

Ahmed, E.M., 1975. Flavor enhancing potential of selected orange oil and essence components and their relationship to product quality. U.S. Dept. of Agric., Agric. Res. Serv., Contract No. 12-14-100-10337(72), June 23, 1970 – March 23, 1973.

Ahmed, E.M., Dennison, R.A. and Shaw, P.E., 1978. Effect of selected oil and essence volatile components on flavor quality of pumpout orange juice. *J. Agric. Food Chem.,* 26: 368 – 372.

Albach, R.F. and Redman, G.H., 1969. Composition and inheritance of flavanones in citrus fruit. *Phytochemistry,* 8: 127 – 143.

Arctander, S., 1969. Perfume and Flavor Chemicals, Vol I. S. Arctander, Montclair, NJ.

Attaway, J.A. and Oberbacher, M.F., 1968. Studies on the aroma of intact Hamlin oranges. *J. Food Sci.,* 33: 287 – 289.

Baker, R.A., 1980. The role of pectin in citrus quality and nutrition. In: S. Nagy and J.A. Attaway (Editors), Citrus Nutrition and Quality, American Chemical Society Symposium No. 143. Washington DC, pp. 109 – 128.

Barrett, H.C. and Rhodes, A.M., 1976. A numerical taxonomic study of affinity relationships in cultivated *Citrus* and its close relatives. *Sys. Bot.,* 1: 105 – 136.

Berry, R.E. and Tatum, J.H., 1986. Bitterness and immature flavor in grapefruit: analyses and improvement of quality. *J. Food Sci.,* 51: 1368 – 1369.

Bissett, O.W., 1958. Processing freeze-damaged oranges. *Proc. Fla. State Hortic. Soc.,* 71: 29 – 31.

Black, C.A., 1968. Soil – Plant Relationships, 2nd edition. John Wiley, New York.

Bowden, R.P., 1968. Processing quality of oranges grown in the near North Coast area of Queensland. *Queensl. J. Agric. Anim. Sci.,* 25: 93 – 119.

Campbell, C.W., 1979. Tahiti lime production in Florida. University of Florida Bulletin 187, Gainesville, 45 pp.

Castle, W.S. and Phillips, R.L., 1980. Performance of Marsh grapefruit and Valencia orange trees on eighteen rootstocks in a closely spaced planting. *J. Am. Soc. Hortic. Sci.,* 105: 496 – 499.

Chamblee, T.S., Clark, B.C. Jr., Radford, T. and Iacobucci, G.A., 1985. Identification of new constituents in cold pressed lime oil. *J. Chromatogr.,* 330: 141 – 151.

Chapman, H.C., 1968. The mineral nutrition of citrus. In: W. Reuther, L.D. Batchelor and H.J. Webber (Editors), The Citrus Industry, Vol. II. University of California Press, Berkely, pp. 127 – 289.

Childers, N.F., 1975. Modern Fruit Science. Horticultural Publications, Rutgers Univ., New Brunswick, NJ.

Citrus Fruit Laws, 1986 edition, Chapter 601, Florida Citrus Code. State of Florida, Department of

Citrus, Lakeland, FL.

Clark, B.C. Jr., Chamblee, T.S. and Iacobucci, G.A., 1987. HPLC isolation of the sesquiterpene hydrocarbon germacrene B from lime oil and its characterization as a flavor impact constituent. *J. Agric. Food Chem.,* 35: 514–518.

Clements, R.L., 1964. Organic acids in citrus fruits. I. Varietal differences. *J. Food Sci.,* 29: 276–286.

Cook, J.A., Horanic, G.E. and Gardner, F.E., 1952. Citrus rootstock trials. *Proc. Fla. State Hortic. Soc.,* 65: 69–77.

Cooper, W.C. and Chapot, H., 1977. Fruit production – with special emphasis on fruit for processing. In: S. Nagy, P.E. Shaw and M.K. Veldhuis (Editors), Citrus Science and Technology, Vol. 2. AVI Publishing Co., Westport, CT, pp. 1–127.

Curl A.L. and Veldhuis, M.K., 1948. The composition of the sugars in Florida Valencia orange juice. *Fruit Prod. J.,* 27: 342–343, 361.

Davis, W.B., 1932. Deposits in the juice sacs of citrus fruits. *Am. J. Bot.,* 19: 101–105.

Demole, E., Enggist, P. and Ohloff, G., 1982. 1-*p*-Menthene-8-thiol: A powerful flavor impact constituent of grapefruit juice (*Citrus paradisi* Macfayden). *Helv. Chim. Acta,* 65: 1785–1794.

Embleton, T.W., Reitz, H.J. and Jones, W.W., 1973. Citrus fertilization. In: W. Reuther (Editor), The Citrus Industry, Vol. 3. University of California Press, Berkely, pp. 122–182.

Fellers, P.J., 1980. Sensory evaluation of citrus products. In: S. Nagy and J.A. Attaway (Editors), Citrus Nutrition and Quality. American Chemical Society, Washington, DC, pp. 319–340.

Fellers, P.J., 1985. Sensory quality of citrus. In: H.E. Pattee (Editor), Evaluation of Quality of Fruits and Vegetables. AVI Publishing Co., Westport, CT, pp. 83–128.

Florida Department of Citrus, 1975. Official Rules Affecting the Florida Citrus Industry. State of Florida, Department of Citrus, Lakeland, FL, Jan. 1, 1975.

Grierson, W. and Ben-Yehoshha, S., 1986. Storage of citrus fruits. In: W. Wardowski, S. Nagy and W. Grierson (Editors), Fresh Citrus Fruits. AVI Publishing Co., Inc., Westport, CT, pp. 479–507.

Guadagni, D.G., Maier, V.P. and Turnbaugh, J.G., 1973. Effects of some citrus constituents on taste thresholds for naringin and limonin bitterness. *J. Sci. Food Agric.,* 4: 1277–1288.

Guadagni, D.G., Maier, V.P. and Turnbaugh, J.G., 1974. Effect of subthreshold concentrations of limonin, naringin and sweetness on bitterness perception. *J. Sci. Food Agric.,* 25: 1349–1354.

Hagen, R.E., Dunlap, W.J. and Wender, S.H., 1966. Seasonal variation of naringin and certain other flavanone glycosides in juice sacs of Texas Ruby Red grapefruit. *J. Food Sci.,* 31: 542–547.

Harding, P.L., 1947. Quality in citrus fruits: seasonal changes in relation to consumer acceptance of oranges, grapefruit, Temple oranges and tangerines. *Proc. Am. Soc. Hortic. Sci.,* 49: 107–115.

Harding, P.L., 1964. Forecasting quality and pounds-solids in Florida oranges. U.S. Agric. Mark. Serv. AMS-533.

Harding, P.L., Winston, J.R. and Fisher, D.F., 1940. Seasonal Changes in Florida oranges. U.S. Dept. Agric. Tech. Bull. 753.

Hodgson, R.W., 1967. Horticultural varieties of citrus. In: W. Reuther, H.J. Webber and L.B. Batchelor (Editors), The Citrus Industry, Vol. 1. University of California Press, Davis, pp. 431–587.

Hodgson, R.W. and Eggers, E.R., 1938. Rootstock influence on the composition of citrus fruits. *Calif. Citrogr.,* 23: 499, 531.

Horowitz, R.M. and Gentili, B., 1977. Flavanoid constituents of citrus. In: S. Nagy, P.E. Shaw and M.K. Veldhuis (Editors), Citrus Science and Technology, Vol. 1. AVI Publishing Co., Inc., Westport, CT, pp. 397–426.

Huet, R., 1969. The aroma of citrus juices. *Fruits,* 23: 453–471.

Hughes, G., 1750. The Natural History of Barbados. Publisher unknown, London.

Hunter, G.L.K. and Brogden, W.B., Jr., 1965. Analysis of the terpene and sesquiterpene hydrocarbons in some citrus oils. *J. Food Sci.,* 30: 383–387.

Hutcheson, D.J., 1977. Influence of rootstock on the performance of Valencia orange. *Proc. Int. Soc. Citriculture,* Vol. 2, pp. 523–525.

Jones, W.W., 1961. Environmental and cultural factors influencing the chemical composition and physical characters. In: W.B. Sinclair (Editor), The Orange: Its Biochemistry and Physiology. University of California Press, Riverside, pp. 25–55.

Jones, W.W. and Embleton, T.W., 1960. Nitrogen-grade-packout relations in Valencias. *Calif. Citrogr.,* 45: 241–242.

122

Jones, W.W. and Parker, E.R., 1949. Effects of nitrogen, phosphorus and potassium fertilizers and of organic materials on the composition of Washington Navel orange juice. *Proc. Am. Soc. Hortic. Sci.,* 53: 91 – 102.

Kader, A., 1985. Postharvest handling systems: subtropical fruits. In: A. Kader (Editor), Postharvest Technology of Horticultural Crops. University of California Special Publ. 3311, Berkeley, pp. 152 – 156.

Kealey, K.S. and Kinsella, J.E., 1978. Orange juice quality with an emphasis on flavor components. *CRC Crit. Rev. Food Sci. Nutr.,* 11: 1 – 40.

Kefford, J.F. and Chandler, B.V., 1961. Influence of rootstocks on the composition of oranges, with special reference to the bitter principles. *Aust. J. Agric. Res.,* 12: 56 – 68.

Kefford, J.F. and Chandler, B.V., 1970. The Chemical Constituents of Citrus Fruits. Academic Press, New York.

Ketchie, D.O. and Ballard, A.L., 1968. Environments which cause heat injury to Valencia oranges. *Proc. Am. Soc. Hortic. Sci.,* 93: 166 – 172.

Kilburn, R.W., 1958. The taste of citrus juice. I. relationship between Brix, acid and pH. *Proc Fla. State Hortic. Soc.,* 71: 251 – 254.

Koo, R.C.J., 1979a. The influence of N, K and irrigation on tree size and fruit production of 'Valencia' orange. *Proc. Fla. State Hortic. Soc.,* 92: 10 – 13.

Koo, R.C.J., 1979b. Fertilization and nutrition of citrus. In: T. Yamada (Editor), *Boletin Tecnico,* 5: 99 – 122. Instituto International da Potassa (Suica), 13.400 Piracicaba, SP, Brazil.

Koo, R.C.J. 1985. Potassium nutrition of citrus. In: Potassium in Agriculture. ASA-CSSA-SSSA, Madison, WI, pp. 1077 – 1086.

Koo, R.C.J. and Reese, R.L., 1977. Influence of nitrogen, potassium and irrigation on citrus fruit quality. *Proc. Int. Soc. Citric.,* 1: 34 – 38.

Kugler, E. and Kovats, E., 1963. Information on mandarin peel oil. *Helv. Chim. Acta,* 46: 1480 – 1513.

Maier, V.P., Bennett, R.D. and Hasegawa, S., 1977. Limonin and other limonoids. In: S. Nagy, P.E. Shaw and M.K. Veldhuis (Editors), Citrus Science and Technology, Vol. 1. AVI Publishing Co., Inc., Westport, CT, pp. 355 – 396.

Marsh, G.L., 1953. Bitterness in navel orange juice. *Food Technol.,* 7: 145 – 150.

Marsh, G.L. and Cameron, S.H., 1950. A rootstock – Navel bitterness relationship. *Calif. Citrogr.,* 35: 458, 477.

Moshonas, M.G. and Shaw, P.E., 1972. Analysis of flavor constituents from lemon and lime essence. *J. Agric. Food Chem.,* 20: 1029 – 1030.

Moshonas, M.G. and Shaw, P.E., 1977. Effects of abscission agents on composition and flavor of cold-pressed orange peel oil. *J. Agric. Food Chem.,* 25: 1151 – 1153.

Moshonas, M.G. and Shaw, P.E., 1978. Compounds new to orange oil from fruit treated with abscission chemicals. *J. Agric. Food Chem.,* 26: 1288 – 1290.

Moshonas, M.G. and Shaw, P.E., 1980. Citrus essential oils: effects of abscission chemicals and evaluation of flavors and aromas. In: T. Swain and R. Kleinman (Editors), the Resource Potential in Phytochemistry. Recent Advances in Phytochemistry, Vol. 14. Plenum Press, New York, pp. 181 – 203.

Moshonas, M.G., Shaw, P.E. and Sims, D.A., 1976. Abscission agents effects on orange juice flavor. *J. Food Sci.,* 41: 809 – 811.

Mussinan, C.J., Mookherjee, B.D. and Malcolm, G.I., 1981. Isolation and identification of fresh lemon juice. In: B.D. Mookherjee and C.J. Mussinan (Editors), Essential Oils. Allured Publishing Corp., Wheaton, IL, pp. 199 – 228.

Nagy, S. and Shaw, P.E., 1980. Processing of grapefruit. In: P.E. Nelson and D.K. Tressler (Editors), Fruit and Vegetable Juice Processing Technology, 3rd Edition. AVI Publishing Co., Inc., Westport, CT, p. 125.

Norman, S., Craft, C.C. and Davis, P.L., 1967. Volatiles from injured and uninjured Valencia oranges at different temperatures. *J. Food Sci.,* 32: 656 – 659.

Nunez, A.J., Maarse, H. and Bemelmans, J.M.H., 1985. Volatile flavour components of grapefruit juice (*Citrus paradisi* Macfadyen). *J. Sci. Food Agric.,* 36: 757 – 763.

Ochse, J.J., Soule, M.J., Dijkman, M.J. and Wehlburg, C., 1961. Tropical and subtropical Agriculture, Vol. 1., Macmillan, New York.

Ohta, H., Tonohara, K., Watanabe, A., Iino, K. and Kimura, S., 1982. Flavor specificities of Satsuma mandarin juice extracted by a new-type screw press extraction system. *Agric. Biol. Chem.,* 46: 1385 – 1386.

Pennzig, O., 1887. Botanical studies on the acid herbs and related plants. Tip. Erdi Botta. Rome, Ministern di Agricolturae Commercio. *Ann. Agric.,* No. 116: 3 – 10.

Reitz, H.J. and Embleton, T.W., 1986. Production practices that influence fresh fruit quality. In: W.F. Wardowdki, S. Nagy and W. Grierson (Editors), Fresh Citrus Fruits. AVI Publishing Co., Westport, CT, pp. 49 – 78.

Reitz, H.J., and Sites, J.W., 1948. Relation between positions on the tree and analysis of citrus fruit with special reference to sampling and meeting internal grades. *Proc. Fla. State Hortic. Soc.,* 54: 80 – 90.

Reuther, W., Rasmussen, G.K., Hilgeman, R.H., Cahoon, G.A. and Cooper, W.C., 1969. A comparison of maturation and composition of Valencia oranges in some major subtropical zones of the United States. *J. Am. Soc. Hortic. Sci.,* 94: 144 – 157.

Rice, R.G., Keller, G.J. and Beavens, E.A., 1952. Flavor fortification of California frozen orange concentrate. *Food Technol.,* 6: 35 – 39.

Roberts, J.A. and Gaddum, L.W., 1937. Composition of citrus fruit juices. *Ind. Eng. Chem.,* 29: 574 – 575.

Schreier, P., 1981. Changes of flavour compounds during the processing of fruit juices. *Proc. Long Ashton Symp.,* 7: 355 – 371.

Schreier, P., Drawert, F., Junker, A. and Mick, W., 1977. The quantitative composition of natural and technologically changed aromas of plants. II. Aroma compounds in oranges and their changes during juice processing. *Z. Lebensm.-Unters.-Forsch.,* 164: 188 – 193.

Scora, R.W. and Newman, J.E., 1967. A phenological study of the essential oils of the peel of Valencia oranges. *Agric. Meteorol.,* 4: 11 – 26.

Scott, W.C., 1970. Limonin in Florida citrus fruits. *Proc. Fla. State Hortic. Soc.,* 83: 270 – 277.

Scott, W.C., Kew, T.J. and Veldhuis, M.K., 1965. Composition of orange juice cloud. *J. Food Sci.,* 30: 833 – 837.

Shaw, P.E., 1977a. Essential oils. In: S. Nagy, P.E. Shaw and M.K. Veldhuis (Editors), Citrus Science and Technology, Vol. 1. AVI Publishing Co., Inc., Westport, CT, pp. 427 – 462.

Shaw, P.E., 1977b. Aqueous essences. In: S. Nagy, P.E. Shaw and M.K. Veldhuis (Editors), Citrus Science and Technology, Vol. 1. AVI Publishing Co., Inc., Westport, CT, pp. 463 – 478.

Shaw, P.E. and Wilson, C.W. III, 1980. Importance of selected volatile components to natural orange, grapefruit, tangerine and mandarin flavours. In: S. Nagy and J.A. Attaway (Editors), Citrus Nutrition and Quality. ACS Symposium Series 143. American Chemical Society, Washington, DC, pp. 167 – 190.

Shaw, P.E. and Wilson, C.W., III, 1982. Volatile sulfides in headspace gases of fresh and processed citrus juices. *J. Agric. Food Chem.,* 30: 685 – 688.

Shaw, P.E., Ahmed, E.M. and Dennison, R.A., 1977. Orange juice flavour: contribution of certain volatile components as evaluated by sensory panels. *Proc. Int. Soc. Citriculture,* 3: 804 – 807.

Shaw, P.E., Wilson, C.W., III and Hansen, R.W., 1987. HPLC determination of trace levels of succinic acid in orange juice from freeze-damaged and undamaged fruit. *J. Sci. Food Agric.,* 41: 153 – 158.

Sinclair, W.B., 1961. Principle juice constituents. In: W.B. Sinclair (Editor), The Orange. University of California Press, Berkeley, pp. 131 – 160.

Sinclair, W.B., 1972. The Grapefruit. University of California Press, Berkeley.

Sinclair, W.B., 1984. The Lemon. University of California Press, Oakland.

Sinclair, W.B. and Bartholomew, E.T., 1947. Compositional factors affecting the edible quality of oranges. *Proc. Am. Soc. Hortic. Sci.,* 50: 177 – 186.

Sites, J.W., 1944. Sourness in grapefruit in relation to seasonal variation and nutritional programs. *Proc. Fla. State Hortic. Soc.,* 57: 122 – 132.

Sites, J.W. and Reitz, H.J., 1949. The variation in individual Valencia oranges from different locations of the tree as a guide to sampling methods and spotpicking for quality. I. Soluble solids in the juice. *Proc. Am. Soc. Hortic. Sci.,* 54: 1 – 10.

Sites, J.W. and Reitz, H.J., 1950. Part II. Titratable acid and soluble solids/titratable acid ratio of juice. *Proc. Am. Soc. Hortic. Sci.,* 55: 73 – 80.

124

Smith, P.F., 1966. Citrus nutrition. In: N.F. Childers (Editor), Fruit Nutrition. Somerset Press, Somerville, NJ, pp. 174 – 207.

Swisher, H.E. and Swisher, L.H., 1980. Lemon and lime juices. In: P.E. Nelson and D.K. Tressler (Editors), Fruit and Vegetable Juice Processing Technology, 3rd edition. AVI Publishing Co., Inc., Westport, CT, pp. 144 – 179.

Syvertsen, J.P. and Albrigo, L.G., 1980. Some effects of grapefruit tree canopy position on microclimate, water relations, fruit yield and juice quality. *J. Am. Soc. Hortic. Sci.,* 105: 454 – 459.

Tatum, J.H. and Berry, R.E., 1987. Conifirin – an astringent glycoside in citrus. *Phytochemistry,* 26: 322 – 323.

Wilson, C.W., III and Shaw, P.E., 1981. Importance of thymol, methyl-*N*-methyl anthranilate and monoterpene hydrocarbons to the aroma and flavor of mandarin cold-pressed oils. *J. Agric. Food Chem.,* 29: 494 – 496.

Wolford, R.W., Kesterson, J.W. and Attaway, J.A., 1971. Physicochemical properties of citrus essential oils from Florida. *J. Agric. Food Chem.,* 19: 1097 – 1105.

Yajima, I., Yanai, T., Nakamura, M., Sakakibara, H. and Hayashi, K., 1979. Compositions of the volatiles of peel oil and juice from Citrus unshiu. *Agric. Biol. Chem.,* 43: 259 – 264.

Yokoyama, H., Gold, S., DeBenedict, C. and Cartu, B., 1986. Bioregulation of essential oils of lemon. *Food Technol.,* 40: 111 – 113.

Young, R., 1986. Fresh fruit cultivars. In: W.F. Wardowski, S. Nagy and W. Grierson (Editors), Fresh Citrus Fruits. AVI Publishing Co., Westport, CT, pp. 101 – 126.

Chapter IV

THE FLAVOUR OF BERRIES

E. HONKANEN and T. HIRVI

Technical Research Centre of Finland, Food Research Laboratory, Espoo (Finland)

1. INTRODUCTION

The berry is a simple fruit, formed by enlargement of the ovary of the pistil, that remains fleshy and succulent when mature and contains immersed seeds. According to botanical classification, such widely divergent crops as blueberries, grapes, cranberries, gooseberries, tomatoes, melons and bananas are all true berries. In common usage, however, the term berry is loosely extended to include a wide group of perennial herbaceous and bushy species that bear edible fruits, such as blackberry, blueberry, cranberry, elderberry, gooseberry, currants and strawberry. The edible berries belong to several botanical families, including *Rosaceae* (strawberry, raspberry), *Ericaceae* (cranberry, blueberry), *Saxifragaceae* (currants, gooseberry) and *Vitaceae* (grapes). The berries are often characterized by delicate and unique aromas, and have therefore been used by mankind since antiquity. Furthermore, most berries are rich in vitamins and minerals.

The volatile flavour compounds of berries have been studied extensively during the recent decades. Since the classical studies by Coppens and Hoejenbos (1939) on raspberry volatiles, instrumental methods have been improved enormously. The introduction of glass capillary gas chromatography combined with mass spectrometry has made it possible to identify several hundred compounds in a single gas chromatographic run.

This chapter will concentrate on the flavours of the following berries: strawberry, raspberry, blackberry, cloudberry, arctic bramble, cranberry, blueberry, currants and sea buckthorn. The methods for isolation, concentration and identification of berry flavour compounds will also be reviewed.

2. ISOLATION AND CONCENTRATION OF VOLATILES OF BERRIES

The commonly used isolation methods include several types of solvent extractions and distillations, and concentrations of headspace volatiles. The choice of isolation and concentration methods depends on many parameters. The odorous compounds of berries are generally present in low concentrations, although differences as great as between μg/kg and mg/kg exist between the amounts of certain individual components (Flath and Forrey, 1977). The chemical and physical properties of the different volatiles vary, and this may influence the selection of isolation method.

Each isolation method causes alterations in the overall aroma composition, and usually only approximate quantitative determinations of the volatiles can be made. Furthermore, the relative proportions of the individual aroma compounds vary even between different berries of the same species and certainly between cultivars, and the formation of new compounds (artefacts) before and even during analyses is possible. The changes can be both enzymic and non-enzymic (Anjou and von Sydow, 1967a; Croteau and Fagerson, 1976; Rapp, 1982; Schreier, 1984). A number of different sampling methods has been developed (Bemelmans, 1978; Charalambous, 1978; Jennings, 1980; Kolb, 1980; Bemelmans and Schaefer, 1981; Schaefer, 1981; Sugisawa, 1981; Cronin, 1982; Nursten, 1982; Jennings and Rapp, 1983). The enrichment of volatiles from berries has been carried out mainly by liquid extraction, by vacuum distillation, by simultaneous steam distillation – extraction, and by headspace methods.

Direct solvent extraction has been used as the isolation method for volatiles of various berries, including: cloudberry (Pyysalo and Honkanen, 1977), cranberry (Hirvi et al., 1981), cultivated raspberry and Finnish wild raspberry (Honkanen et al., 1980; Guichard, 1982), hybrids between wild raspberry and arctic bramble (Pyysalo, 1976b), strawberry (Tressl et al., 1969; Tressl and Drawert, 1971; Drawert et al., 1973; Mussinan and Walradt, 1975; Staudt et al., 1975; Blakesley et al., 1979; Pyysalo et al., 1979; Pickenhagen et al., 1981; Schreier and Drawert, 1981; Hirvi and Honkanen, 1982; Hirvi, 1983), bilberry, high-bush blueberry and bog blueberry (Hirvi and Honkanen, 1983a,b), sea buckthorn (Hirvi and Honkanen, 1984), black chokeberry (Hirvi and Honkanen, 1985) and red currants (Schreier et al., 1977).

Weurman (1969) and Teranishi et al. (1981) have listed the solvents used for aroma research. Schultz et al. (1967) studied the extraction efficiency of various solvents. Diethyl ether and liquid carbon dioxide gave the highest recoveries. In most cases, the press juices of berries have been extracted with a mixture of redistilled pentane – diethyl ether. Pentane is used to minimize the extraction of water. Extraction is often performed in a Kutscher – Stäudel extraction apparatus. Diethyl ether is the main solvent, because it has been shown to be effective and easy to evaporate after the extraction.

The most frequently used distillation method is steam distillation at atmospheric pressure or in vacuum. Steam distillation at normal pressure is a common method for isolating essential oils of spices, but it is too harsh for the isolation of berry volatiles because of the formation of artefacts by thermal degradation or hydrolysis. The formation of artefacts can be markedly reduced by the use of vacuum distillation. Many studies of berry volatiles have been carried out using vacuum distillation: blueberry (Parliment and Kolor, 1975), arctic bramble (Kallio, 1976a,b), cloudberry (Honkanen and Pyysalo, 1976), a hybrid between raspberry and arctic bramble (Pyysalo, 1976b), strawberry

(Pyysalo *et al.*, 1979; Schreier, 1980) and a hybrid between *Rubus stellatus* and *R. arcticus* (Kallio *et al.*, 1980).

A combined steam distillation – extraction method can be used if the volatile constituents can be distilled without degradation. This type of method was first reported by Likens and Nickerson (1964), and Nickerson and Likens (1966). A large number of modifications of the Likens – Nickerson apparatus has been developed (Maarse and Kepner, 1970; Römer and Renner, 1974; Groenen *et al.*, 1976; Schultz *et al.*, 1977; Teranishi *et al.*, 1977; Godefroot *et al.*, 1981). Schultz *et al.* (1977) used the name simultaneous distillation – extraction (SDE) to describe this system. Using SDE, volatiles can be concentrated from a dilute solution with a small amount of solvent. Usually the Likens – Nickerson apparatus is used at atmospheric pressure, but it can be used at reduced pressure to minimize artefact formation. SDE has been applied to the concentration of a wide range of samples (Bemelmans, 1978; Bemelmans and Schaefer, 1981), including various berries, e.g. elder (Davidek *et al.*, 1982) and rabbit-eye blueberries (Horvat *et al.*, 1983). Because steam distillation is a rather harsh method and many volatiles of berries, particularly those of strawberries (Hirvi *et al.*, 1980) and raspberries (Honkanen *et al.*, 1980), are sensitive to degradation, SDE has not, in general, been used in the investigation of berry volatiles. SDE is, furthermore, a rather complex method, and in some cases only those compounds distilled with water can be collected, with consequent distortion of the aroma profile.

A simultaneous distillation – adsorption method (SDA) has been applied in the microanalysis of volatiles of some biological materials (Sugisawa and Hirose, 1981). The adsorbed volatiles are desorbed by solvent extraction. This method has not gained wide application in investigations of berry volatiles.

Extraction of berries or distillates produces such dilute solutions that concentration is necessary before instrumental determinations. Bemelmans and Schaefer (1981) have reviewed concentration methods for volatiles. The most generally used procedures are distillation, freeze concentration, zone melting, and adsorption on charcoal or porous polymers. Junk *et al.* (1974) performed concentration by fractional distillation using Vigreaux and three cavity Snyder columns. The Vigreaux column has been applied in the concentration of cloudberry (Pyysalo and Honkanen, 1977), raspberry (Honkanen *et al.*, 1980), cranberry (Hirvi *et al.*, 1981), cowberry (Anjou and von Sydow, 1967a, 1969), strawberry (Pyysalo *et al.*, 1979; Hirvi and Honkanen, 1982; Hirvi, 1983), blueberry (Hirvi and Honkanen, 1983a,b) and black chokeberry (Hirvi and Honkanen, 1985) volatiles.

Solvent-free extracts can be prepared by carbon dioxide extraction (Schultz and Randall, 1970; Jennings, 1979). Because this method does not require evaporation of the excess solvent, the concentrations of low boiling point components can also be assayed. Honkanen and Karvonen (1969) developed a carbon dioxide distillation apparatus for the separation and collection of volatiles from fats and oils. Sugisawa (1981) has reviewed liquid carbon dioxide extraction methods. The use of carbon dioxide has recently increased, particularly in the extraction of spices (Stahl *et al.*, 1978; Stahl and Schütz, 1978; Calame and Steiner, 1982; Stahl and Gerard, 1982; Roselius *et al.*, 1982; Wüst *et al.*, 1982; Ravid *et al.*, 1983). The utilization of carbon dioxide will undoubtedly also increase in the isolation of berry volatiles.

One important alternative sampling method is the isolation of volatiles from the vapour phase. As well as in aroma analysis, this method is also used in medical research and in environmental studies. The simplest technique is the direct analysis of equilibrium headspace vapour (a static headspace). This allows the investigation of only very low boiling compounds. However, the higher-boiling compounds may be significant for the odour properties. Some of these compounds can be detected using a dynamic headspace method. Several techniques have been developed for concentration of volatiles present in the headspace. Three main methods of collection are available: liquid adsorption, solid adsorption and the cryogenic method (Schaefer, 1981).

In all applications, a bulk gas (nitrogen, air, carbon dioxide, helium or argon) is passed through the headspace of a sample to a collection system. The most frequently used collection system is solid adsorption. The relevant properties of the available adsorbents have been investigated in many different studies (Weurman, 1969; Sakodynskii *et al.*, 1974; Butler and Burke, 1976; Versino *et al.*, 1976; Jennings and Filsoof, 1977; Murray, 1977; Rix *et al.*, 1977; Boyko *et al.*, 1978; Sydor and Pietrzyk, 1978; Withycombe *et al.*, 1978; Brown and Purnell, 1979; Cole, 1980; Jennings, 1980; Lewis and Williams, 1980; Singleton and Pattee, 1980; Barnes *et al.*, 1981; Novak *et al.*, 1981; Schaefer, 1981; Sugisawa, 1981; Cronin, 1982; Jennings and Rapp, 1983). Tenax GC and various Porapaks and Chromosorbs are synthetic porous polymers which are frequently used in adsorption traps. Selection of the adsorbent depends on the properties and concentrations of the compounds of interest and on the purity of the sample. Desorption of volatiles from polymers is usually performed by heating, followed by isolation in a cold trap. Volatiles can also be desorbed by solvent extraction. In the cryogenic method, compounds are usually desorbed by heating the trap, which is attached to a gas chromatographic injector.

Headspace concentration techniques have been used in flavour investigations of berries, including bilberry (von Sydow *et al.*, 1970), blackcurrant (von Sydow and Karlsson, 1971a) and strawberry (Yamashita *et al.*, 1975, 1976, 1977; Dirinck *et al.*, 1977, 1981; Hirvi and Honkanen, 1982; Hirvi, 1983). According to these studies, it is possible to determine only very highly volatile components using the headspace technique. Of the strawberry volatiles, 2,5-dimethyl-4-methoxy-2*H*-furan-3-one is the least volatile compound that has been identified by this method. The character impact compounds of wild strawberry, myrtenol, eugenol, methyl anthranilate and methyl *N*-formylanthranilate, are too high-boiling to be detected using headspace concentration. On the other hand, readily volatile compounds may be evaporated during solvent extraction or steam distillation, and escape identification. According to Hirvi and Honkanen (1982), about 40 compounds in a solvent extract could not be detected by the headspace method. The advantage of this concentration method over solvent extraction is that very small samples of the berries can be used. Because only very small samples are needed in headspace injection, this technique is obviously very useful when investigating the aroma compounds of new berry cultivars.

3. IDENTIFICATION AND QUANTIFICATION OF AROMA COMPOUNDS OF BERRIES

Gas-liquid chromatography (GLC) is the most frequently used method for the separation of aroma compounds of berries. The mixture of volatiles is usually very complex, consisting of compounds of different chemical classes and molecular weights, and a pre-separation stage is needed. Many reviews of methods for the separation and identification of aroma compounds have been published (Jennings, 1980, 1981, 1982; Kubeczka, 1981; Maarse and Belz, 1981; Teranishi et al., 1981). Capillary gas chromatography offers a practical method for the separation of complex mixtures, combining a rather short time of analysis with high separation ability. At present, wall-coated open tubular glass or flexible fused silica capillary columns are commonly used in berry flavour research. The bonded phase, fused silica columns resist phase stripping, and are therefore superior in on-column and splitless injection (Schreier, 1984). Many volatiles of berries are enantiomers, and in the future, polysiloxane-based chiral phases will obviously be used for the separation of stereoisomeric compounds (Frank et al., 1978; König, 1984).

Survey of the aroma literature shows that polar columns and splitless injection systems are the most favoured for studies in this field. Jennings and Shibamoto (1980) recorded the retention indices of volatile compounds according to the Kováts Index system (on Carbowax 20M and OV 101 phases). Most of the compounds listed are typical berry aroma compounds.

The flame ionization detector (FID) is nowadays widely used as the GLC detector, because it has a high sensitivity to different organic compounds. Specific detectors, e.g. nitrogen-phosporus detectors (NPD) and flame photometric detectors (FPD), are superior in the analysis of nitrogen or sulphur compounds, respectively. However, the necessity for specific detectors is rather unusual in berry aroma research, because in most cases the character impact compounds consist only of hydrogen, carbon and oxygen. In strawberries, certain sulphur-containing (Dirinck et al., 1981) and nitrogen-containing (Pyysalo et al., 1979) compounds have also been found.

Mass spectrometry as GLC-detector is a powerful tool for the identification and structure elucidation of aroma compounds of berries. Jennings and Shibamoto (1980) have compiled the mass spectra of about 700 aroma volatiles.

In the elucidation and confirmation of the structure of an unknown compound, additional necessary information can be obtained by nuclear magnetic resonance (NMR), and UV- and IR-spectroscopic measurements. Recently, IR-spectrophotometers have been connected to gas chromatographs. Some food flavours have been studied by capillary GC/FT-IR (Fourier transform IR) (Schreier, 1984).

Quantification of the identified berry volatiles is mainly performed by comparing the peak areas in an FID chromatogram. Quantitative estimation of some substances is also possible using mass spectrometric selected ion monitoring (SIM) as GLC-detector. By monitoring ions of one mass, or a few specific masses, instead of the whole mass spectrum, very high sensitivity can be attained. Furthermore, great specificity can be achieved with SIM. The SIM-technique is strictly a quantitative method, applicable to the quadrupole mass filter and to most magnetic sector mass spectrometers. However, the former system is more practical, cheaper and easier to use than the magnetic sector mass spectrometers.

Selected ion monitoring has already found wide application in the fields of biochemistry, medicine and environmental science. Only a few flavour studies of berries have hitherto been carried out using the SIM-technique, but it will become more general in aroma research in the future. The quantities of characteristic flavour compounds in some strawberry and blueberry cross-breedings have been estimated by SIM (Hirvi, 1983; Hirvi and Honkanen, 1983b). Before the mass fragmentographic determinations, the volatile compounds of the strawberry varieties and blueberry species had already been studied in detail. Of the compounds identified, the most important character impact compounds, i.e. those having the highest odour values or the highest concentrations among the volatiles, were selected for monitoring. This technique is so sensitive that it is possible to determine the characteristic compounds of strawberry even in samples from a single plant, without any enrichment of volatiles. According to these studies, the correlation between the sensory parameters of odour and the concentrations of the most important compounds measured with SIM is significant.

Obviously, SIM would be suitable for distinguishing wild strawberry from cultivated strawberry, because the characteristic compounds of wild strawberry (*Fragaria vesca* L), methyl anthranilate and methyl *N*-formylanthranilate, have never been found in cultivated strawberry varieties. These compounds have ions in their mass spectra which are suitable for monitoring purposes. The aromas of bilberry and blueberry are so different that with the SIM-technique it would be possible to distinguish the berries, juices or jams of bilberry from those of other *Vaccinium* berries. Using the specific and sensitive SIM-technique, it is possible to follow very easily the inheritance of the characteristic flavour components of the progenitors of cross-breedings. Thus, SIM could be used by botanists for estimation of the aromas of new cultivars in plant breeding experiments, without the large samples required for sensory evaluations. Another advantage of SIM is the ease of preparation of concentrates prior to this technique.

4. RELATIONSHIP BETWEEN INSTRUMENTAL AND SENSORY ANALYSES OF AROMA
   COMPOUNDS OF BERRIES

In many cases, sensory evaluation has been used together with instrumental identification of volatile compounds in assessing the quality of aroma and the importance of berry volatiles. Odour description of gas chromatographic eluates can be used for determination of the compounds responsible for the odour of berries. Using this method it is possible to obtain valuable information concerning the character and strength of odorous components. Numerous papers have been published in this connection, and large numbers of modified techniques have been developed (Forss, 1981; van Gemert, 1981). Some standard procedures have been described for odour assessment of gas chromatographic eluates (von Sydow and Anjou, 1969; von Sydow et al., 1970; Qvist et al., 1976; van Gemert, 1981), but many difficulties in standardization have been reported.

In many cases, a small number of gas chromatographic peaks, 'character impact compounds', may correlate very closely with typical berry flavour (Jennings, 1969). For example, von Sydow and Anjou (1969) and von Sydow et al. (1970) showed that only three compounds contribute to the characteristic odour of bilberry. Clear correlations have also been observed between the character of odour and the overall impression of odour of fresh strawberries, on the one hand, and the concentrations of ethyl butanoate, ethyl hexanoate, trans-hex-2-enal, 2,5-dimethyl-4-methoxy-2H-furan-3-one and linalool in the berries, on the other. The 'odour values' of the major compounds identified in some strawberry cultivars are presented in Table IV.I.

The odour value, introduced by Mulders (1973), has often been used for determination of the relative contributions of individual components. The odour value of a compound is the ratio of its concentration to its odour threshold value. Some authors have criticized the use of odour values, claiming that artefacts and some complicating features may cause errors in the determinations of threshold values (Rothe, 1976; Frijters, 1978). In addition, factors such as the carrier medium of the volatiles (Rothe, 1975), interactions and synergism between different compounds (Salo et al., 1972; Salo, 1975; Williams, 1978a; Laing and Panhuber, 1979; Dravnieks et al., 1981) and the different increase of strength of odour when the compounds are present above their threshold concentration (Patte et al., 1975; Frijters, 1978; Williams and Ismail, 1981), may all influence the contributions of different compounds to the total flavour. However, in spite of these limitations, odour values may provide much valuable information (Kallio, 1976a; Pyysalo, 1976a; Hirvi and Honkanen, 1982; Hirvi, 1983). For example, in Table IV.I the sums of the odour values of the most important compounds are clearly highest in the berries of the strawberry cultivars Senga Sengana, Kristina, and Lihama × Senga Sengana. However, the overall highest concentration was in the cultivar Annelie, which is a hybrid between cultivated and wild strawberry (Trajkovski, 1978). According to Fig. IV.1, (Tuorila-Ollikainen et al., 1983) these cultivars

132

TABLE IV.I

Odour thresholds, and odour values, of the major compounds identified in different strawberry cultivars (Hirvi and Honkanen, 1982; Hirvi, 1983)

Compound	Odour threshold in water (mg/l)	Odour value								
		Senga Sengana (SS)	Red Gauntlet (RG)	Kristina	Zephyr	Ostara	Lihama × SS	Ydun × SS	SS × RG	Annelie
Ethyl hexanoate	0.003 (Pyysalo et al., 1977a)	133	13	100	33	13	27	50	27	960
Ethyl butanoate	0.001 (Flath et al., 1967)	100	50	100	10	–	100	166	67	1100
trans-Hex-2-enal	0.017 (Buttery et al., 1971)	41	29	58	29	29	58	50	18	88
2,5-Dimethyl-4-methoxy-2H-furan-3-one	0.01 (Pyysalo et al., 1979)	120	40	120	80	110	170	40	30	80
Linalool	0.006 (Buttery et al., 1971)	50	–	5	–	–	10	–	5	8
		444	132	383	152	152	365	306	147	2256

have also been described as the best with regard to their character of odour and overall impression of odour. By contrast, the hybrid Senga Sengana × Red Gauntlet is completely different from Senga Sengana, and very similar to its other progenitor. The odour values of the major compounds are about equal in Red Gauntlet and in the hybrid Senga Sengana × Red Gauntlet. As can be seen in Fig. IV.1, the hybrid Annelie forms an interesting exception in the 'family portrait', being atypical for its odour but giving a strong and pleasant overall impression of odour and taste. In the field of sensory research of berry flavours,

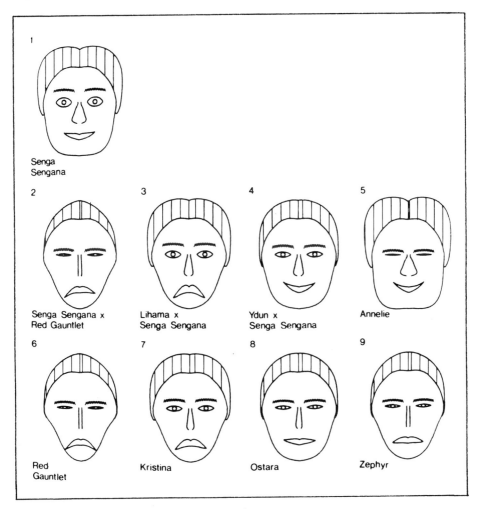

Fig. IV.1. Faces illustrating eight sensory properties evaluated in nine strawberry cultivars. Properties evaluated: size of the nose = intensity of odour; eye size and vertical position of eyebrow = character of odour; upper hair line = overall impression of odour; curvature of mouth = sweetness; face line = overall impression of taste; size of mouth = sourness; symmetry of the right-hand side of the nose = off-odour; symmetry of the right-hand side of the curvature of the mouth = off-taste.

the method of facial presentation could be used for simplicity and clarity of research results.

Relationships between instrumental gas chromatographic and sensory analyses have been studied, and correlations have frequently been found. Kallio (1976a) reported a good correlation between the content of 2,5-dimethyl-4-methoxy-2*H*-furan-3-one and odour evaluations of the berries of arctic bramble (*Rubus arcticus* coll.). Some dependence of the overall impression of taste on the content of the above-mentioned methoxyfuranone is also evident in the berries of arctic bramble (Kallio *et al.,* 1984). In the wild raspberries, ethyl 5-hydroxyoctanoate and ethyl 5-hydroxydecanoate, together with the δ-lactones, probably make a great contribution to the aroma (Honkanen *et al.,* 1980). Pyysalo (1976a) and Pyysalo *et al.* (1977a) studied the gas chromatographic eluates of cloudberries and concentrations of 4-vinylphenol and 5-methoxy-4-vinylphenol. Of the *Vaccinium* berries, the aroma of cranberries has been characterized by the presence of several aromatic compounds, but no character impact compound has been found (Anjou and von Sydow, 1967b; Croteau and Fagersson, 1976; Hirvi *et al.,* 1981). In American blueberries, hydroxy-citronellol has been considered as blueberry-like. On the other hand, methyl and ethyl esters of 2- and 3- hydroxy-3-methylbutanoic acids have been found to be the typical compounds of bilberry. The odours of the pure synthetic hydroxy esters have been described by sensory panelists as bilberry-like. These esters may therefore contribute to the typical aroma of bilberry (Hirvi and Honkanen, 1983a). The unique aroma of sea buckthorn consists of numerous 3-methylbutyl esters. Several different esters apparently contribute to the typical aroma of this fruit, because their measured concentrations are well above their threshold values (Hirvi and Honkanen, 1984).

Besides simple comparison, more sophisticated techniques such as stepwise discrimination, multiple discrimination and various ranking techniques between sets of analytical and sensory data have also been used. Williams (1978a,b,c) reviewed briefly the various methods which have been used to give both descriptive and hedonic meaning to chemical data. Correlations between sensory and instrumental assessments of strawberries have been studied by correlograms and regression analysis (Hirvi, 1983; Tuorila-Ollikainen *et al.,* 1983).

5. STRAWBERRIES

Strawberries belong to the genus *Fragaria* (*Rosaceae*). Along with many other genera of plants, the genus *Fragaria* probably has its origin in the mountain regions of the Himalayas and Southeast Asia (Darrow, 1966; Scott and Lawrence, 1975). It is possible to demonstrate a relationship between three residing Asian species, *F. daltoniana, F. nubicola* and *F. nilgerrensis*, and the species growing in Europe, Asia and America (Fig. IV.2).

During the period of distribution, the continent of Eurasia was connected to North America. Over this long period, the chromosome number of the original

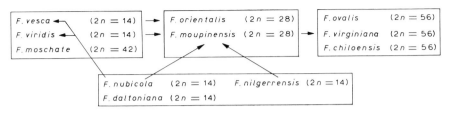

Fig. IV.2. Proposed relationships between wild *Fragaria* species.

diploid species ($2n = 14$) was multiplied by 2, 3 or 4, so that American *Fragaria* species are octaploid ($2n = 56$) (Fig. IV.2).

Nowadays, *F. vesca* L. has the widest area of distribution, including Europe, Asia and North America. Some variant forms of *F. vesca* are also known, for example *F. vesca f. alba*, which has white or pale yellow berries, and *F. vesca f. semperflorens*, which has a chromosome number of $2n = 56$ and is best known by the name 'Rügen'. Other wild *Fragaria* species, for example *F. moschata* Duch. and *F. viridis* Duch., are native to Europe; *F. nubicola* Lindl. ex Lacaita, *F. nilgerrensis* Schlecht., *F. orientalis* Losinsk., *F. moupinensis* Franch. and *F. daltoniana* J. Gay to Asia; *F. ovalis* Lehm. and *F. virginiana* Duch. to North America; and *F. chiloensis* Duch. to South America (Chile). Cultivated strawberries (*F.* × *ananassa*) originate from the progenies of crossing between *F. virginiana* and *F. chiloensis*, which were introduced to Europe (France and England) in the 16th and 17th centuries, and several thousands of cultivars have been developed. Today there is a demand for new strawberry cultivars with the pleasant herbaceous aroma typical of wild strawberries. In the hope of obtaining such new cultivars, *F. vesca f. semperflorens* and *F. virginiana*, which both have the same chromosome number ($2n = 56$) as *F.* × *ananassa*, have been cross-bred with cultivated strawberry varieties such as 'Valentine' and 'Senga Sengana', respectively (Brooks and Olmo, 1968; Trajkovski, 1978).

The volatile flavour compounds of cultivated strawberries have been extensively studied by many authors during the past 25 years. Winter and Willhalm (1964) identified over 60 compounds in the cultivar 'Surprise des Halles', including *trans*-hex-2-en-1-ol, 2-ethylhexan-1-ol, ethyl butanoate, hexyl acetate, *trans*-hex-2-en-1-yl acetate, ethyl acetoacetate and α-terpineol. Later, Willhalm *et al.* (1966) reported that the main volatile acids of strawberries consisted of 2-methylpropanoic, 2-methylbutanoic, and hexanoic acids.

McFadden *et al.* (1965) reported the identification of *cis*- and *trans*-hex-3-en-1-yl hexanoates, *trans*-hex-2-en-1-yl hexanoate, linalool and several other compounds not reported previously as constituents of strawberry aroma.

Tressl *et al.* (1969) identified about 200 compounds in the berries of the cultivar 'Revata'. The main components included previously unidentified compounds such as octyl butanoate, octyl 2-methylbutanoate, octyl hexanoate and γ-dodecalactone. Stoltz *et al.* (1970) identified 1,2-dihydro-1,1,6-trimethyl-

136

TABLE IV.II

Compounds identified in cultivated strawberries (*Fragaria* × *ananassa*)

Compound	Ref.[a]
Terpenes	
Limonene	7, 9, 15
α-Pinene	7, 9
β-Pinene	9
Linalool	3, 9, 11, 12, 15
Nerolidol	12
α-Terpineol	3, 11, 12, 15
Borneol	17, 18
Isofenchyl alcohol	17, 18
Linalool oxides	11, 12
Alcohols	
Methanol	1
Ethanol	1, 3, 11
Propan-1-ol	16, 17, 18
Propan-2-ol	16, 18
2-Methylpropan-1-ol	15
Butan-1-ol	1, 3, 12
Butan-2-ol	16, 17, 18
2-Methylbutan-1-ol	3
3-Methylbutan-1-ol	3, 11, 12, 15
2-Methylbutan-2-ol	16, 17, 18
Pentan-1-ol	12, 15
Pentan-2-ol	12
Pentan-3-ol	16
Pent-1-en-3-ol	1
Hexan-1-ol	1, 9, 10, 11, 12, 15
Hexan-2-ol	15
Hexan-3-ol	16
trans-Hex-2-en-1-ol	1, 3, 11, 12
cis-Hex-3-en-1-ol	3, 11, 12
Hex-1-en-3-ol	16
Heptan-1-ol	12
Heptan-2-ol	1, 9, 10, 11, 12
Heptan-3-ol	16
Octan-1-ol	9, 12
Octan-2-ol	16
Octan-3-ol	12
Oct-3-en-1-ol	16
Oct-1-en-3-ol	16
Nonan-1-ol	12
Nonan-2-ol	9, 10, 11, 12
Non-1-en-3-ol	16
Decan-1-ol	16
Decan-2-ol	16
Undecan-2-ol	9, 16
Dodecan-1-ol	16
Dodecan-2-ol	16
Tridecan-2-ol	9, 10
Pentadecan-2-ol	9, 10

TABLE IV.II (*continued*)

Compound	Ref.[a]
Carbonyl compounds	
Acetaldehyde	1, 16, 20
Propanal	1, 16
Propenal	1
Butanal	16
But-2-enal	1
Pent-2-enal	1
Hexanal	1, 11, 12, 16
trans-Hex-2-enal	1, 7, 12, 15, 16, 11
cis-Hex-3-enal	1, 15, 16
Heptanal	17, 18
Hept-2-enal	15
Oct-2-enal	15
Nonanal	16
Decanal	12
Propanone	1, 16
Butanone	16
Methylbutanone	17, 18, 20
Diacetyl (Butanedione)	1, 16, 15
Pentan-2-one	1, 16
Pentan-3-one	16
Pent-3-en-2-one	15
Hexan-2-one	16
Heptan-2-one	1, 12, 11, 16
Octan-2-one	16
Nonan-2-one	12, 11, 16
Decan-2-one	16
Undecan-2-one	16
Acids	
Formic acid	16
Acetic acid	11, 16
Propanoic acid	6, 11, 16
2-Methylpropanoic acid	6, 11, 16
Butanoic acid	6, 11, 16
2-Methylbutanoic acid	6, 11, 16
3-Methylbutanoic acid	6
2-Methylbut-2-enoic acid	6
Pentanoic acid	6, 16
4-Methylpentanoic acid	6
2-Methylpent-2-enoic acid	6
2-Methylpent-3-enoic acid	6
Hexanoic acid	6, 11, 16
Hex-2-enoic acid	6, 16
5-Methylhexanoic acid	6, 16
3-Hydroxyhexanoic acid	6
Heptanoic acid	6, 16
Octanoic acid	6, 11, 16
Oct-2-enoic acid	6
3-Hydroxyoctanoic acid	6
Nonanoic acid	6, 16

(*continued*)

138

TABLE IV.II (*continued*)

Compound	Ref.[a]
Non-3-enoic acid	6
Decanoic acid	6, 11, 16
Dec-2-enoic acid	16
Undecanoic acid	6
Dodecanoic acid	6, 11, 16
Tridecanoic acid	16
Tetradecanoic acid	11, 16
Tetradec-2-enoic acid	16
Pentadecanoic acid	16
Hexadecanoic acid	6, 11, 16
Hexadec-9-enoic acid	16
Heptadecanoic acid	16
Octadec-9-enoic acid	16
Octadeca-9,12-dienoic acid	16
Octadeca-9,12,15-trienoic acid	16
Nonadecanoic acid	16
Eicosanoic acid	16
Esters	
Methyl formate	18
Ethyl formate	16
Butyl formate	16
3-Methylbutyl formate	16
Hexyl formate	11
Methyl acetate	16, 20
Ethyl acetate	1, 7, 11, 12, 15, 16, 20
Propyl acetate	1, 16
Isopropyl acetate	7, 16
Butyl acetate	1, 7, 9, 10, 11, 12, 15, 16
2-Methylpropyl acetate	16
2-Methylbutyl acetate	7
3-Methylbut-2-enyl acetate	11
Pentyl acetate	16
1-Methylbutyl acetate	16
3-Methylbutyl acetate	1, 12, 15, 16
Hexyl acetate	1, 7, 9, 10, 11, 12, 16
1-Methylpentyl acetate	16
trans-Hex-2-enyl acetate	1, 9, 11, 12, 15, 16
cis-Hex-3-enyl acetate	9, 11, 12, 16
Hex-1-en-3-yl acetate	16
1-Methylhexyl acetate	16
Hept-1-en-3-yl acetate	16
Octyl acetate	9, 10, 12, 16
Decyl acetate	9, 10, 12
Methyl propanoate	7
Ethyl propanoate	1, 16, 20
cis-Hex-3-enyl propanoate	16
Methyl 1-methyl propanoate	7, 16, 20
Ethylmethyl propanoate	16, 20
Methyl butanoate	1, 7, 11, 12, 15, 16, 20
Ethyl butanoate	1, 7, 12, 11, 15, 16

TABLE IV.II (*continued*)

Compound	Ref.[a]
Propyl butanoate	7, 16
Isopropyl butanoate	1, 16, 17, 18
Butyl butanoate	12, 16
2-Methylpropyl butanoate	16
Pentyl butanoate	16
1-Methylbutyl butanoate	16
3-Methylbutyl butanoate	12, 16
Pent-3-enyl butanoate	12, 16
Hexyl butanoate	1, 9, 10, 11, 12, 16
trans-Hex-2-enyl butanoate	1, 12
cis-Hex-3-enyl butanoate	12
1-Methylhexyl butanoate	12, 16
Octyl butanoate	12, 16
1-Methyloctyl butanoate	16
Decyl butanoate	12
Ethyl but-2-enoate	1, 12, 17, 18
Methyl 2-methylbutanoate	7, 12, 16, 17, 18
Ethyl 2-methylbutanoate	7, 12, 16, 17, 18
Isopropyl 2-methylbutanoate	7
Butyl 2-methylbutanoate	12, 16, 17, 18
2-Methylpropyl 2-methylbutanoate	16
2-Methylbutyl 2-methylbutanoate	16
3-Methylbutyl 2-methylbutanoate	16
Hexyl 2-methylbutanoate	16
Octyl 2-methylbutanoate	12, 16
Ethyl 3-methylbutanoate	12, 15
Butyl 3-methylbutanoate	12
Methyl 3-hydroxybutanoate	12
Ethyl 3-oxobutanoate	17, 18
Ethyl pentanoate	15, 16
Methyl 4-methylpentanoate	7
Methyl hexanoate	1, 9, 10, 11, 12, 15, 16
Ethyl hexanoate	1, 7, 9, 10, 11, 12, 15, 16
Butyl hexanoate	9, 12, 16
Pentyl hexanoate	16
3-Methylbutyl hexanoate	16
1-Methylbutyl hexanoate	16
Hexyl hexanoate	9, 10, 12, 16
Hex-2-enyl hexanoate	12
trans-Hex-3-enyl hexanoate	17, 18
1-Methylhexyl hexanoate	16
Octyl hexanoate	12, 16
Decyl hexanoate	9, 10
Ethyl *trans*-hex-2-enoate	12
Methyl 3-hydroxyhexanoate	12
Ethyl 3-hydroxyhexanoate	12
Methyl heptanoate	16
Ethyl heptanoate	15, 16
Methyl octanoate	9, 10, 12, 15, 16
Ethyl octanoate	9, 10, 12, 15, 16
Isopropyl octanoate	16

(*continued*)

TABLE IV.II (*continued*)

Compound	Ref.[a]
Butyl octanoate	16
3-Methylbutyl octanoate	16
Hexyl octanoate	12, 16
cis-Hex-3-enyl octanoate	16
Methyl nonanoate	16
2-Methylpropyl nonanoate	16
3-Methylbutyl nonanoate	16
Methyl decanoate	9, 10, 11, 12, 16
Ethyl decanoate	9, 10, 15, 16
Isopropyl decanoate	16
Hexyl decanoate	9, 10
Methyl dodecanoate	11, 16
Ethyl dodecanoate	16
Methyl hexadecanoate	16
Methyl octadecanoate	16
Methyl octadec-9-enoate	16
Methyl octadeca-9,12,15-trienoate	16
Lactones	
γ-Hexalactone	12, 16
γ-Octalactone	16
γ-Decalactone	9, 10, 12, 16
γ-Dodecalactone	9, 10, 12, 16
δ-Hexalactone	11
δ-Heptalactone	16
δ-Octalactone	11, 16
Acetals	
Dimethoxymethane	17, 18, 20
Diethoxymethane	20
1,1-Dimethoxyethane	17, 18, 20
1-Ethoxy-1-methoxyethane	15, 20
1-Butoxy-1-methoxyethane	17, 18
1-Methoxy-1-pentoxyethane	17, 18
1,1-Diethoxyethane	17, 18, 20
1-Ethoxy-1-propoxyethane	15, 17, 18, 20
1-Butoxy-1-ethoxyethane	17, 18
1-Ethoxy-1-pentoxyethane	17, 18
1-Ethoxy-1-hexoxyethane	17, 18
1-Ethoxy-1-(hex-3-enoxy)ethane	17, 18
1,1-Dihexoxyethane	17, 18
1,1-Diethoxypentane	17, 18
Furans	
2-Furfural (2-furaldehyde)	17, 18
2-Furancarboxylic acid	6
2,5-Dimethyl-4-hydroxy-2*H*-furan-3-one	12, 19
2,5-Dimethyl-4-methoxy-2*H*-furan-3-one	8, 12, 19
Aromatic compounds	
Benzyl alcohol	1, 11, 12, 16, 17, 18

TABLE IV.II (*continued*)

Compound	Ref.[a]
2-Phenylethanol	11, 12, 16, 17, 18
2-(4-Hydroxyphenyl)ethanol	17, 18
trans-Cinnamyl alcohol	11
Benzaldehyde	11, 12, 15, 16, 17, 18
Acetophenone	17
Benzoic acid	6, 16, 18
4-Methylbenzoic acid	6
2-Hydroxybenzoic acid	17, 18
Phenylacetic acid	6
3-Phenylpropanoic acid	6
trans-Cinnamic acid	6, 11, 16
Benzyl acetate	9, 10, 11, 12, 15, 16, 17, 18
2-Phenethyl acetate	12, 16, 18
Methyl salicylate	17, 18
Ethyl salicylate	17, 18
Methyl cinnamate	9, 10, 11, 12, 16, 18
Ethyl cinnamate	9, 10, 15, 18
4-Vinylphenol	11
2-Methoxy-4-vinylphenol	11
Eugenol	11
Sulphur compounds	
Methanethiol	21
Ethylthioethane	16
Ethyldithioethane	21

[a] 1. Winter and Willhalm, 1964; 2. Willhalm *et al.,* 1966; 3. McFadden *et al.*, 1965; 4. Tressl *et al.*, 1969; 5. Stoltz *et al.*, 1970; 6. Mussinan and Walradt, 1975; 7. Dirinck *et al.*, 1977; 8. Sundt, 1970; 9. Drawert *et al.*, 1973; 10. Staudt *et al.*, 1975; 11. Pyysalo *et al.*, 1979; 12. Schreier, 1980; 13. Hirvi and Honkanen, 1982; 14, Hirvi, 1983; 15. Shen *et al.*, 1980; 16. Drawert, 1970; 17. Nursten and Williams, 1967; 18. Gierscher and Baumann, 1968; 19. Re *et al.*, 1973; 20. Teranishi *et al.*, 1963; 21. Kemp *et al.*, 1973.

naphthalene in strawberry oil, and Mussinan and Walradt (1975) found over 20 volatile acids which had not previously been reported in strawberries. Dirinck *et al.* (1977) used headspace concentration on Tenax to evaluate the aroma quality of cultivated strawberries. They identified about 30 volatiles eluting before undecane using a SE-30 capillary column.

The more than 60 esters, 30 alcohols and about 30 carbonyl compounds identified in these studies do, of course, contribute to the estery and green notes of the odour of strawberries, but they can hardly be considered to include a 'character-impact' compound or compounds. Sundt (1970), however, reported the identification of 2,5-dimethyl-4-hydroxy-2*H*-furan-3-one in strawberries, and this compound may well be considered the most important aroma constituent of strawberries so far reported. Schreier (1980) compared the volatile constituents of three fresh and deep-frozen cultivated strawberry varieties, 'Senga

Sengana', 'Senga Litessa' and 'Senga Gourmella', and identified 74 volatile components, of which 43 were esters.

These results for cultivated strawberries are summarized in Table IV.II.

Drawert *et al.* (1973) and Staudt *et al.* (1975) compared the volatile constituents of some wild *Fragaria* species (*F. vesca, F. moschata, F. chiloensis* and *F. virginiana*) with those of the strawberry cultivar 'Revata'. They found that *F. vesca* and *F. moschata* have the same main flavour compounds. Typical for both these berries were high concentrations of alkan-2-ones and alkan-2-ols, especially tridecan-2-one and tridecan-2-ol. On the other hand, *F. chiloensis* contained neither alkan-2-ones nor alkan-2-ols. The presence of ethyl esters of C_6-C_{12} acids is characteristic for *F. chiloensis. F. virginiana* has the strongest aroma of all these four species. It also gives alkan-2-ones, and methyl and ethyl esters of C_6-C_{12} acids. The high concentrations of methyl and ethyl cinnamates and γ-decalactone are also characteristic of *F. virginiana*. The volatile constituents of *F. nilgerrensis* differ clearly from those of *F. virginiana*. Typical compounds are octyl and decyl acetates, methyl benzoate and benzyl acetate.

Pyysalo *et al.* (1979) compared quantitatively the volatile constituents of wild strawberry (*F. vesca*) with those of the cultivar 'Senga Sengana', and identified a total of 87 volatile components. They found that the main volatile compound in the press juice of *F. vesca* was 2,5-dimethyl-4-methoxy-2*H*-furan-3-one (10.7% of the total volatiles). This compound was first reported as a constituent of canned Alfonso mango by Hunter *et al.* (1974), and later as a component of arctic bramble (*Rubus arcticus*) by Kallio and Honkanen (1974). The cultivated berries of *F.* × *ananassa* var. Senga Sengana were also shown to contain this compound, although at a considerably lower concentration. On the other hand, the corresponding hydroxy compound, 2,5-dimethyl-4-hydroxy-2*H*-furan-3-one, which had previously been identified in strawberries (Sundt, 1970), was not found in either the wild or cultivated berries studied (Pyysalo *et al.*, 1979). Other major neutral volatiles identified in the wild berries included *trans*-hex-2-en-1-ol (4.1% of total volatiles), methyl butanoate (3.1%), hexan-1-ol (3.0%), pentadecan-2-one (2.8%), benzyl alcohol (2.7%), ethyl butanoate (1.8%), heptan-2-one (1.8%), pentadecan-2-ol (1.7%) and eugenol (1.6%). Typical for the flavour of *F. vesca* were also methyl anthranilate (methyl 2-aminobenzoate), methyl nicotinate (methyl pyridine-3-carboxylate), carvyl acetate, verbenone, citronellol and myrtenol, which had not previously been identified as strawberry volatiles.

The volatile constituents of wild strawberries are summarized in Table IV.III.

Hirvi and Honkanen (1982) studied the volatiles of two new strawberry cultivars, 'Annelie' and 'Alaska Pioneer', obtained by backcrossing cultivated strawberries with wild strawberries, *F. vesca semperflorens* 'Rügen' and *F. virginiana*. The flavours of the berries of both cultivars resembled that of the fresh wild strawberry more closely than that of the cultivated strawberry. The main volatile compound of wild strawberry (*F. vesca*), 2,5-dimethyl-4-methoxy-2*H*-furan-3-one, was also an important constituent of the aromas of both berries. The concentrations of this compound were 0.8 and 0.45 mg/kg in 'Annelie'

TABLE IV.III

Compounds identified in some wild *Fragaria* species

Compound	Ref.[a]				
	F. vesca	F. moschata	F. chiloensis	F. virginiana	F. nilgerrensis
Acids					
Acetic acid	3				
Propanoic acid	3				
Methylpropanoic acid	3				
Butanoic acid	3				
Hexanoic acid	3				
Octanoic acid	3				
Decanoic acid	3				
Dodecanoic acid	3				
Tetradecanoic acid	3				
trans-Cinnamic acid	3				
Alcohols					
Ethanol	3				
3-Methylbutan-1-ol	3				
3-Methylbutan-2-ol	3				
Pentan-1-ol	3				
Hexan-1-ol	1 – 3	1, 2	1, 2		
Heptan-1-ol	3				
Heptan-2-ol	1 – 3	1, 2			
Octan-1-ol	1 – 3	1, 2	1, 2		
2-Ethylhexan-1-ol	3				
Nonan-2-ol	1 – 3	1, 2			
Decan-1-ol	3				
Undecan-2-ol	1 – 3	1, 2			
Tridecan-2-ol	1 – 3	1, 2			
Pentadecan-2-ol	1 – 3	1, 2			
trans-Hex-2-en-1-ol	3				
cis-Hex-3-en-1-ol	3				
6-Methylhept-5-en-2-ol	3				
Linalool	3				
α-Terpineol	3				
Citronellol	3				
Myrtenol	3				
Benzyl alcohol	3				
2-Phenylethanol	3				
3-Phenylpropanol	3				
Cinnamyl alcohol	3				
Carbonyl compounds					
Hexanal	3				
Heptan-2-one	1 – 3	1, 2			
Nonan-2-one	1 – 3				
Undecan-2-one	1 – 3	1, 2		1, 2	
Tridecan-2-one	1 – 3	1, 2		1, 2	
Pentadecan-2-one	1 – 3	1, 2			
Heptadecan-2-one	1, 2	1, 2			

(*continued*)

TABLE IV.III (*continued*)

Compound	Ref.[a]				
	F. vesca	*F. moschata*	*F. chiloensis*	*F. virginiana*	*F. nilgerrensis*
Acetoin (3-hydroxy-butanone)	3				
Verbenone	3				
Benzaldehyde	3				
Vanillin (4-hydroxy-3-methoxybenzaldehyde)	3				
Esters					
Methyl butanoate	3				
Methyl hexanoate	1 – 3	1, 2	1, 2	1, 2	1, 2
Methyl octanoate	1 – 3	1, 2	1, 2	1, 2	
Methyl decanoate	1 – 3	1, 2	1, 2	1, 2	
Methyl dodecanoate	1 – 3	1, 2			
Methyl benzoate	1, 2	1, 2			1, 2
Methyl salicylate	3				
Methyl cinnamate	1 – 3	1, 2	1, 2	1, 2	1, 2
Methyl nicotinate (methyl pyridine-3-carboxylate)	3				
Methyl anthranilate (methyl 2-aminobenzoate)	3				
Methyl *N*-formyl-anthranilate	3				
Ethyl acetate	3				
Ethyl butanoate	3				
Ethyl hexanoate	1 – 3	1, 2	1, 2	1, 2	
Ethyl octanoate	1, 2	1, 2	1, 2	1, 2	1, 2
Ethyl decanoate	1, 2		1, 2	1, 2	1, 2
Ethyl dodecanoate	1, 2		1, 2		
Ethyl but-2-enoate	3				
Ethyl acetoacetate	3				
Ethyl cinnamate	1, 2	1, 2	1, 2	1, 2	1, 2
Butyl formate	3				
Butyl acetate	1 – 3	1, 2	1, 2	1, 2	
Butyl hexanoate				1, 2	
3-Methylbut-2-enyl acetate	3				
Hexyl formate	3				
Hexyl acetate	1 – 3	1, 2	1, 2	1, 2	
Hexyl butanoate	3			1, 2	
Hexyl hexanoate				1, 2	
Hexyl decanoate				1, 2	
trans-Hex-2-enyl acetate	3				
cis-Hex-3-enyl acetate	1 – 3	1, 2	1, 2		
Octyl acetate	1 – 3	1, 2	1, 2	1, 2	1, 2
Decyl acetate	1 – 3	1, 2			1, 2
Decyl butanoate	3				
Decyl hexanoate				1, 2	
Benzyl acetate	1 – 3	1, 2			1, 2
Carvyl acetate	3				

TABLE IV.III (*continued*)

Compound	Ref.[a]				
	F. vesca	F. moschata	F. chiloensis	F. virginiana	F. nilgerrensis
Miscellaneous					
Linalool oxides	3				
γ-Hexalactone	3				
γ-Heptalactone	3		1, 2		
γ-Octalactone	3				
γ-Decalactone	1, 2	1, 2	1, 2	1, 2	1, 2
γ-Dodecalactone	1, 2		1, 2	1, 2	1, 2
δ-Hexalactone	3				
δ-Octalactone	3				
2,5-Dimethyl-4-methoxy-2H-furan-3-one	3				
4-Vinylphenol	3				
Eugenol	3				

[a] 1. Drawert *et al.*, 1973; 2. Staudt *et al.*, 1975; 3. Pyysalo *et al.*, 1979.

and 'Alaska Pioneer' cultivars, respectively. These values are somewhat lower than those reported for *F. vesca* (1.7 mg/kg) by Pyysalo *et al.* (1979). Only traces of the corresponding hydroxy compound, 2,5-dimethyl-4-hydroxy-2H-furan-3-one, were found in both berries. In general, the aroma of both berries consists of the same compounds, the concentrations of which are, however, higher in the cultivar 'Annelie'. Table IV.IV shows the 'odour values' of the major compounds in these cultivars compared with those of *F. vesca* and *F.* × *ananassa* cultivar 'Senga Sengana'.

Hirvi (1983) has studied the aromas of eight fresh and deep-frozen strawberry cultivars (Senga Sengana, Red Gauntlet, Kristina, Zephyr, Ostara and three new cultivars Lihama × Senga Sengana, Ydun × Senga Sengana and Senga Sengana × Red Gauntlet) by mass spectrometric and sensory evaluation. The concentrations of 17 major components of Senga Sengana were determined in each cultivar. Intensity of odour, character of odour, overall impresson of odour, sweetness, overall impression of taste, sourness, off-odours and off-tastes were evaluated using a graphical scale method. The concentrations of the most important aroma compounds, ethyl hexanoate, ethyl butanoate, *trans*-hex-2-enal, 2,5-dimethyl-4-methoxy-2H-furan-3-one and linalool, were highest among the cultivars Senga Sengana, Kristina and Lihama × Senga Sengana. These three were rated best with regard to character of odour and overall impression of odour. Clear correlations were also obtained between titratable acids and sourness.

Yamashita *et al.* (1976) investigated the formation of volatile alcohols, aldehydes and esters in strawberries. They found that acetaldehyde, propanal, 2-methylpropanal, butanal, 3-methylbutanal, pentanal and hexanal were reduced to their corresponding alcohols by incubation with strawberry fruit. The alcohols formed were then converted to their acetate, propanoate, butanoate, 3-methylbutanoate and hexanoate esters during the incubation. In all, seven alcohols and 54 esters were produced during the incubation.

Skrede (1980) investigated 12 strawberry cultivars for potential use by the jam processing industry. Chemical, physical and sensory analyses were performed with fresh and thawed fruit, and with jams made from the fruits. It was concluded that the cultivars Jonsok, Totem and Bounty appeared suitable for industrial jam production when compared with the cultivar Senga Sengana now in use. The cultivars Chanil, Fanil and Sivetta were not acceptable.

Schreier and Drawert (1981) investigated the volatile substances of strawberry wine made from the cultivar Senga Sengana. They found that by comparison of the concentrations of genuine fruit aroma components, such as 2,5-dimethyl-4-methoxy-2H-furan-3-one, methyl cinnamate and γ-decalactone, it could be possible to demonstrate unpermitted additions of aroma compounds.

TABLE IV.IV

Odour thresholds and odour values of the major compounds identified in some different strawberry cultivars (Hirvi and Honkanen, 1982)

Compound	Odour threshold in water (mg/kg)	Odour value			
		Annelie	Alaska Pioneer	Senga Sengana	Wild strawberry (*F. vesca* L.)
Ethyl hexanoate	0.003	9600	660	4000	
Ethyl butanoate	0.001	1100	100	1200	300
trans-Hex-2-enal	0.017	88	59	28	2.9
2,5-Dimethyl-4-methoxy-2H-furan-3-one	0.01	80	45	30	170
Linalool	0.006	8.0		25	
Benzaldehyde	0.035	5.7			
Methyl salicylate	0.04	5.0			
Hexanal	0.02		13	5.0	
γ-Octalactone	0.088		1.7		1.0
Vanillin (4-hydroxy-3-methoxybenzaldehyde)	0.02	1.0	1.0		1.0
Methyl anthranilate (methyl 2-amino-benzoate)	0.003	6.7			100
Myrtenol	0.007				86
Carvyl acetate	0.0015				27
Eugenol	0.03				8.1
Heptan-2-one	0.14			1.5	2.1
Nonan-2-one	0.038				3.9

6. BERRIES OF THE GENUS *RUBUS*

The genus *Rubus* (*Rosaceae*) has a very wide global distribution, appearing as several different species. The berries are aggregates of small drupelets and are yellow, orange, red, blue or nearly black in colour, generally with delicate aromas.

6.1 Raspberries

Wild and cultivated raspberries (*Rubus idaeus* L.) are economically the most important species in the genus *Rubus*. In addition to raspberry, blackberry (boysen berry) (*R. fruticosus* Coll., syn. *R. ursinus, R. laciniata* L.) and its crossings with raspberry are widely cultivated in central and southern Europe and in America. Some other closely related species are also known, such as yellow raspberry (*R. idaeus f. chlorocarpus*), *R. xanthocarpus*, dewberry (*R. caesius,* L.) and black raspberry (*R. occidentalis,* L.) with yellow, orange, blue or black coloured berries, respectively. These species are, however, of little or no economic importance.

Volatile components of cultivated raspberries have been investigated extensively during recent decades. In the studies of Schinz and Seidel (1957), Winter and Sundt (1962), Sundt and Winter (1962) and Bohnsack (1967a,b), numerous aroma compounds were isolated and characterized by classical methods. In more recent work, Winter and Enggist (1971) identified 39 neutral volatiles. These and other studies are summarized in Table IV.V. 4-(4-Hydroxy-phenyl)butan-2-one (raspberry ketone) together with α- and β-ionones have been found to be the character impact compounds of the aroma of raspberries.

It is well known that the flavour of fresh wild raspberries is much stronger and more pleasant than that of cultivated raspberries. The aroma of wild raspberries is, however, very sensitive to heat and storage. Even deep freezing destroys the pleasant aroma nuances of the fresh berries. For this reason, Honkanen *et al.* (1980) investigated the possible qualitative and quantitative differences between volatiles of fresh wild and cultivated raspberries, and also the chemical and enzymic reactions which occur during the processing or storage of the berries. A total of 75 components was identified in fresh wild berries, corresponding to about 64 mg/kg of raspberry oil. More than 40 compounds not previously reported as raspberry volatiles were detected (see Table IV.VI). These included 2,5-dimethyl-4-hydroxy-2*H*-furan-3-one, 2,5-dimethyl-4-meth-oxy-2*H*-furan-3-one and 11 terpenes. Two of the identified esters, ethyl 5-hydroxyoctanoate and ethyl 5-hydroxydecanoate, have not previously been identified in natural products. They were also identified in cultivated raspberries in somewhat lower concentrations. These compounds are probably precursors of the corresponding δ-lactones, which are also present in raspberry oil. The esters are very unstable, losing ethanol on standing or heating (probably also enzymically). The concentration of 4-(4-hydroxyphenyl)butan-2-one (raspberry ketone) is about 3 times higher in wild raspberries (3.1 mg/kg) than in the

TABLE IV.V

Compounds identified in cultivated raspberries (*Rubus idaeus* L.)

Compound	Ref.[a]
Acids	
Formic acid	14
Acetic acid	1, 8, 10, 13, 14
Propanoic acid	8, 10, 13
2-Methylpropanoic acid	8, 10, 13, 14
Butanoic acid	8, 10, 13, 14
3-Methylbutanoic acid	8, 10, 13, 14
Pentanoic acid	8, 10, 13, 14
Hexanoic acid	8, 10, 13, 14
Heptanoic acid	8
Octanoic acid	8, 10, 13, 14
Nonanoic acid	8
Decanoic acid	8, 14
Hex-2-enoic acid	10, 13, 14
Hex-3-enoic acid	10, 13, 14
Alcohols	
Methanol	4
Ethanol	4, 8, 12, 13, 14
Propan-1-ol	8
Butan-1-ol	4, 8, 13, 14
trans-But-2-en-1-ol	4
2-Methylpropan-1-ol	14
Pentan-1-ol	4, 7
3-Methylbutan-1-ol	7, 8, 13, 14
Pent-1-en-3-ol	4
3-Methylbut-2-en-1-ol	4, 8, 12, 13
3-Methylbut-3-en-1-ol	4
Hexan-1-ol	1, 3, 14, 7, 8, 12, 13
cis-Hex-3-en-1-ol	6, 4, 7, 8, 13, 14
Butan-2-ol	2
trans-Pent-2-en-1-ol	7
Heptan-1-ol	7
Heptan-2-ol	14
Octan-1-ol	7
cis-Oct-2-en-1-ol	7
Nonan-1-ol	7
Carbonyl compounds	
Acetaldehyde	3, 14
Propanal	3, 14
Propenal	3
3-Methylbut-2-enal	2
Methylpropanal	14
Pent-2-enal	3
Hexanal	7, 14
trans-Hex-2-enal	3, 7, 14
cis-Hex-3-enal	3, 7, 14
Acetone	3, 12, 14

TABLE IV.V (*continued*)

Compound	Ref.[a]
Butanedione	3, 5, 8, 14
Pentan-2-one	3
Acetoin (3-hydroxybutanone)	3, 7, 8
Nonan-2-one	7
Heptan-2-one	14
Esters	
Methyl acetate	13
Ethyl acetate	8, 12, 13, 14
Butyl acetate	13
Pentyl acetate	2
Ethyl 5-hydroxyoctanoate	9
Ethyl 5-hydroxydecanoate	9
3-Methylbutyl acetate	13, 14
Ethyl hexanoate	14
Ethyl octanoate	14
cis-Hex-3-enyl acetate	7, 14
Hexyl acetate	13
Terpenes	
α-Ionone	3, 7, 8, 14
β-Ionone	3, 5, 7, 8, 14
Menthone	3, 14
Geraniol	4, 7, 8, 11, 14
Menthol	2
Linalool	7, 8, 14
4-Terpineol	7, 14
α-Terpineol	14
Nerol	7
Theaspirane	7
β-Dihydroionone	7, 8
Epoxy-β-ionone	7
p-Menth-2-en-1-ol	7
Geranial	7
Neral	7
Piperitone	7
Camphene	7
Damascenone	7
Myrtenol	14
Aromatic compounds	
Benzaldehyde	3, 9, 14
Benzyl alcohol	7, 8, 14
2-Phenylethanol	1, 5, 12, 14
4-Methoxybenzaldehyde	3
Acetophenone	7
Ethyl salicylate	14
Phenylacetaldehyde	14
Methyl benzoate	14
4-(4-Hydroxyphenyl)butan-2-one	1, 3, 9, 14
Benzoic acid	14

(*continued*)

150

TABLE IV.V (*continued*)

Compound	Ref.[a]
Coumarin	14
3-Phenylpropanoic acid	14
Heterocyclic compounds	
2-Furfural (2-furaldehyde)	3
5-Methyl-2-furfural	3
γ-Hexalactone	7, 8
γ-Octalactone	7, 8, 14
δ-Hexalactone	9
δ-Octalactone	9
δ-Decalactone	1, 7
δ-Dodecalactone	9
γ-Hex-2-enolide	2

[a] 1. Schinz and Seidel, 1957; 2. Nursten and Williams, 1967; 3. Winter and Sundt, 1962; 4. Sundt and Winter, 1962; 5. Bohnsack, 1967a; 6. Bohnsack, 1967b; 7. Winter and Enggist, 1971; 8. Pyysalo, 1976a; 9. Honkanen *et al.*, 1980; 10. Winter *et al.*, 1962; 11. Sundt and Winter, 1960; 12. Herrmann, 1963; 13. Schwob and Dupaigne, 1965; 14. Broderick, 1976.

cultivar Preussen; in the cultivar Ottawa, only traces of this ketone were detected (Honkanen *et al.*, 1980). The furanone compounds mentioned above occur in very low concentrations and are also very unstable. These compounds have been identified previously, both in strawberries (Sundt, 1970; Pyysalo *et al.*, 1979) and in arctic bramble (Kallio and Honkanen, 1974). The concentrations of α- and β-ionones in the raspberry cultivars (Preussen and Ottawa) were 1.5 – 2 times higher than in the wild raspberries. With the exception of the ionones, the concentrations of individual components present in both wild and cultivated berries were in general 3 – 4 times higher in wild raspberries. Furthermore, the numerous compounds found only in wild berries may also contribute to the distinct difference between the aromas of wild and cultivated raspberries. It seems to be a general phenomenon that increase in berry size, obtained by breeding, hybridization, or fertilization, inevitably leads to deterioration of the aroma of the berries. Climatic conditions and even geographical situation can also have a major influence on the aroma of some berries. For example, the berries grown in the northern countries are in general more highly flavoured than those of the same species grown further south. This may be due to the stronger solar radiation in the north during the summer.

6.2 Blackberries

Although the aroma of raspberries has been studied very extensively, only little attention has been paid to the flavour of blackberries (*Rubus fruticosus*, Coll.). Until recently, only three papers had been published during the past 20 years (Scanlan *et al.*, 1970; Houchen *et al.*, 1972; Gulan *et al.*, 1973), all dealing

TABLE IV.VI

Compounds identified in wild raspberries (*Rubus idaeus* L.) (Honkanen *et al.*, 1980)

Compound	Compound
Terpene hydrocarbons	3-Methylbut-2-en-1-yl acetate
α-Pinene	*cis*-Hex-3-en-1-yl formate
α-Phellandrene	*cis*-Hex-3-en-1-yl acetate
Limonene	Ethyl 5-hydroxyoctanoate
α-Terpinolene	Ethyl 5-hydroxydecanoate
Sabinene	
trans-Caryophyllene	Carbonyl compounds
Humulene	Hexanal
α-Elemene	*trans*-Hex-2-enal
	6-Methylhept-5-en-2-one
Alcohols	Acetoin (3-hydroxybutanone)
Ethanol	Piperitone
2-Methylpropan-1-ol	α-Ionone
2-Methylbutan-1-ol	β-Ionone
Pentan-1-ol	
3-Methylbut-2-en-1-ol	Aromatic compounds
Hexan-1-ol	Xylene
trans-Hex-3-en-1-ol	*p*-Cymene
cis-Hex-3-en-1-ol	Benzaldehyde
trans-Hex-2-en-1-ol	Benzyl acetate
Linalool	Benzyl alcohol
4-Terpineol	2-Phenylethanol
α-Terpineol	2-Methoxy-4-vinylphenol
cis-Sabinol	2-Methoxy-5-vinylphenol
Nerol	3,4-Dimethoxybenzaldehyde
Geraniol	Eugenol
Menthanol	4-Vinylphenol
	trans-Cinnamyl alcohol
Acids	Benzoic acid
Acetic acid	Vanillin (4-hydroxy-3-methoxybenzaldehyde)
Propanoic acid	Gingerone
Butanoic acid	4-(4-Hydroxyphenyl)butan-2-one (raspberry ketone)
2-Methylbutanoic acid	
3-Methylbut-2-enoic acid	Heterocyclic compounds
3-Methylbut-3-enoic acid	Linalool oxides
Hexanoic acid	2,5-Dimethyl-4-methoxy-2*H*-furan-3-one
Octanoic acid	5-Methyl-4-hydroxy-2*H*-furan-3-one
Decanoic acid	2,5-Dimethyl-4-hydroxy-2*H*-furan-3-one
Tetradecanoic acid	γ-Hexalactone
Hexadecanoic acid	γ-Octalactone
	δ-Hexalactone
Esters	δ-Octalactone
Ethyl acetate	δ-Decalactone
3-Methylbut-2-en-1-yl formate	δ-Dodecalactone

with the flavour of commercial blackberry essence (var. Thornless Evergreen). Twenty-three compounds were identified, of which only 3,4-dimethoxyallyl-benzene was considered to contribute to the characteristic flavour of blackberries.

Georgilopoulos and Gallois (1987a) recently studied the aroma compounds of fresh blackberries (var. Thornless Evergreen) in detail. They identified 245 compounds, including 212 for the first time. The main compounds identified were heptan-2-ol (43.06%), p-cymen-8-ol (3.72%), heptan-2-one (3.32%), hexan-1-ol (3.05%), α-terpineol (2.38%), pulegone (2.05%), octan-1-ol (1.83%), isoborneol (1.76%), myrtenol (1.28%), 4-terpineol (1.21%), carvone (1.14%), elemicine (1.12%) and nonan-1-ol (1.01%).

Table IV.VII lists all the compounds identified to date as blackberry volatiles.

The aroma composition of blackberry is thus quite different from that of raspberry. Only traces of β-ionone have been found in blackberry oil, and α-ionone and raspberry ketone are completely absent, whereas all three compounds are characteristic of raspberry flavour. The main blackberry compound, heptan-2-ol, has previously been identified also in arctic bramble (*R. arcticus*) by Kallio (1976a) as an important flavour compound, and it possesses an intense fruity taste with herbaceous nuances. It is interesting to note that the same ethyl esters of 5-hydroxyacids previously detected in raspberries by Honkanen *et al.* (1980) are also present in blackberries in trace amounts. Despite the large numbers of volatiles identified, according to the authors (Georgilopoulos and Gallois, 1987a) no single compound could be described as blackberry-like. Attempts to reproduce the original aroma by mixing appropriate amounts of pure compounds chosen from among the most abundant, resulted in an odour somewhat reminiscent of blackberries, but lacking the delicate aroma of the natural extract.

In a more recent publication, Georgilopoulos and Gallois (1987b) have investigated the volatile flavour compounds in heated blackberry juices. They found that the concentrations of some aldehydes, lactones and furan compounds were increased very markedly on heating.

6.3 Cloudberries

The cloudberry (*R. chamaemorus*, L.) is common throughout the cool temperate zone of the northern hemisphere, i.e. in the subarctic regions and the coniferous forest zone. The species also occurs in some mountain ranges in central Europe and central Asia (Resvoll, 1925, 1928; Hulten, 1971; Taylor, 1971). It is common in Scandinavia, especially in Finland except in the southernmost areas. In Finland, cloudberry grows in mires, principally in fjeld Lapland. The berries resemble raspberries, but are larger and yellow or orange in colour. The odour of cloudberry is very mild and not as distinctive as that of raspberry. The taste, however, is very pleasant, and therefore cloudberries are well suited for the flavouring of ice-cream, yoghurt, bakery products, etc. In Finland, large quantities of cloudberries are also frozen for use in whole-berry form.

TABLE IV.VII

Compounds identified in blackberries (*Rubus fruticosus* Coll.)

Compound	Ref.[a]
Hydrocarbons	
Benzene	4
Toluene	4
o-Xylene	4
m-Xylene	4
p-Xylene	4
Styrene	4
α-Pinene	4
β-Pinene	4
Camphene	4
α-Phellandrene	4
p-Cymene	4
Limonene	4
γ-Terpinene	4
α,*p*-Dimethylstyrene	4
Car-2-ene	4
α-Terpinolene	4
α-Cubebene	4
δ-Cadinene	4
Esters	
Ethyl acetate	1, 4
Diethyl carbonate	4
Methyl propanoate	4
Methyl 2-methylpropanoate	4
Methyl methylpropenoate	4
Ethyl propanoate	4
Methyl butanoate	4
Methyl but-2-enoate	4
Methyl 3-methylbutanoate	4
Butyl acetate	4
Methyl pentanoate	4
Ethyl 2-methylbutanoate	4
Ethyl 3-methylbutanoate	4
3-Methylbutyl acetate	4
Pentyl acetate	4
Methyl hexanoate	4
Dimethyl malonate	4
Butyl butanoate	4
trans-Hex-2-enyl acetate	4
trans-Hex-3-enyl acetate	4
Ethyl hexanoate	4
Hexyl acetate	4
Methyl heptanoate	4
Dimethyl succinate	4
Butyl pentanoate	4
Diethyl malonate	4
Methyl octanoate	4
Diethyl succinate	4

(*continued*)

154

TABLE IV.VII (*continued*)

Compound	Ref.[a]
Ethyl benzoate	4
Ethyl octanoate	4
Octyl acetate	4
Methyl nonanoate	4
Ethyl salicylate	4
Hexyl 2-methylpropanoate	4
Methyl decanoate	4
Ethyl decanoate	4
Ethyl cinnamate	4
Methyl dodecanoate	4
Ethyl 5-hydroxyhexanoate	4
Ethyl dodecanoate	4
Methyl tridecanoate	4
Trimethyl citrate	4
Ethyl 5-hydroxyoctanoate	4
Methyl tetradecanoate	4
Ethyl 5-hydroxydecanoate	4
Benzyl benzoate	4
Ethyl tetradecanoate	4
Methyl pentadecanoate	4
Ethyl pentadecanoate	4
Methyl hexadecanoate	4
Methyl hexadec-9-enoate	4
Ethyl hexadecanoate	4
Methyl heptadecanoate	4
Methyl octadec-9-enoate	4
Methyl octadeca-9,12,15-trienoate	4
Methyl eicosa-11,14,17-trienoate	4

Aldehydes

Compound	Ref.
Butanal	4
3-Methylbutanal	4
Pentanal	4
Hexanal	4
trans-Hex-2-enal	4
Heptanal	4
Benzaldehyde	4
Hepta-*trans*-2,*cis*-4-dienal	4
Octanal	4
Hepta-*trans*-2,*trans*-4-dienal	4
Phenylacetaldehyde	4
trans-Oct-2-enal	4
Nonanal	4
Octa-*trans*-2,*trans*-4-dienal	4
trans-Non-2-enal	4
Myrtenal	4
Decanal	4
Nona-*trans*-2,*trans*-4-dienal	4
trans-Dec-2-enal	4
Cinnamaldehyde	4
Perillaldehyde	4

TABLE IV.VII (*continued*)

Compound	Ref.[a]
Undecanal	4
Deca-*trans*-2,*trans*-4-dienal	4
trans-Undec-2-enal	4
Dodecanal	4
Tridecanal	4
3-(4-Isopropylphenyl)-2-methylpropanal (cyclamenaldehyde)	4
Tetradecanal	4
Pentadecanal	4
Ketones	
Butenone	4
Butanedione	4
Butanone	4
Pentan-2-one	4
2-Methylpent-1-en-3-one	4
Acetoin (3-hydroxybutanone)	4
Pent-3-en-2-one	4
4-Methylpentan-2-one	4
Heptan-2-one	1 – 4
Octan-3-one	4
Octane-2,3-dione	4
6-Methylhept-5-en-2-one	4
Pulegone	4
Acetophenone	4
Octa-*trans*-3,*trans*-5-dien-2-one	4
Camphor	4
4-Methylacetophenone	4
Verbenone	4
Carvone	1, 4
Piperitone	4
Carvenone	4
Isopiperitone	4
Undecan-2-one	4
Damascenone	4
Dodecan-2-one	4
α-Ionone	3
β-Ionone	4
Dihydro-β-ionone	4
Tridecan-2-one	4
trans-Geranylacetone	4
4-Hydroxy-3-methoxypropiophenone (propiovanillone)	4
2-Oxacyclotetradecanone	4
Hexahydrofarnesylacetone	4
2-Oxacycloheptadecanone	4
Acetals	
1,1-Dimethoxyethane	1, 4
1,2-Dimethoxyethane	4
1-Ethoxy-1-methoxyethane	4
1,1-Dimethoxypropane	4
1,1-Diethoxyethane	1, 4

(*continued*)

TABLE IV.VII (*continued*)

Compound	Ref.[a]
1,3-Dimethoxypropane	4
1-Ethoxy-1-propoxyethane	4
1,1-Dimethoxyheptane	4
1,1-Dimethoxyoctane	4
1,1-Dimethoxynonane	4
1,1-Dimethoxydecane	4
Phenols and phenol ethers	
Eugenol	3, 4
Methyleugenol	1, 4
Vanillin (4-hydroxy-3-methoxybenzaldehyde)	4
Isoeugenol	4
Methylisoeugenol	4
4-Allyl-2,6-dimethoxyphenol	4
1-Allyl-3,4,5-trimethoxybenzene (elemicine)	3, 4
Alcohols	
Ethanol	1, 4
Pent-1-en-3-ol	4
3-Methylbutan-1-ol	4
Pentan-1-ol	4
3-Methylbut-2-en-1-ol	4
cis-Hex-3-en-1-ol	4
trans-Hex-2-en-1-ol	4
Hexan-1-ol	1, 4
Heptan-1-ol	4
Heptan-2-ol	1, 4
2-Butoxyethanol	4
Oct-1-en-3-ol	4
6-Methylhept-5-en-2-ol	4
6-Methylheptan-1-ol	4
2-Ethylhexan-1-ol	4
Benzyl alcohol	4
Octan-1-ol	4
2-Phenylpropan-2-ol	4
Linalool	1, 4
2-Phenylethanol	3, 4
trans-Pinocarveol	4
Borneol	1, 4
Isoborneol	4
Nonan-1-ol	4
4-Terpineol	1, 4
p-Cymen-7-ol	4
p-Cymen-8-ol	3, 4
α-Terpineol	1, 4
Myrtenol	4
cis-Piperitol	4
3-Phenylpropan-1-ol	4
Geraniol	4
Decan-1-ol	4
p-Mentha-1,4-dien-7-ol	4

TABLE IV.VII *(continued)*

Compound	Ref.[a]
p-Mentha-1,5-dien-7-ol	4
Perillyl alcohol	4
Thymol	4
cis-Cinnamyl alcohol	4
Decan-1-ol	4
Undecan-1-ol	4
2-(3-Isopropylphenyl)propan-1-ol	4
Cadinol	4
Lactones	
γ-Butyrolactone	4
γ-Hexalactone	4
γ-Octalactone	4
γ-Nonalactone	4
γ-Decalactone	4
γ-Undecalactone	4
γ-Dodecalactone	4
δ-Octalactone	4
δ-Decalactone	4
δ-Undecalactone	4
δ-Dodecalactone	4
Furans	
2-Methylfuran	4
Tetrahydrofuran	4
2-Methyltetrahydrofuran	4
3-Methyltetrahydrofuran	4
2,4-Dimethyltetrahydrofuran	4
3-Methyl-3*H*-dihydrofuran	4
3,4-Dimethylfuran-2,5-dione	4
Dibenzofuran	4
Ethers	
3-Methoxyhexane	4
3-Methoxyoctane	4
8-Ethoxy-*p*-cymene	4
Miscellaneous	
3,4-Dimethyl-1,3-dioxane	4
2,4,5-Trimethyl-1,3-dioxolane	4
Pyridine	4
2,5-Dimethoxy-4-methoxy-2*H*-furan-3-one	4
2-Methoxy-2-hydroxy-1-phenylethane	4
Phenylacetonitrile	4
Benzothiazole	4
Quinoline	4
cis-Theaspirane	4
trans-Theaspirane	4
Dimethylphenylcarbinol butanoate	4
Dihydroactinidiolide (5,6,7,7a-tetrahydro-4,4,7a-trimethyl-4*H*-benzo-furan-2-one)	4

(continued)

TABLE IV.VII *(continued)*

Compound	Ref.[a]
cis-Megastigma-5,8-dien-4-one	4
Tonalide (2-acetyl-3,5,5,6,8,8-hexamethyl-5,6,7,8-tetrahydro-naphthalene)	4
Acids	
Acetic acid	4
Hexanoic acid	4
Octanoic acid	4
Decanoic acid	4
Undecanoic acid	4
Dodecanoic acid	4
Tridecanoic acid	4
Tetradecanoic acid	4
Hexadecanoic acid	4

[a] 1. Scanlan *et al.*, 1970; 2. Houchen *et al.*, 1972; 3. Gulan *et al.*, 1973; 4. Georgilopoulos and Gallois, 1987a.

Cloudberries have become valuable raw material for the Finnish liqueur, 'Lakka'. The natural growing areas of cloudberry have been deplorably reduced during the past few decades. Although cultivation has proved difficult and not yet profitable, good results have been obtained by the fertilization of its natural growing areas. Honkanen and Pyysalo (1976) and Pyysalo and Honkanen (1977) studied the qualitative and quantitative aroma composition of fresh and heated cloudberries. In these studies a total of 80 compounds were identified and these are listed in Table IV.VIII.

The aroma of cloudberries is characterized by the presence of several benzene derivatives. The main compound is benzyl alcohol, which constitutes about 30% of the essential oil. According to Pyysalo *et al.* (1977b), ethyl hexanoate, 4-vinylphenol, 2-methoxy-5-vinylphenol and 2-methylbutanoic acid are, however, the most important compounds for the aroma of cloudberry. 4-Vinylphenol and 2-methoxy-5-vinylphenol are formed enzymically by decarboxylation of 4-hydroxycinnamic acid and 3-hydroxy-4-methoxycinnamic acid, respectively. Both compounds have very low odour threshold values, and their odours were described as 'fruity' or 'fresh'. The concentrations of these phenols increase 4 to 5 times if the cloudberries are heated for 1 h at 100°C, and they may contribute to the delicate aroma of cooked cloudberry jam.

6.3.1   Formation of the major volatiles in cloudberries

As mentioned above, an important feature of the aroma of cloudberries is the high content of numerous aryl compounds among the volatiles. The neutral volatile fraction includes more than 50 aromatic components, amounting to about 80% of all the neutral volatiles identified (Pyysalo and Honkanen, 1977). Benzoic acid is the main volatile acid in cloudberries.

TABLE IV.VIII

Compounds identified in the press juice of cloudberries (*Rubus chamaemorus* L)

Compound	Ref.[a]
Acids	
Acetic acid	1, 2
Propanoic acid	1, 2
2-Methylpropanoic acid	1, 2
Butanoic acid	1, 2
2-Methylbutanoic acid	1, 2
Pentanoic acid	2
Hexanoic acid	1, 2
Octanoic acid	1, 2
2-Methyloctanoic acid	1, 2
Nonanoic acid	2
Decanoic acid	2
Dodecanoic acid	2
Tetradecanoic acid	2
Hexadecanoic acid	2
Benzoic acid	1, 2
cis-Cinnamic acid	1, 2
trans-Cinnamic acid	1, 2
4-Hydroxycinnamic acid	1
3-Phenylpropanoic acid	2
Alcohols	
Methanol	1
Ethanol	1, 2
Butan-2-ol	1
2-Methylpropan-1-ol	1
2-Methylbutan-1-ol	1, 2
Pentan-1-ol	1
Pentan-2-ol	1
Hexan-1-ol	1, 2
Heptan-2-ol	1
Pent-3-en-1-ol	1, 2
3-Methylbut-2-en-1-ol	1
trans-Hex-2-en-1-ol	1, 2
trans-Hex-3-en-1-ol	1, 2
cis-Hex-3-en-1-ol	1, 2
Oct-1-en-3-ol	1, 2
trans-Oct-2-en-1-ol	1, 2
Carbonyl compounds	
Acetaldehyde	1
Acetone	1
Butanedione	1, 2
Acetoin (3-hydroxybutanone)	1, 2
Hexanal	1, 2
Pentan-2-one	1
Heptan-2-one	1
6-Methylhept-5-en-2-one	1, 2

(*continued*)

TABLE IV.VIII (*continued*)

Compound	Ref.[a]
Esters	
Ethyl acetate	1, 2
Butyl formate	1, 2
Methyl hexanoate	1, 2
Ethyl hexanoate	1, 2
Terpenes	
cis-, *trans*-Linalool oxides	1, 2
Linalool	1, 2
Linalyl acetate	1, 2
α-Terpineol	1, 2
Geraniol	2
Aromatic compounds	
Benzyl alcohol	1, 2
1-Phenylethanol	1, 2
2-Phenylethanol	1, 2
3-Phenylpropanol	1, 2
cis-Cinnamyl alcohol	1, 2
trans-Cinnamyl alcohol	1, 2
Phenol	1, 2
4-Ethylphenol	1, 2
2-Methoxyphenol	1, 2
4-Vinylphenol	1, 2
2-Methoxy-5-vinylphenol	1, 2
4-Allylphenol	1, 2
Eugenol	1, 2
4-Methoxybenzaldehyde	1, 2
Vanillin (4-hydroxy-3-methoxybenzaldehyde)	2
Cinnamaldehyde	1, 2
Acetophenone	1, 2
4-Methylacetophenone	1
4-Hydroxyacetophenone	2
4-Methoxyacetophenone	1, 2
4-Hydroxy-3-methoxyacetophenone (acetovanillone)	2
Methyl benzoate	1, 2
Ethyl benzoate	1, 2
Methyl 3-hydroxybenzoate	2
Methyl 4-hydroxybenzoate	2
Benzyl acetate	1, 2
2-Phenethyl formate	1, 2
Methyl vanillate	2
Methyl cinnamate	1, 2
Heterocyclic compounds	
2-Furfural (2-furaldehyde)	1, 2
5-Methyl-2-furfural	1, 2
5-(Hydroxymethyl)-2-furfural	2
Coumarin	1, 2
3,5-Dihydroxy-2-methyl-4*H*-pyran-2-one	2

[a] 1. Honkanen and Pyysalo, 1976; 2. Pyysalo and Honkanen, 1977.

Both *trans* and *cis* cinnamic acids are also abundant in cloudberries, and
many neutral volatiles are formed from these acids via cinnamyl-CoA (Ebel and
Griesenbach, 1973; Mansell *et al.*, 1973). The concentration of *cis*-cinnamyl
alcohol in cloudberries is only 1/20 of the *trans* alcohol, whereas the corresponding acids are almost equal in concentration. *Para* as well as *ortho* hydroxyla-
tion of cinnamic acid has been well established in several higher plants (Neish,
1964; Potts, 1973; Russel, 1971), yielding *p*- and *o*-coumaric acids, respectively.
3-Hydroxylation and subsequent 3-*O*-methylation of *p*-coumaric acid to yield
ferulic acid was studied by Tressl and Drawert (1973) using ^{14}C-labelled

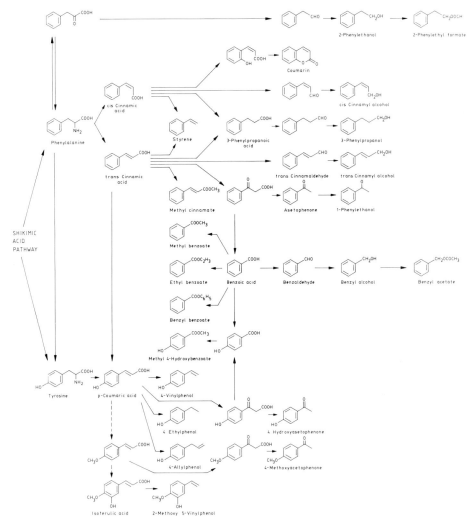

Fig. IV.3. Suggested derivation of the major aromatic compounds identified in cloudberries. Only named compounds have been identified unambiguously in the berries (Pyysalo, 1976a).

substances and plant tissue extracts. In cloudberries, however, ferulic acid is only a minor component, if present at all (Pyysalo, 1976a), whereas isoferulic and *p*-coumaric acids are the most abundant of the substituted cinnamic acids. Isoferulic acid could be formed by 4-*O*-methylation and subsequent 3-hydroxylation of *p*-coumaric acid. The enzymic decarboxylation of the cinnamic acids identified in cloudberries leads to formation of styrene, 4-vinylphenol and 2-methoxy-5-vinylphenol. The proposed enzymic pathways of the major aromatic compounds in cloudberries are shown in Fig. IV.3.

6.4 Berries of Arctic Bramble

The arctic bramble (*R. arcticus*, L.) occurs in the circumpolar subarctic zone of the northern hemisphere. Two closely related species, formerly named *R. stellatus* Sm. and *R. acaulis* Michx, growing in the Aleutian Islands, Canada and Alaska, have been shown to be subspecies of *R. arcticus* by Hulten (1971), and must therefore be named as *R. arcticus* ssp. *stellatus* and *R. arcticus* ssp. *acaulis*, respectively. The berries of arctic bramble are dark red or reddish brown in colour, and have an extremely pleasant, strong and unique aroma which is clearly different from that of other berries or fruits. The berries are therefore valuable, and are a much sought after raw material for the foodstuff industry. In Finland the best known product of the berries is the liqueur, 'Mesimarja'.

The natural growing areas of arctic bramble in Finland have also been reduced during recent decades, as in the case of cloudberry. The best growing areas are in Central Finland. Because the annual berry yield varies greatly, attempts have been made to cultivate them. Many difficulties, however, are connected with the cultivation. The berry yield is, in general, much lower than, for instance, in the case of strawberries. Furthermore, picking of the berries is very tedious and the harvest time is long.

In order to obtain more productive cultivars, arctic bramble has been crossbred with raspberry (Vaarama, 1948; Rousi, 1965; Hiirsalmi, 1973) and with *R. arcticus* ssp. *stellatus*. These cultivars are indeed more productive, but the typical and pleasant aroma of arctic bramble is only partly inherited by the progenies (Pyysalo, 1976b).

The aroma of the berries of arctic bramble has been studied by Kallio and Honkanen (1974), Kallio (1975a, 1976b)), and recently by Kallio *et al.* (1984). More than 70 compounds have been identified (see Table IV.IX).

Eleven compounds have been shown to be most important for the aroma of arctic bramble (Kallio, 1976a; Pyysalo *et al.,* 1977), namely 2,5-dimethyl-4-methoxy-2*H*-furan-3-one, 2,5-dimethyl-4-hydroxy-2*H*-furan-3-one, linalool, heptan-2-ol, 3-methylbut-2-en-1-ol, *cis*-hex-3-en-1-ol, 3-methylbutanoic acid, 2-phenylethanol, eugenol, vanillin and 1-methylbut-2-en-1-yl acetate (Kallio, 1975b). Kallio *et al.* (1984) also studied the volatile compounds of some selected wild *R. arcticus* ssp. *arcticus* strains and those of a crossing between *R. arcticus* ssp. *arcticus* and *R. arcticus* ssp. *stellatus*. (see Table IV.X).

TABLE IV.IX

Compounds identified in arctic bramble (*Rubus arcticus* L.) and some subspecies (*R. stellatus* and *R. acaulis*)

	Ref.[a]		
	R. arcticus	*R. stellatus*	*R. acaulis*
Alcohols			
Ethanol	1 – 6	3, 5	3, 5
Propan-1-ol	1 – 4, 6	3, 5	3, 5
Butan-1-ol	1 – 4, 6	3, 5	3, 5
trans-Pent-3-en-1-ol	1 – 4, 6	3, 5	3, 5
Heptan-2-ol	1 – 6	3, 5	3, 5
3-Methylbut-2-en-1-ol	1 – 6	3, 5	3, 5
Hexan-1-ol	1 – 4, 6	3	3
cis-Hex-3-en-1-ol	1 – 6	3, 5	3, 5
3-Methylbutan-1-ol	1 – 4, 6	3	3
trans-Hex-3-en-1-ol	1 – 2, 4, 6		
Linalool	1 – 6	3, 5	3, 5
α-Terpineol	1 – 3, 6	3	
Nerol	1 – 2, 4, 6		
Geraniol	1 – 2, 4, 6		
Acids			
Acetic acid	1 – 4, 6	3	3
Propanoic acid	1, 2, 4, 6		
Methylpropanoic acid	1 – 4, 6	3	3
Butanoic acid	1 – 4, 6	3	3
3-Methylbutanoic acid	1 – 6	3, 5	3, 5
3-Methylbut-2-enoic acid	1 – 2, 4, 6		
3-Methylbut-3-enoic acid	1 – 2, 4, 6		
Hexanoic acid	6		
Octanoic acid	6		
Nonanoic acid	6		
Decanoic acid	6		
Esters			
Ethyl acetate	6	3, 5	3, 5
cis-Hex-3-en-1-yl acetate	1 – 4, 6	3	3
2-Methylbut-2-enyl acetate	1 – 6	3, 5	3, 5
Carbonyl compounds			
Acetaldehyde	2		
Acetone	1 – 6	3, 5	3, 5
Pentanal	6		
Acetoin (3-hydroxybutanone)	1 – 4, 6	3, 5	3, 5
Diacetyl (butanedione)	1 – 4, 6	3	3
Heptan-2-one	1 – 4, 6	3	3
2-Methylbut-2-enal	3, 6	3	3
trans-Hex-2-enal	6		
Nonan-2-one	6		
Dihydro-β-ionone	6		

(*continued*)

TABLE IV.IX (*continued*)

	Ref.[a]		
	R. arcticus	R. stellatus	R. acaulis
Heterocyclic compounds			
Linalool oxides	1 – 4, 6	3	3
2,5-Dimethyl-4-hydroxy-2*H*-furan-3-one	5	5	5
2,5-Dimethyl-4-methoxy-2*H*-furan-3-one	1 – 6	3, 5	3, 5
Ethyl nicotinate (ethyl pyridine-3-carboxylate)	1 – 4, 6	3	3
Aromatic compounds			
Eugenol	1 – 2, 4 – 6	5	5
Eugenol methylether	1 – 4, 6	3	3
Benzyl alcohol	1 – 4, 6	3	3
2-Phenylethanol	1 – 6	3, 5	3, 5
Ethyl benzoate	1 – 4, 6	3	3
Benzyl acetate	1 – 4, 6	3	3
2-Phenethyl acetate	1 – 2, 6		
Phenol	1 – 2, 6		
2-Methoxyphenol	1 – 2, 6		
Vanillin (4-hydroxy-3-methoxybenzaldehyde)	1 – 2, 5, 6	5	5

1. Kallio and Honkanen, 1974; 2. Kallio, 1975a; 3. Kallio, 1975b; 4. Kallio, 1976a; 5. Kallio *et al.*, 1984; 6. Pyysalo, 1976b.

The concentrations of 2,5-dimethyl-4-methoxy-2*H*-furan-3-one in the selected strains and in the crossing were found to be unusually high (8 – 20 mg/kg or 30% of total volatiles). This compound was first identified in canned Alphonso mango by Hunter *et al.* (1974), in strawberries by Pyysalo *et al.* (1979), and at trace levels in wild raspberries by Honkanen *et al.* (1980). The concentration of 2,5-dimethyl-4-methoxy-2*H*-furan-3-one in wild strawberries (*F. vesca*) is about 1.7 mg/kg, or only 1/10 to 1/5 of that found in arctic bramble. The content of 2,5-dimethyl-4-hydroxy-2*H*-furan-3-one (5 – 14 mg/kg) is also very high. Because the latter compound has a considerably lower threshold concentration (0.00003 mg/kg) than the methoxyfuranone (0.01 mg/kg) (Honkanen *et al.*, 1980), the hydroxy compound may contribute much more to the typical aroma of arctic bramble. Both of these compounds have very strong, sweet and pleasant odours. The concentration of heptan-2-ol (0.9 – 3.4 mg/kg) is also high. It is interesting to note that, in blackberry (*R. fructicosus*) oil, this compound constitutes about 43% of the total volatiles, whereas the percentage of 2,5-dimethyl-4-methoxy-2*H*-furan-3-one is only 0.07% (Georgiopoulos and Gallois, 1987a).

Pyysalo (1976a,b) studied the volatiles of some hybrids between raspberry and arctic bramble. The main aroma compound of arctic bramble, 2,5-dimethyl-4-methoxy-2*H*-furan-3-one, was inherited by the crossings, but only in traces. The highest concentrations detected in two selections were 0.002

TABLE IV.X

Contents (mg/kg) of 11 aroma compounds in the press juices of some selected *Rubus arcticus* ssp. *arcticus* strains, *R. arcticus* ssp. *stellatus*, and of a hybrid between *R. arcticus* ssp. *arcticus* × *R. arcticus* ssp. *stellatus* (Kallio et al., 1984)

Compound	*R. arcticus* ssp. *arcticus* strains						*R. arcticus* ssp. *stellatus*	*R. arcticus* ssp. *arcticus* × *R. arcticus* ssp. *stellatus*
	'Pima 1'	'Pima 2'	'Mespi 1'	'Mespi 2'	Wild str.	Mesma		
2,5-Dimethyl-4-methoxy-2*H*-furan-3-one	12.0	3.8	20.0	11.0	3.1	8.1	13.0	2.8
2,5-Dimethyl-4-hydroxy-2*H*-furan-3-one	6.3	4.9	11.0	8.1	10.0	14.0	0.75	4.8
Linalool	0.30	0.07	0.65	0.45	0.70	0.25	0.03	0.08
Heptan-2-ol	2.7	1.6	0.90	1.0	0.65	3.4		1.9
3-Methylbut-2-en-1-ol	0.95	0.07	0.95	0.40	0.80	1.5	0.02	0.06
cis-Hex-3-en-1-ol	0.15	0.10	0.25	0.30	0.04	0.30		0.06
3-Methylbutanoic acid	0.14	0.03	0.15	0.10	1.5	0.20	0.30	1.1
2-Phenylethanol	0.75	0.80	0.40	0.55	0.10	0.55	tr	0.10
Eugenol	tr	tr	tr	tr	0.01	tr	0.01	tr
Vanillin (4-hydroxy-3-methoxybenzaldehyde)	0.02	0.01	0.02	0.03	0.05	0.02	0.04	0.03
3-Methylbut-2-enyl acetate	0.15	0.02	0.2	0.05	0.45	0.20		

tr = trace.

and 0.008 mg/kg, whereas the concentration of the furanone in arctic bramble varies from 8 to 20 mg/kg, as mentioned above, or about 100 – 500 times more than that found in these hybrids. At the same time, the content of ionones, the characteristic compounds in the aroma of raspberry, also decreased to a considerably lower level than in the raspberry parent. The resulting berries therefore have only a rather weak odour. Their taste is, however, sharply acid, which can be considered to be an advantage in comparison with the flat flavour characteristic of many raspberry cultivars.

7. FRUITS OF SEA BUCKTHORN

Sea buckthorn (*Hippophae rhamnoides* L., *Elaeagnaceae*) has a very wide natural distribution throughout Europe and Asia, and several local subspecies with somewhat different growth habitats have been described (Rousi, 1971; Franke and Müller, 1983). The only species growing wild in Scandinavia, as well as in England and Scotland, is correctly named as *H. rhamnoides* ssp. *rhamnoides*. *H. rhamnoides* ssp. *fluviatilis* is native to the Alps, the Apennines and the Pyrenees, *H. rhamnoides* ssp. *carpatica* to the Caucasus and Armenia, *H. rhamnoides* ssp. *turkestanica* to Kirgiz and Sinhiang, *H. rhamnoides* ssp. *mongolica* to Mongolia, *H. rhamnoides* ssp. *yannannensis* to Yannan and Tibet and *H. rhamnoides* ssp. *gyantsensis* to Tibet and Lhasa. Two closely related species, *H. salicifolia* D. Don and *H. tibetana* Schlecht, are native to the Himalayas. All these species have very decorative orange-coloured ellipsoidal fruits, rich in vitamins C and E, carotene and an oil. The name arose from Hippos (horse), because the ancient Romans made a preparation from the fruits for the grooming of horses. The fruits of the sea buckthorn have probably been used for centuries as a folk medicine for the treatment of several diseases. Sea buckthorn fruits have a sourish taste and a unique aroma, not comparable to that of any other common berries and fruits. Hirvi and Honkanen (1984) studied the volatile compounds of the fruits of sea buckthorn, and a total of 60 compounds was identified, corresponding to about 36 mg/kg of the total volatile oil (see Table IV.XI).

The volatile oil was characterized by the presence of several aliphatic esters, such as ethyl, 3-methylbutyl and *cis*-hex-3-en-1-yl esters. The most important compounds include ethyl hexanoate, 3-methylbutyl 3-methylbutanoate, 3-methylbutanoic acid, 3-methylbutyl hexanoate, 3-methylbutyl benzoate, 3-methylbutyl 2-methylbutanoate and 3-methylbutyl octanoate. It is interesting to note that 3-methylbutyl acetate, which is the commonest 3-methylbutyl ester in many fruits (for example in banana), was completely absent in the fruit of sea buckthorn, although a nearly complete homologous series of other 3-methylbutyl esters was present. All these compounds may contribute to the typical aroma of the fruit, because their measured concentrations were well above their odour threshold values. The odour values, obtained by dividing the concentrations of the compounds by their threshold concentrations, are presented for the most important flavour compounds in Table IV.XII.

TABLE IV.XI

Compounds identified in the fruits of sea buckthorn (*Hippophae rhamnoides* L) (Hirvi and Honkanen, 1984)

Compound	Compound	Compound
Terpene hydrocarbons	Pentanoic acid	3-Methylbutyl 3-methyl-
Limonene	Hexanoic acid	butanoate
Sabinene	Octanoic acid	3-Methylbutyl pentanoate
	Decanoic acid	3-Methylbutyl hexanoate
Alcohols and phenols	Phenylacetic acid	3-Methylbutyl heptanoate
cis-Hex-3-en-1-ol		3-Methylbutyl octanoate
trans-Hex-2-en-1-ol	Esters	3-Methylbutyl decanoate
Linalool	Methyl salicylate	3-Methylbutyl dodecanoate
α-Terpineol	Ethyl hexanoate	3-Methylbutyl tetradecanoate
Benzyl alcohol	Ethyl octanoate	3-Methylbutyl hexadecanoate
2-Phenylethanol	Ethyl decanoate	3-Methylbutyl furoate
Eugenol	Ethyl dodecanoate	3-Methylbutyl benzoate
2-Methoxy-5-vinylphenol	Ethyl tetradecanoate	3-Methylbut-3-enyl 3-methyl-
	Ethyl hexadecanoate	butanoate
Carbonyl compounds	Ethyl 3-hydroxy-3-methyl-	*cis*-Hex-3-en-1-yl acetate
Butanedione	butanoate	*cis*-Hex-3-en-1-yl butanoate
Benzaldehyde	Ethyl benzoate	*cis*-Hex-3-en-1-yl 3-methyl-
Phenylacetaldehyde	Ethyl cinnamate	butanoate
4-(4-Hydroxyphenyl)butan-	Propyl 2-methylbutanoate	*cis*-Hex-3-en-1-yl pentanoate
2-one	Propyl hexanoate	*cis*-Hex-3-en-1-yl hexanoate
	1-Methylpropyl benzoate	Benzyl pentanoate
Acids	2-Methylpropyl hexanoate	Benzyl benzoate
Acetic acid	3-Methylbutyl formate	2-Phenethyl pentanoate
Butanoic acid	3-Methylbutyl 2-methyl-	2-Phenethyl hexanoate
3-Methylbutanoic acid	butanoate	2-Phenethyl benzoate

TABLE IV.XII

Odour values of the most important compounds identified in the fruits of sea buckthorn (Hirvi and Honkanen, 1984)

Compound	Odour value	Threshold value in water (mg/kg)
Ethyl hexanoate	3700	0.0003
3-Methylbutyl 3-methylbutanoate	310	0.02
3-Methylbutanoic acid	38	0.15
3-Methylbutyl benzoate	22	0.25
3-Methylbutyl hexanoate	20	0.32
3-Methylbutyl octanoate	14	0.07
3-Methylbutyl 2-methylbutanoate	8	0.14
Ethyl benzoate	5	0.1

The concentrations of terpene and aromatic compounds were surprisingly low in the fruits of sea buckthorn. Only small amounts of linalool and α-terpineol, together with benzyl alcohol, 2-phenylethanol and eugenol, were detected. Sea buckthorn fruits have been used for making juices and jams, especially for health food products. In Finland, a liqueur (Tyrni) is made from these fruits.

8. CURRANTS

Currants belong to the genus *Ribes* (*Saxifragaceae*). About 150 species are known, growing principally in the northern hemisphere with a very wide natural distribution. The most important species are blackcurrants (*R. nigrum*), and red and white currants (*R. rubrum* and *R. rubrum f. alba*, respectively), which are native to north and central Europe and to some parts of Asia. Some other species, for example *R. niveum, R. americanum, R. sativum* and *R. hirtellum*, grow in America, and *R. alpinum, R. auraum, R. gordonianum* and *R. divaricatum* in both Europe and Asia.

8.1 Blackcurrants

Blackcurrants are extensively used in the preparation of jams, jellies, juices, preserves and wines. Hundreds of different cultivars have been developed. During recent years, cultivars which are erect in stature and hence suitable for mechanical harvesting have become popular.

Blackcurrents have a strong and herbaceous aroma, quite different from that of red and white currants. Several papers concerning the volatile constituents of blackcurrants have been published during recent decades: Andersson and von Sydow (1964, 1966a,b), Nursten and Williams (1969a,b), von Sydow and Karlsson (1971), Williams and Tucknott (1972), and Karlsson-Ekström and von Sydow (1973a,b). A total of about 120 compounds has been identified in these studies (see Table IV.XIII).

According to Andersson and von Sydow (1964) the higher boiling (b.p. > 150°C) fraction of the aroma oil of blackcurrants amounted to about 9 mg/kg of the fresh weight, consisting mainly of several terpene compounds, including the following (percentage amounts in parentheses): car-3-ene (25.9%), caryophyllene (11.6%), α-pinene (7.0%), β-phellandrene (5.3%), terpinolene (4.9%), limonene (3.6%), citronellyl acetate (2.5%), terpinen-4-ol (3.3%), *trans*-β-ocimene (2.9%), *cis*-β-ocimene (2.6%), *p*-cymen-8-ol (2.9%) and α-terpineol (1.4%).

Several lower boiling aliphatic compounds, including alcohols and esters, have also been identified by Nursten and Williams (1969a,b); e.g. hexan-1-ol, *trans*-hex-2-en-1-ol, 3-methylbutyl acetate, ethyl butanoate, methyl hexanoate and ethyl hexanoate. Together with the terpene compounds mentioned above, these may contribute to the typical aroma of blackcurrants.

TABLE IV.XIII

Compounds identified in blackcurrants (*Ribes nigrum* L)

Compound	Ref.[a]
Terpenes	
α-Pinene	1, 4, 5, 8
β-Pinene	4
cis-β-Ocimene	1, 4, 5
trans-β-Ocimene	1, 4, 8
Limonene	1, 4, 8
m-Cymene	1
p-Cymene	1, 4, 8
Myrcene	1, 4, 7, 8, 9
Car-3-ene	1, 3, 4, 5, 7
Car-4-ene	4
Terpinolene	1, 3, 4, 5, 7
α-Terpinene	4, 5, 8
γ-Terpinene	3, 4, 8, 9
α-Thujene	4
α-Fenchene	4
Camphene	8
p-Mentha-1,8-diene	4
α-Phellandrene	4, 5, 8, 9
β-Phellandrene	1, 4, 8
1-Methyl-4-isopropenylbenzene	4, 5, 7, 8, 9
Caryophyllene	1, 3, 4, 9
Humulene	4
γ-Elemene	4
δ-Cadinene	4
1,8-Cineol	4, 8, 9
4-Terpineol	4, 8, 9
α-Terpineol	1, 3, 4, 8, 9
p-Cymen-8-ol	1, 4, 8
Mentha-4,6-dien-8-ol	4
Citronellol	4, 8
Carveol	4
Linalool	8
Geraniol	4, 8
p-Menthan-9-ol	4
Piperitone	4
Citronellyl acetate	1, 3, 4
α-Terpinyl acetate	4
Linalool oxides	4
Rose oxide (4-methyl-2-(2-methylprop-1-enyl)tetrahydro-2*H*-pyran)	4
Carbonyl compounds	
Acetaldehyde	4, 5, 7, 8
Propanal	4, 5, 7
2-Methylbutanal	4, 5, 7
3-Methylbutanal	4
3-Methylbut-2-enal	4
Pentanal	4, 7
Hexanal	2, 4, 5

(continued)

TABLE IV.XIII (*continued*)

Compound	Ref.[a]
trans-Hex-2-enal	4, 7, 8, 9
Nonanal	4, 5, 7
Decanal	4
Acetone	2
Butanone	2
Pentan-2-one	8
3-Methylbutan-2-one	8
Hexan-2-one	9
Hexan-3-one	4
Octan-3-one	4
Alcohols	
Methanol	4, 8
Ethanol	2, 4, 8, 9
Propan-1-ol	2, 8
Butan-1-ol	2, 8, 9
2-Methylpropan-1-ol	2, 8, 9
Pentan-2-ol	8
Pent-1-en-3-ol	4, 8
2-Methylbut-3-en-2-ol	2, 4, 5, 7, 8
3-Methylbut-2-en-1-ol	2
3-Methylbutan-1-ol	2, 9
Hexan-1-ol	4, 8, 9
Hexan-2-ol	8
cis-Hex-2-en-1-ol	4, 8
trans-Hex-2-en-1-ol	8, 9
Octan-4-ol	8
Oct-1-en-3-ol	1, 8
Esters	
Methyl formate	4
Methyl acetate	2, 4, 8, 9
Ethyl acetate	2, 4, 7, 8
Methyl butanoate	8
Ethyl butanoate	4, 9
Isopropyl acetate	8
Methyl hexanoate	8, 9
Ethyl hexanoate	9
2-Methylpropyl butanoate	8
3-Methylbutyl acetate	4, 8, 9
Butyl acetate	8
Pentyl acetate	8
Hexyl acetate	4
Aromatic compounds	
Benzaldehyde	4, 8
Methyl benzoate	4, 8, 9
Ethyl benzoate	1, 4, 8
Methyl salicylate	1, 8
4-Methylacetophenone	4

TABLE IV.XIII (*continued*)

Compound	Ref.[a]
Heterocyclic compounds	
Furan	5
2-Methylfuran	4
2-Pentylfuran	4, 5
Furfural (2-furaldehyde)	4, 5

1. Andersson and von Sydow, 1964; 2. Andersson and von Sydow, 1966a; 3. Andersson and von Sydow, 1966b; 4. von Sydow and Karlsson, 1971a; 5. von Sydow and Karlsson, 1971b; 6. Karlsson-Ekström and von Sydow, 1973a; 7. Karlsson-Ekström and von Sydow, 1973b; 8. Nursten and Williams, 1969a; 9. Nursten and Williams, 1969b.

TABLE IV.XIV

Quantitative data for some major components in the essential oils of six cultivars of blackcurrant (mg/kg in berries) (Andersson and von Sydow, 1966b)

	Essential oil	Car-3-ene	γ-Terpinene	Terpino-lene	4-Terpineol	Citronellyl acetate	Caryo-phyllene
Brodtorp	13	2.3	0.1	0.75	0.14	0.19	1.6
Wellington XXX	10	1.4	0.35	0.50	1.2	0.12	0.5
Silvergiertes Zwarte	11.5	1.5	0.34	0.50	1.2	0.14	0.6
Cotswold Cross	31	< 0.60	2.3	0.50	8.0	0.40	3.3
Wellington XXX × Brodtorp	10.5	0.9	0.39	0.35	1.5	0.20	1.3
Cotswold Cross × Brodtorp	12	< 0.2	0.42	< 0.10	1.5	0.13	1.5

Andersson and von Sydow (1966a,b) compared the aroma of six blackcurrant cultivars and found qualitative and quantitative differences between the cultivars studied. 'Cotswold Cross', for example, contained roughly three times as much essential oil as the other cultivars. The results are summarized in Table IV.XIV.

The essential oil of blackcurrant leaves was studied by Andersson and von Sydow (1964) and a total of 17 compounds was identified. Qualitatively, the aroma composition of the leaves is rather similar to that of the berries. Quantitatively, the main difference is that there is a larger relative amount of monoterpene hydrocarbons in the berries than in the leaves. Rigaud *et al.* (1986) isolated an interesting sulphur compound, 4-methoxy-2-methylbutane-2-thiol, from the essential oil of blackcurrant buds. This compound has the typical catty note of blackcurrant buds and leaves. It is possible that it may also be responsible for the typical odour of blackcurrants, although it has not yet been identified in the berries (Latrasse *et al.*, 1982). Nishimura *et al.* (1987) identified *cis* and *trans* 2-hydroxymethylbut-2-enonitrile and 3-hydroxy-2-methylbutyronitrile in blackcurrant buds.

Marriott (1986) has studied the biogenesis of the terpene fraction of blackcurrant aroma. Changes in concentrations of monoterpene hydrocarbons, alcohols and esters occurred mainly during the period of sugar accumulation, when a sharp decrease in the level of terpene hydrocarbons and an increase in monoterpene alcohols and esters was observed. The presence of monoterpene and aromatic glycosides was also tentatively identified in blackcurrants for the first time.

8.2 Red currants

Only one investigation concerning the volatile compounds of red currants has appeared in the literature (Schreier *et al.*, 1977). In this study, the qualitative and quantitative differences between the flavour compositions in fruits and juices (fresh, UHT-heated and pasteurized juices) from red currants were compared (see Table IV.XV).

A total of 38 compounds was identified in the press juice of fresh red currants. The concentrations of the individual components were very low, only of the order of 0.002 – 0.070 mg/kg. The main compounds included butan-1-ol, *trans*-pent-2-enal, 3-methylbutan-1-ol, oct-1-en-3-ol, terpinen-4-ol, 3-methylbutanal, myrcene, β-pinene, caryophyllene and *trans*-non-2-enal. During juice

TABLE IV.XV

Compounds identified in red currants (*Ribes rubrum* L) (Schreier *et al.*, 1977)

Compound	Compound	Compound
Hydrocarbons	Carbonyl compounds	Alcohols
Toluene	3-Methylbutanal	2-Methylpropan-1-ol
1-Methyl-4-isopropenyl-	Pentanal	Butan-1-ol
benzene	*trans*-Pent-2-enal	Butan-2-ol
p-Xylene	Hexanal	3-Methylbutan-1-ol
Butylbenzene	*cis*-Hex-3-enal	Pentan-1-ol
β-Pinene	*trans*-Hex-2-enal	Pentan-2-ol
Myrcene	Octanal	Pentan-3-ol
Limonene	Nonanal	Pent-1-en-3-ol
β-Phellandrene	*trans*-Non-2-enal	Hexan-1-ol
γ-Terpinene	Decanal	*cis*-Hex-3-en-1-ol
β-Caryophyllene	*trans*-Dec-2-enal	*trans*-Hex-2-en-1-ol
α-Humulene	Benzaldehyde	Oct-1-en-3-ol
δ-Cadinene		4-Terpineol
		Miscellaneous
		2-Pentylfuran

processing, hexanal and *cis*-hex-3-enal are formed as intermediates, which are rapidly reduced enzymically to the corresponding alcohols. *cis*-Hex-3-en-1-ol is isomerized to *trans*-hex-2-en-1-ol, which is the main compound in the processed juices.

9. BERRIES OF THE GENUS *VACCINIUM*

The genus *Vaccinium* (*Ericaceae*) includes several species having red or blue edible berries. In Europe, the most important berries are cowberries (lingonberry, mountain-cranberry), *V. vitis ideae* L., bilberries (whortleberries), *V. myrtillus* L., and European cranberries, *V. oxycoccus* L., whereas in North America the cultivated American cranberry (*V. macrocarpon* Ait.) and some high-bush blueberry varieties are of great importance. Some other wild *Vaccinium* species are known, for example in Europe, bog blueberry (*V. uliginosum* L.) and in North America the bush-like *V. corymbosum* L., the dwarf shrub-like *V. lamarckii* Camp., rabbit eye blueberry (*V. ashei* Reade), and the bushy *V. australe* Small.

Cowberry is the most important wild berry-bearing plant in northern Europe. Cowberry crops vary greatly from year to year. Attempts have therefore been made to domesticate the cowberry (Lehmushovi and Säkö, 1975). Many interesting results have been obtained in experiments with fertilizers, substrates and mulches. Cowberry is a uniquely Eurasian species, with no closely related species in North America.

The continents of Eurasia and America have been separated for a very long time, and therefore their plant species have diversified in the course of evolution in different directions. However, it is interesting to note that, for instance, bog blueberry is still genetically so near to some American blueberries that crossing is possible, leading to intergenetic hybrids. The enzymic systems in related species producing similar spectra of aroma compounds have also remained very similar. Successful crosses have been achieved between bog blueberry and high-bush blueberry. These *Vaccinium* species have the same tetraploid chromosome number, $2n = 48$. American high-bush blueberries have many distinctive characteristics, such as vigorous bushy growth, high fruit yield and large berry size, making them more suitable for cultivation than the north European *Vaccinium* species. However, none of the American blueberry varieties could be recommended for practical cultivation under climatic conditions prevailing in the Nordic latitudes. These varieties have rather poor winter hardiness and are susceptible to infection of the shoots by the fungus *Fusicoccum putrefaciens* Shear (Hårdh, 1959), causing blueberry cancer. On the other hand, the European bog blueberry has somewhat smaller berries than high-bush blueberry, and the berries have almost no aroma. Attempts have been made by Hiirsalmi and Rousi (Rousi, 1963, 1966; Hiirsalmi, 1977a,b; Hiirsalmi and Säkö, 1977) to cross high-bush blueberry with the bog blueberry in order to obtain more hardy varieties with attractive aroma and taste characteristics. Some of these crossings are now in practical cultivation in Finland (Hiirsalmi and Lehmuskoski, 1982).

European cranberries are native to the temperate regions of the northern hemisphere. Cranberries grown in bogs are harvested by hand either in the autumn or in the spring after thawing of the snow. The cultivation of American cranberries is, however, very easy. The berries are collected with mechanical harvesters and the productivity is very high. American cranberry has considerably larger berries than the European species, but the European cranberries have a much stronger and more herbaceous aroma. The European species is therefore a very valuable raw material in the manufacture of foods and alcoholic drinks.

9.1 Cowberries

The aroma of cowberry (lingonberry) (*Vaccinium vitis idaea* L.) is characterized by the presence of several aromatic compounds, together with α-terpineol and 2-methylbutanoic acid. Anjou and von Sydow (1967a, 1969) identified 44 compounds corresponding to about 95% of the total volatiles. Fifteen of these were aliphatic alcohols, eight aliphatic aldehydes and ketones, five terpene derivatives, seven aromatic compounds and nine were other compounds. The content of 2-methylbutanoic acid recorded by Anjou and von Sydow (1969) was 48% of the total volatile oil. 2-Methylbutanoic acid and the aromatic compounds are probably the most important components of cowberry aroma. Of the alcohols, 2-methylbut-3-en-2-ol is the major component in the press residue of cowberries. This compound, along with some other unsaturated alcohols, is probably important for the overall aroma. Of the many aliphatic aldehydes, hepta-2,4-dienal may be important for the unique aroma of cowberry. Of the terpene derivates, α-terpineol is a relatively important monoterpene alcohol in cowberries. The compounds identified in cowberries are presented in Table IV.XVI.

The studies of cowberry aroma were carried out over 20 years ago. Using modern analytical facilities, many new, important compounds would almost certainly be identified in the aroma of cowberries.

9.2 Cranberries

The aromas of American cranberries (*Vaccinium macrocarpon* Ait.) and European cranberries (*V. oxycoccus* L.) have been studied in detail, and according to van Straten (1981) over 70 volatile compounds have been identified in cranberries. The aromas of cranberries are characterized by the presence of several aromatic compounds, together with α-terpineol.

The aroma of American cranberries has been studied by Croteau and Fagerson (1968), Anjou and von Sydow (1967a) and Hirvi *et al.* (1981). Comparison of the results obtained in these studies is rather difficult, however, because different samples, isolation methods and concentration techniques were used. Anjou and von Sydow (1967b) investigated extracts of the press residue of

TABLE IV.XVI

Compounds identified in cowberries (*Vaccinium vitis idaea* L), American cranberries (*V. macrocarpon*, Ait.) and European cranberries (*V. oxycoccus* L.)

Compound	Ref.[a]		
	V. vitis idaea	*V. macrocarpon*	*V. oxycoccus*
Terpenes			
α-Pinene		2, 3, 4	4
β-Pinene		3	
α-Terpinene		2	4
Limonene		2, 3, 4	4
Terpinolene			4
Myrcene		2, 3	
Car-2-ene		2	
Camphene		2	
p-Mentha-1,8-diene			4
p-Cymene		4	4
Thujene			4
α,*p*-Dimethylstyrene			4
Linalool	1	2, 3, 4	4
α-Terpineol	1	2, 3, 4	4
4-Terpineol	1	2	
Nerol		2, 3	
1,8-Cineol	1	2, 4	4
trans-Linalool oxide	1	2, 4	4
cis-Linalool oxide		2, 4	4
p-Mentha-1,8-diol		4	4
α-Cedrene		2	
trans,trans-Farnesol		2	
Cedrenol		2	
Pimaradiene		2	
Kaurene		2	
Manoyloxide		2	
Alcohols			
Ethanol	1		
Propan-1-ol	1		
2-Methylpropan-1-ol	1		
Butan-1-ol	1	4	4
2-Methylbutan-1-ol	1	4	4
2-Methylbut-3-en-2-ol	1	2, 3	
2-Methylbut-2-en-1-ol	1		
3-Methylbut-3-en-2-ol		2	
Pentan-1-ol	1	2, 3, 4	4
Pentan-2-ol	1	2, 3	
Pentan-3-ol		2	
Pent-1-en-3-ol	1		
Hexan-1-ol	1	2, 3, 4	4
Hexan-3-ol	1	2	
cis-Hex-3-en-1-ol	1	2, 4	4
trans-Hex-2-en-1-ol	1	4	4
Heptan-3-ol		2	

(*continued*)

TABLE IV.XVI (*continued*)

Compound	Ref.[a]		
	V. vitis idaea	*V. macrocarpon*	*V. oxycoccus*
Octan-1-ol		2, 3	ıxm
Octan-2-ol		2	
Octan-3-ol		2	
Oct-1-en-3-ol		2, 3, 4	4
Nonan-1-ol		2, 3	
Decan-1-ol		2, 3	
Undecan-1-ol		2	
Octadecan-1-ol		3	
2-Methylcyclohexanol		2	
3-Methylcyclohexanol		2	
Carbonyl compounds			
Acetaldehyde		3, 4	4
Pentanal		2, 3	
3-Methylbutanal		2	
Hexanal		2, 3	
Heptanal		2	
trans-Hept-2-enal		2	
Hepta-*trans*-2,*trans*-4-dienal	1	2	
Octanal		2, 3	
trans-Oct-2-enal		2	
Nonanal		2, 3	
trans-Non-2-enal		2	
Decanal		2, 3	
trans-Dec-2-enal		2	
Deca-2,4-dienal		2	
Undecanal		2	
Dodecanal		2	
Acetone	1		
Butanone	1		
Diacetyl (butanedione)	1	2, 3, 4	4
Acetoin (3-hydroxybutanone)	1	4	4
Pent-3-en-2-one	1		
Heptan-2-one	1		
Octan-2-one		2	
6-Methylhept-5-en-2-one	1	2, 4	4
Tridecan-2-one		2	
Pentadecan-2-one		2	
Octadecan-2-one		2	
Carvone			4
Acids			
Acetic acid		4	4
Propanoic acid		4	4
2-Methylpropanoic acid	1		4
Butanoic acid		4	4
2-Methylbutanoic acid	1	2, 3, 4	4
Pentanoic acid			4
Hexanoic acid		4	4
(Benzoic acid)		3, 4	

TABLE IV.XVI (*continued*)

Compound	Ref.[a]		
	V. vitis idaea	*V. macrocarpon*	*V. oxycoccus*
Aromatic compounds			
Benzyl alcohol	1	2, 3, 4	4
1-Phenylethanol			4
2-Phenylethanol		2, 3, 4	4
3-Phenylpropan-1-ol		4	4
Cuminyl alcohol			4
trans-Cinnamyl alcohol		4	4
2-(4-Hydroxyphenyl)ethanol		4	4
2-(4-Methoxyphenyl)ethanol			4
Phenol		2, 4	4
4-Ethylphenol		2, 4	4
Thymol		2, 4	4
Cresol		2	
Carvacrol		2, 4	4
Eugenol		2	4
o-Hydroxybiphenyl		2, 3	
4-Ethyl-2-methoxyphenol		2	
Benzaldehyde	1	2, 3, 4	4
4-Methoxybenzaldehyde		3	4
Salicylaldehyde			4
3-Phenylpropanal			4
trans-Cinnamylaldehyde		4	4
Vanillin (4-hydroxy-3-methoxybenzal- dehyde)			4
Acetophenone	1	2, 3, 4	4
4-Hydroxyacetophenone			4
4-Methoxyacetophenone			4
4-(4-Hydroxyphenyl)butan-2-one			4
4-(4-Hydroxy-3-methoxyphenyl)butan-2-one			4
Methyl benzoate	1	2, 3, 4	4
Ethyl benzoate		2, 3, 4	4
Benzyl benzoate		2, 3, 4	4
Methyl cinnamate			4
Ethyl cinnamate		2	
Benzyl formate	1	2, 3	4
Benzyl acetate	1	3, 4	4
Methyl salicylate	1	2	
2-Phenethyl formate			4
Benzyl ethyl ether		3	
Benzothiazole		2	
Dibenzofuran		2	
Other compounds			
Ethyl acetate	1	2, 3	
Methyl butanoate		2	
Ethyl butanoate		2	
Ethyl 2-methylbutanoate	1		
Isopropyl butanoate		2	
Hexyl butanoate		2	

(*continued*)

TABLE IV.XVI (*continued*)

Compound	Ref.[a]		
	V. vitis idaea	*V. macrocarpon*	*V. oxycoccus*
Methyl heptanoate		3	
2-Methoxyethyl acetate	1		
2-Ethoxyethyl acetate	1		
2-Butylfuran		2	
2-Pentylfuran		2	
2-Furfural (2-furaldehyde)	1	2, 3	
2,2,4-Trimethyl-1,3-dioxolane	1		
1,1-Diethoxyhexane	1		
2-Hexylthiophene		2	
2-Heptylthiophene		2	
Isophorone (3,5,5-trimethylcyclohex-2-enone)		2	
Indene		2	
Octadec-1-ene		2	
Dihydroactinidiolide (5,6,7,7a-tetrahydro-4,4,7a-trimethyl-4*H*-benzofuran-2-one)			4
γ-Octalactone			4
γ-Decalactone			4
δ-Octalactone			4
δ-Octalactone			4

[a] 1. Anjou and von Sydow, 1969; 2. Anjou and von Sydow, 1967b; 3. Croteau and Fagerson, 1968; 4. Hirvi *et al.*, 1981.

American cranberry, whereas Croteau and Fagerson (1968) and Hirvi *et al.* (1981) studied the extract of the press juice. According to Anjou and von Sydow (1967b) the amount of total volatiles in American cranberry is 1.1 mg/kg, and according to Croteau and Fagerson (1968), 11 mg/kg. Hirvi *et al.* (1981) obtained a value of 3.7 mg/kg for American cranberry. The total amount of volatiles in European cranberries is 4–5 times higher (16 mg/kg) than in American cranberries. According to Anjou and von Sydow (1967b) and Croteau and Fagerson (1968), α-terpineol is the main volatile component in American cranberry. On the other hand, Hirvi *et al.* (1981) obtained benzyl alcohol as the main component in both cranberries. Although there are large quantitative differences between these three studies, the qualitative compositions of both cranberry species were in general very similar. Some compounds reported as missing, or as only minor components, in American cranberry are typical of European cranberry, including *p*-mentha-1,8-diene, *p*-menthane-1,8-diol (terpin hydrate), carvenone, 2-(4-hydroxyphenyl)ethanol (tyrosol), 4-(4-hydroxyphenyl)butan-2-one (raspberry ketone) and 4-(4-hydroxy-3-methoxyphenyl)-butan-2-one (gingerone). Terpin hydrate has not been identified in any other berries or fruits (Hirvi *et al.*, 1981), but gingerone has been identified in wild raspberry (Honkanen *et al.*, 1980) and tyrosol in strawberries (Nursten and

Williams, 1967). Terpin hydrate and tyrosol have a bitter taste. Their taste threshold concentrations are, however, at a higher level than their concentrations in both cranberries. Of great importance is that 2-methylbutanoic acid is missing, or present in only very small amounts, in American cranberries. Like 2-methylbutanoic acid, α-terpineol also has high specific odour intensity. Some similarities between the flavours of European cranberries and cowberries may be explained by the similar concentrations of these compounds.

Deep-freezing and time of the harvesting have very little effect on the aroma composition of European cranberries. Often, European cranberries are collected in spring after melting of the snow. For comparison, the amounts of the aroma compounds found in the press juices of European and American cranberries are presented in Table IV.XVI.

Wang *et al.* (1978) studied polyphenolic compounds in cranberries. The astringency of cranberries may be partly due to the presence of high molecular weight polymeric polyphenols.

9.3 Bilberries and Blueberries

The aromas of bilberry and high-bush blueberry have been investigated by von Sydow and Anjou (1969), von Sydow *et al.* (1970), Parliment and Kolor (1975, 1976), Hirvi and Honkanen (1983a,b) and Horvat *et al.* (1983). In these studies, over 100 volatile compounds have been identified in bilberry and blueberries. von Sydow and Anjou (1969) suggested that the character impact compounds of bilberry are *trans*-hex-2-en-1-ol, ethyl 3-methylbutanoate and ethyl 2-methylbutanoate. von Sydow and Anjou (1969) also found considerable amounts (4.7% of the total volatiles) of 2-phenethyl benzoate in bilberries. However, this compound has not been detected in any other studies of bilberries. Parliment and Kolor (1975), and Parliment and Scarpellino (1977) stressed the importance of *trans*-hex-2-enal, *trans*-hex-2-en-1-ol and linalool to the characteristic flavour of blueberry. In those works, however, vacuum steam distillation was used in isolation and a packed column in gas chromatographic fractionation of volatiles, so that many important compounds may have escaped identification.

According to Hirvi and Honkanen (1983a,b), the total amounts of volatiles in bilberry and blueberry juices are below 2 mg/kg, and most of the components are present below their threshold concentrations. The aroma of bog blueberry was found to be particularly weak. No components have been found in this species which were not also present in the aroma concentrates of bilberry and high-bush blueberry. According to Hirvi and Honkanen (1983a), the aroma compositions of all three berries have many similarities. The aroma compositions of bilberry, bog blueberry and high-bush blueberry are presented in Table IV.XVII. It can be seen that the typical compounds of high-bush blueberry are terpenes: linalool, geraniol, citronellol, hydroxycitronellol, farnesol and farnesyl acetate. None, or only traces, of these components has been detected in other *Vaccinium* berries. The odour of an aqueous solution of hydroxy-

TABLE IV.XVII

Compounds identified in bilberry (*Vaccinium myrtillus* L.), bog blueberry (*V. uliginosum* L.), rabbit-eye blueberry (*V. ashei*) and high-bush blueberry (*V. corymbosum*)

Compound	Ref.[a]			
	V. myrtillus	*V. uliginosum*	*V. ashei*	*V. corymbosum*
Terpenes				
α-Pinene	2			1
β-Pinene	2			
γ-Terpinene			4	
Myrcene	2			1
Car-3-ene	2			
Limonene	1, 2	1	4	1, 3
Terpinolene	2			
Camphor	2			
Geranial	2			
Linalool	1, 2	1	4	1, 3
α-Terpineol	1, 2	1	4	1
γ-Terpineol	2		4	
Carveol			4	
Myrcenol		1		
Myrtenol	2			
Nerol	1	1	4	1, 3
Geraniol	1, 2		4	1, 3
Citronellol	1			1
Pinocarveol	2			
Borneol	2			
p-Menth-1-en-9-ol	2			
Perilla alcohol	2			
p-Mentha-1,4-dien-7-ol	2			
1,8-Cineol	1, 2			1
Hydroxycitronellol				1
Farnesol				1
cis-Caran-3-ol			4	
α-Bergamotene	2			
Caryophyllene	2			1
α-Cadinene	2			
δ-Cadinene	2			
Palustrol	2			
Ledol	2			
Dehydroabietane	2			
Cineralone			4	
α-Cedrene			4	
Sabinol			4	
Geranyl formate			4	
Linalyl acetate			4	
Farnesyl acetate				1
Alcohols				
Ethanol			4	3
Butan-1-ol	1			1
Butan-2-ol	1			1

TABLE IV.XVII (*continued*)

Compound	Ref.[a]			
	V. myrtillus	*V. uliginosum*	*V. ashei*	*V. corymbosum*
2-Methylbutan-1-ol	1, 2			
3-Methylbutan-1-ol	2	1		
Pentan-1-ol	1, 2	1		1
Pentan-2-ol	2			
Pentan-3-ol	2			
Pent-1-en-3-ol	2		4	1, 3
Pent-2-en-1-ol	2			1, 3
2-Methylbut-3-en-2-ol	2			
3-Methylbut-3-en-2-ol	2			
Hexan-1-ol	1, 2	1	4	1, 3
2-Ethylhexan-1-ol			4	3
trans-Hex-2-en-1-ol	2	1	4	3
cis-Hex-3-en-1-ol	1, 2	1		1, 3
trans-Hex-3-en-1-ol	1, 2			1, 3
Heptan-1-ol	2		4	3
Octan-1-ol	1, 2			1
trans-Oct-2-en-1-ol	2			1
Oct-1-en-3-ol	2			
Nonan-1-ol	2			1, 3
Decan-1-ol	2			
Undecan-1-ol	2			
Carbonyl compounds				
Acetaldehyde			4	
Pentanal	2			
Hexanal	2	1	4	1, 3
trans-Hex-2-enal	1	1	4	1, 3
Heptanal	2			
trans-Hept-2-enal	2			
Hepta-2,4-dienal	2			
Octanal	2			
trans-Oct-2-enal	2			
Nonanal	1, 2			
trans-Non-2-enal	2			
Decanal	2			
trans-Dec-2-enal	2			
Deca-*trans*-2,*cis*-4-dienal	2			
Deca-*trans*-2,*trans*-4-dienal	2			
Undecanal	2			
Undec-2-enal	2			
Dodecanal	2			
Tridecanal	2			
Tetradecanal	2			
Pentadecanal	2			
Hexadecanal	2			
Acetone			4	
Acetoin (3-hydroxy-butanone)	1			

(*continued*)

TABLE IV.XVII (*continued*)

| Compound | Ref.[a] | | | |
	V. myrtillus	*V. uliginosum*	*V. ashei*	*V. corymbosum*
Pentan-2-one			4	
6-Methylhept-5-en-2-one	2			
6,10-Dimethylundecan-2-one		2		
6,10,14-Trimethylpentadecan-2-one	2			
Undecan-2-one			4	
Tridecan-2-one			4	
Acids				
Acetic acid	1	1		1
Butanoic acid	1	1		1
2-Methylbutanoic acid	1	1		1
Pentanoic acid	1			1
Hexanoic acid	1	1		1
trans-Hex-2-enoic acid	1			
Esters				
Ethyl acetate			4	3
Ethyl 2-methylbutanoate	2			
Ethyl 3-methylbutanoate				3
Hexyl acetate				1
trans-Hex-2-enyl acetate	1, 2			
cis-Hex-3-enyl acetate	2			
trans-Hex-2-enyl butanoate	1			
Methyl dodecanoate	2			
Methyl tetradecanoate	2			
Ethyl tetradecanoate			4	
Methyl hexadecanoate	2			
Ethyl hexadecanoate	2			
Methyl octadecanoate	2			
Methyl 2-hydroxy-3-methylbutanoate	1			
Methyl 3-hydroxy-3-methylbutanoate	1			
Ethyl 2-hydroxy-3-methylbutanoate	1			
Ethyl 3-hydroxy-3-methylbutanoate	1			
Dimethyl octanedioate			4	
Aromatic compounds				
Benzene	2			
Toluene			4	
Xylene	2			
Styrene	2			
α-*p*-Dimethylstyrene	1			1
Naphthalene	2			

TABLE IV.XVII (*continued*)

Compound	Ref.[a]			
	V. myrtillus	*V. uliginosum*	*V. ashei*	*V. corymbosum*
Biphenyl	2			
p-Cymene			4	
Benzyl alcohol	1, 2	1	4	1
2-Phenylethanol	1, 2	1		1
3-Phenylpropan-1-ol	1, 2			1
trans-Cinnamyl alcohol	1, 2	1		1
Guaiacol		1		
Phenol	1	1		1
p-Cresol	1			
Pyrocatechol		1		
Ethylphenol	2			
2-Methoxy-5-vinylphenol	1	1		1
4-Vinylphenol	1	1		1
Thymol	2		4	1
Eugenol	1, 2		4	1
Iso-eugenol	1			1
Methyleugenol	2			
Myristicin				1
Benzaldehyde	1, 2	1		1
Phenylacetaldehyde	1, 2			
trans-Cinnamaldehyde	1, 2			1
Vanillin (4-hydroxy-3-methoxybenzaldehyde)	1	1		1
1-Allyl-4-methoxybenzene	2			
Methyl benzoate	2			
Ethyl benzoate	2			
Pentyl benzoate	2			
cis-Hex-3-enyl benzoate	2			
2-Phenethyl formate	1			
2-Phenethyl acetate	1, 2			
2-Phenethyl benzoate	2			
Methyl salicylate	1			
Piperonal			4	
Butyl phenyl ether			4	
Other compounds				
γ-Butyrolactone	1			
γ-Octalactone	1			
γ-Decalactone	1, 2			
γ-Dodecalactone	1			
δ-Decalactone	1			
2-Pentylfuran	1, 2	1		1
2-Furfural (2-furaldehyde)	2		4	
5-Methyl-2-furfural	2		4	
Coumarin	2			
2,2,6-Trimethylcyclohexanone	2			

(*continued*)

TABLE IV.XVII (*continued*)

Compound	Ref.[a]			
	V. myrtillus	*V. uliginosum*	*V. ashei*	*V. corymbosum*
Dihydroactinidiolide (5,6,7,7a-tetrahydro-4,4,7a-trimethyl-4H-benzofuran-2-one)	2			
β-Ionone			4	
Hydroxyisocarvomenthol	1			1
Dec-2-yne			4	

1. Hirvi and Honkanen, 1983; 2. von Sydow and Anjou, 1969; 3. Parliment and Kolor, 1975; 4. Horvat *et al.*, 1983.

citronellol has been described by sensory panelists as blueberry-like. Because this and many other compounds of high-bush blueberry are present below their threshold concentrations (the threshold concentration of hydroxycitronellol is 5 mg/kg in water), it is possible that synergism may play an important rôle in the overall impression of odour of high-bush blueberries. According to van Straten (1987), hydroxycitronellol has not been found in any other fruits or berries.

The aroma of bilberry differs clearly from that of high-bush blueberry. The most typical compounds of bilberry have been found to be the methyl and ethyl esters of 2- and 3-hydroxy- 3-methylbutanoic acids (Hirvi and Honkanen, 1983a). No traces of these esters have been identified in other *Vaccinium* species. As well as in bilberries, ethyl 2-hydroxy-3-methylbutanoate has been identified in some wines and brandies (Schreier *et al.*, 1976, 1978, 1979; Schreier and Drawert, 1981). The odours of pure synthetic hydroxy esters have been considered as bilberry-like. Together with ethyl 3-methylbutanoate and ethyl 2-methylbutanoate, these esters may therefore contribute to the typical aroma of bilberry described by von Sydow and Anjou (1969), and von Sydow *et al.* (1970).

Horvat *et al.* (1983) studied the flavour of rabbit-eye blueberries (*Vaccinium ashei* Reade, cv Tiftblue) using gas chromatographic and mass spectrometric techniques. These berries are native to southeastern North America and have been improved by breeding for adaptability to commercial production.

Horvat *et al.* (1983) obtained 25 mg/kg of steam volatile oils from rabbit-eye blueberries. They identified a total of 42 compounds. The aroma of rabbit-eye blueberries differs markedly from other blueberries; 29 of the identified compounds have not been reported as constituents of other blueberry volatiles.

The compounds identified in rabbit-eye blueberries are also presented in Table IV.XVII.

In many investigations, benzyl alcohol has been shown to be typical for the berries of most *Vaccinium* species: *V. myrtillus, V. uliginosum,* L. (Hirvi and

Honkanen, 1983), *V. vitis idaea* L. (Anjou and von Sydow, 1967a, 1969), *V. oxycoccus* L. and *V. macrocarpon* Ait. (Hirvi *et al.*, 1981). As an aroma substance benzyl alcohol is, however, of little significance, as its odour is rather weak and vacant.

REFERENCES

Andersson, J. and von Sydow, E., 1964. The aroma of black currants I. *Acta Chem. Scand.*, 18: 1105–1114.
Andersson, J. and von Sydow, E., 1966a. The aroma of black currants II. *Acta Chem. Scand.*, 20: 522–528.
Andersson, J. and von Sydow, E., 1966b. The aroma of black currants III. *Acta Chem. Scand*, 20: 529–535.
Anjou, K. and von Sydow E., 1967a. The aroma of cranberries I. *Vaccinium vitis-idaea* L., *Acta Chem. Scand.*, 21: 945–953.
Anjou, K. and von Sydow, E., 1967b. The aroma of cranberries II. *Vaccinium macrocarpon* Ait. *Acta Chem. Scand.*, 21: 2076–2082.
Anjou, K. and von Sydow, E., 1969. The aroma of cranberries III. Juice of *Vaccinium vitis-idaea* L. *Acta Chem Scand.*, 23: 109–114.
Barnes, R.D., Law, L.M. and MacLeod, A.J., 1981. Comparison of some porous polymers as adsorbents for collection of odour samples and the application of the technique to an environmental malodour. *Analyst*, 106: 412–428.
Bemelmans, J.M.H., 1978. Review of isolation and concentration techniques. In: D.G. Land and H.E. Nursten (Editors), Progress in Flavour Research., Applied Science Publishers Ltd., London, pp. 79–98.
Bemelmans, J.M.H. and Schaefer, J., 1981. Isolation and concentration of volatiles from foods. In: H. Maarse and R. Belz (Editors), Handbuch der Aromaforschung. Isolation, Separation and Identification of Volatile Compounds in Aroma Research. Akademie-Verlag, Berlin, pp. 4–59.
Blakesley, C.N., Loots, J.G., du Plessis, L.M. and de Bruyn, G., 1979. γ-Irradiation of subtropical fruits. 2. Volatile components, lipids and amino acids of mango, papaya and strawberry pulp. *J. Agric. Food Chem.*, 27: 42–48.
Bohnsack, H., 1967a. Über Untersuchungsergebnisse des natürlichen Himbeerfruchtöles III. *Riechst., Aromen, Körperpflegem.*, 17: 357.
Bohnsack, H., 1967b. Beitrag zur Kenntnis der ätherischen Öle, Riech- und Geschmackstoffe XX Mitteilung über Untersuchungsergebnisse des natürlichen Himbeerfruchtöles. III Teil. Die Inhaltstoffe des Himbeertrester – Extraktöles. *Riechst., Aromen, Körperpflegem.*, 17: 514–516.
Boyko, A.L., Morgan, M.E. and Libbey, L.M., 1978. Porous polymers trapping for GC/MS analysis of vegetable flavours. In: G. Charalambous (Editor), Analysis of Foods and Beverages, Headspace Techniques. Academic Press Inc., New York, pp. 57–79.
Broderick, J.J., 1976. Raspberry. A case history. *Int. Flavours Food Addit.*, 7: 27.
Brooks, R.M. and Olmo, H.P., 1968. Register of new fruit and nut varieties Lis. 23. *Am. Hortic. Sci.*, 93: 879.
Brown, R.H. and Purnell, C.J., 1979. Collection and analysis of trace organic vapour pollutants in ambient atmospheres. The performance of a Tenax-GC adsorbent tube. *J. Chromatogr.*, 178: 79–90.
Butler, L.D. and Burke, M.F., 1976. Chromatographic characterization of porous polymers for use as adsorbents in sampling columns. *J. Chromatogr. Sci.*, 14: 117–122.
Buttery, R.G., Seifert, R.M., Guadagni, D.G. and Ling, L.C., 1971. Characterization of additional volatile components of tomato. *J. Agric. Food Chem.*, 19: 524–529.
Calame, J.P. and Steiner, R., 1982. CO_2 extraction in the flavour and perfumery industries. *Chem. Ind. (London)*, 12: 399–402.
Charalambous, G. (Editor), 1978. Analysis of Food and Beverages, Headspace Techniques. Academic Press Inc., New York.

Cole, R.A., 1980. The use of porous polymers for the collection of plant volatiles. *J. Sci. Food Agric.*, 31: 1242 – 1249.

Coppens, A. and Hoejenbos, L., 1939. Investigation of the volatile constituents of raspberry juice (*Rubus idaeus*, L.). *Rev. Trav. Chim. Pays-Bas*, 58: 675 – 679.

Cronin, D.A., 1982. Techniques of analysis of flavours. Chemical methods including sample preparation. In: I.D. Morton and A.J. MacLeod (Editors), Food Flavours Part A. Introduction. Elsevier, Amsterdam, p. 15.

Croteau, R.J. and Fagerson, I.S., 1976. Major volatile components of the juice of American cranberry. *J. Food Sci.*, 33: 386 – 389.

Darrow, G., 1966. The Strawberry. History, Breeding and Physiology. Holt, Rinehard and Winston, New York, Chicago, San Francisco, p. 447.

Davidek, J., Pudil, F., Velsik, J. and Kubelka, V., 1982. Volatile constituents of elder (*Sambucus nigra* L.) II. Berries. *Lebensm.-Wiss. Technol.*, 15: 181 – 182.

Dirinck, O., Schreyen, I. and Schamp, N., 1977. Aroma quality evaluation of tomatoes, apples and strawberries. *J. Agric. Food Chem.*, 25: 759 – 763.

Dirinck, P.J., De Pooter, H.L., Willaert, G.A. and Schamp N.M., 1981. Flavor quality of cultivated strawberries: The role of the sulfur compounds. *J. Agric. Food Chem.*, 29: 316 – 321.

Dravnieks, A., Bock, F.C. and Jarke, F.H., 1981. Sensory structure of odor mixtures. In: H.R. Moskowitz and C.B. Warren (Editors), Odor Quality and Chemical Structure. ACS Symp. Ser. No. 148, Am. Chem. Soc., Washington, DC, pp. 79 – 91.

Drawert, F., 1970. Modern physiochemical methods and their application to the problem of biogenesis of aromatic substances. *Ernaehr. Umsch.*, 17: 392 – 400.

Drawert, F., Tressl, R., Staudt, G. and Köppler, H., 1973. Gaschromatographisch-massenspektrometrische Differenzierung von Erdbeerarten. *Z. Naturforsch.*, 282: 488 – 493.

Ebel, J. and Grisenbach, H., 1973. Reduction of cinnamic acids to cinnamyl alcohols with an enzyme preparation from cell suspension cultures of soybean (*Glycine max*). *FEBS Lett.*, 30: 141 – 143.

Flath, R.A. and Forrey, R.R., 1977. Volatile components of papaya (*Carica papaya* L., Solo variety). *J. Agric. Food Chem.*, 25: 103 – 109.

Flath, R.A., Black, D.R., Guadagni, D.G., McFadden, W.H. and Schultz, 1967. Identification and organoleptic evaluation of compounds in delicious apple essence. *J. Agric. Food Chem.*, 15: 29 – 35.

Forss, D.A., 1981. Sensory characterization. In: R. Teranishi, R. Flath and H. Sugisawa (Editors), Flavor Research Recent Advances. Marcel Dekker Inc., New York, pp. 125 – 174.

Frank, H., Nicholson, G.J. and Bayer, E., 1978. Chirale Polysiloxane zur Trennung von optischen Antipoden. *Angew. Chem.*, 90: 396.

Franke, W. and Müller, H., 1983. Biology of food plants. Quantity and fatty acid composition of fruit pulp and seed oils of sea buckthorn fruits (*Hippohae rhamnoides*, L.). *Angew. Bot.*, 57: 77 – 83.

Frijters, J.E.R., 1978. Some psychophysical notes on the use of the odour unit number. In: D.G. Land and H.E. Nursten (Editors), Progress in Flavour Research. Applied Science Publishers Ltd., London, pp. 47 – 51.

van Gemert, L.J., 1981. Coordination of sensory and instrumental analysis. In: H. Maarse and R. Belz (Editors), Handbuch der Aromaforschung. Isolation, Separation and Identification of Volatile Compounds in Aroma Research. Akademie-Verlag, Berlin, pp. 240 – 258.

Georgilopoulos, D.N. and Gallois, A.N., 1987a. Aroma compounds of fresh blackberries (*Rubus laciniata* L.). *Z. Lebensm.-Unters.-Forsch.*, 184: 374 – 380.

Georgilopoulos, D.N. and Gallois, A.N., 1987b. Volatile flavour compounds in heated blackberry juices. *Z. Lebensm.-Unters.-Forsch.*, 185: 299 – 306.

Gierscher, K. and Baumann, G., 1968. Aromastoffe in Früchten. *Riechst. Aromen, Koerperpflegem.*, 18: 3, 37, 94, 134, 179, 220, 322.

Godefroot, M., Sandra, P. van Verzele, M., 1981. New method for quantitative essential oil analysis. *J. Chromatogr.*, 203: 325 – 335.

Groenen, P.J., Jonk, R.J.G., van Ingen, C. and ten Noever de Rwown, M.C., 1976. Determination of eight volatile nitrosoamines in thirty cured meat products with capillary gas chromatography-high-resolution mass spectrometry: The presence of nitrosodiethylamine and the absence of nitrosopyrrolidine. *IA RC Sci. Publ.*, 14: 321 – 331.

Guichard, E., 1982. Identification des constituants volatils aromatiques de la variété de framboises Lloyd

George. *Sci. Aliment.*, 2: 173.

Gulan, M.P., Veek, M.H., Scanlan, R.A. and Libbey, L.M., 1973. Compounds identified in commercial blackberry essence. *J. Agric. Food Chem.*, 21: 741.

Hårdh, J., 1959. Pensasmustikan viljelyä haittaavista tekijöistä Suomessa. Summary: On factors affecting blueberry culture in Finland. *J. Sci. Agric. Soc. Finland*, 31: 131 – 140.

Herrmann, K., 1963. Die flüchtigen Aromastoffe der Obstarten. *Fruchtsaft-Ind.*, 8: 329 – 339.

Hiirsalmi, H., 1973. Breeding of *Rubus idaeus* × *R. arcticus*. *J. Yugoslav. Pomol.*, 25: 117 – 121.

Hiirsalmi, H., 1977a. Culture and breeding of highbush blueberries in Finland. *Acta Hortic.*, 61: 101 – 110.

Hiirsalmi, H., 1977b. Inheritance of characters in hybrids of *Vaccinium uliginosum* and highbush blueberries. *Ann. Agric. Fenn.*, 16: 7 – 18.

Hiirsalmi, H. and Säkö, J., 1977. Variety trials with the highbush blueberry in Finland. *Ann. Agric. Fenn.*, 12: 190 – 199.

Hiirsalmi, H. and Lehmuskoski, A., 1982. Pensasmustikasta kotimainen lajike. *Puutarha*, 6: 350 – 352.

Hirvi, T., 1983. Mass fragmentographic and sensory analyses in the evaluation of the aroma of some strawberry varieties. *Lebensm.-Wiss. Technol.*, 16: 157 – 161.

Hirvi, T. and Honkanen, E., 1982. The volatiles of two new strawberry cultivars, 'Annelie' and 'Alaska Pioneer', obtained by backcrossing of cultivated strawberries with wild strawberries, *Fragaria vesca*, Rügen and *Fragaria virginiana*. *Z. Lebensm.-Unters.-Forsch.*, 175: 113 – 116.

Hirvi, T. and Honkanen, E., 1983a. The aroma of blueberries. *J. Sci. Food Agric.*, 34: 992 – 998.

Hirvi, T. and Honkanen, E., 1983b. The aroma of some hybrids between high-bush blueberry (*Vaccinium corymbosum*, L.) and bog blueberry (*Vaccinium uliginosum* L.). *Z. Lebensm.-Unters.-Forsch.*, 176: 346 – 349.

Hirvi, T. and Honkanen, E., 1984. The aroma of the fruit of Sea Buckthorn, *Hippophae rhamnoides* L. *Z. Lebensm.-Unters.-Forsch.*, 1979: 387 – 388.

Hirvi, T. and Honkanen, E., 1985. Volatile constituents of black chokeberry, *Aronia melanocarpa*, Ell. *J. Sci. Food Agric.* 36: 808 – 810.

Hirvi, T., Honkanen, E. and Pyysalo, T., 1980. Stability of 2,5-dimethyl-4-hydroxy-3(2H)furanone and 2,5-dimethyl-4-methoxy-3(2H)furanone in aqueous buffer solutions. *Lebensm.-Wiss. Technol.*, 13: 324 – 325.

Hirvi, T., Honkanen, E. and Pyysalo, T., 1981. The aroma of cranberries. *Z. Lebens,.-Unters.-Forsch.*, 172: 365 – 367.

Honkanen, E. and Karvonen, P., 1969. Isolation of volatile flavour compounds from fats and oils by vacuum carbon dioxide distillation. *Acta Chem. Scand.*, 20: 2626 – 2627.

Honkanen, E. and Pyysalo, T., 1976. The aroma of cloudberries (*Rubus chamaemorus* L.). *Z. Lebensm.-Unters.-Forsch.*, 160: 393 – 400.

Honkanen, E., Pyysalo, T. and Hirvi, T., 1980. The aroma of Finnish wild raspberries, *Rubus idaeus*, L. *Z. Lebensm.-Unters.-Forsch.*, 171: 180 – 182.

Horvat, R.J., Seuter, S.D. and Dekazos, E.P., 1983. GLC – MS-analysis of volatile constituents in rabbit eye blueberries. *J. Food Sci.*, 48: 278 – 279.

Houchen, M., Scanlan, R.A., Libbey, L.M. and Bills, D.D, 1972. Possible precursor for 1-methyl-4-isopropenylbenzene in commercial blackberry flavor essence. *J. Agric. Food Chem.*, 20: 170.

Hulten, E., 1971. The circumpolar plants. II Dicotyledons. *K. Sven. Vetenskapsakad. Handl.* 4, Ser. 13.1: 164, 373.

Hunter, G.L.K., Bucek, W.A. and Radford, T., 1974. Volatile components of canned Alphonso mango. *J. Food Sci.*, 39: 900 – 903.

Jennings, W.G., 1969. Chemistry of flavor. *Lebensm.-Wiss. Technol.*, 2: 75 – 77.

Jennings, W.G., 1979. Vapor-phase sampling. *J. High Res. Chrom., Chrom. Comm.*, 2: 221 – 224.

Jennings, W.G. (Editor), 1980. Gas Chromatography with Glass Capillary Columns. Academic Press Inc., New York, pp. 183 – 200.

Jennings, W.G., 1981. Comparisons of Fused Silica and Other Glass Columns in Gas Chromatography. Hüthig-Verlag, Heidelberg.

Jennings, W.G., 1982. Applications of Glass Capillary Gas Chromatography. M. Dekker, New York.

Jennings, W.G. and Filsoof, M., 1977. Comparison of sample preparation techniques for gas chromatographic analysis. *J. Agric. Food Chem.*, 25: 440 – 445.

188

Jennings, W.G. and Rapp, A., 1983. Sample Preparation for Gas Chromatographic Analysis. Hüthig-Verlag, Heidelberg.

Jennings, W.G. and Shibamoto, T. (Editors), 1980. Qualitative Analysis of Flavor and Fragrance Volatiles by Glass Capillary Gas Chromatography. Academic Press, Inc., New York.

Junk, G.A., Richard, J.J., Grieser, M.D., Witiak, D., Witiak, J.L., Arguello, M.D., Vick, R., Svec, H.J., Fritz, J.S. and Calder, G.V., 1974. Use of macroreticular resins in the analysis of water for trace organic contaminants. *J. Chromatogr.*, 99: 745 – 762.

Kallio, H., 1975a. Identification of volatile aroma compounds in arctic bramble (*Rubus arcticus* L.) and their development during ripening of the berry, with special reference to *Rubus stellatus*. S.M. Acad. Diss., University of Turku, Finland.

Kallio, H., 1975b. Chemical constituents of the volatile aroma compounds in *Rubus arcticus* L. subsp. *stellatus* (sm) Boivin, with reference to *Rubus arcticus* L. subsp. *arcticus*. *Rep. Kevo Subarctic Res. Stat.*, 12: 60 – 65.

Kallio, H., 1976a. Development of volatile aroma compounds in arctic bramble, *Rubus arcticus* L. *J. Food Sci.*, 41: 563 – 566.

Kallio, H., 1976b. Identification of vacuum steam distilled aroma compounds in the press juice of arctic bramble *Rubus arcticus* L. *J. Food Sci.*, 41: 555 – 562.

Kallio, H. and Honkanen, E., 1974. An important major aroma compound in arctic bramble, *Rubus arcticus* L. *Proc. Int. Congr. Food Sci. Technol.*, 4th, 1: 84 – 92.

Kallio, H., Laine, M. and Huopalahti, R., 1980. Aroma of the berries of the *Rubus stellatus* Sm × R. *arcticus* L. *Rep. Kevo Subarctic Res. Stat.*, 16: 17 – 21.

Kallio, H., Lapveteläinen, A., Hirvi, T. and Honkanen, E., 1984. Volatiles in relation to aroma in the berries of *Rubus arcticus* Coll. P. Schreier: Analysis of Volatiles. Proc. Int. Workshop Würzburg, Federal Republic of Germany, Sept. 20 – 30, 1983 pp. 433 – 446.

Karlsson-Ekström, G. and von Sydow, E., 1973a. The aroma of blackcurrants VI. *Lebensm.-Wiss. Technol.*, 6: 86 – 89.

Karlsson-Ekström, G. and von Sydow, E., 1973b. The aroma of blackcurrants VI. *Lebensm.-Wiss. Technol.*, 6: 165 – 169.

Kemp, T., Knavel, D. and Stoltz, L.P., 1973. Volatile *Cucumis melo* components: Identification of additional compounds and effects of storage conditions. *Phytochemisty*, 12: 2921 – 2924.

Kolb, B., 1980. Applied Headspace Gas Chromatography. Heyden & Son Ltd., London.

Kubeczka, K.-H., 1981. Application of HPLC for the separation of flavor compounds. In: P. Schreier (Editor), Flavour 81. W. de Gruyter & Co., Berlin, pp. 345 – 359.

König, W.A., 1984. New developments in enantiomer separation by capillary gas chromatography. In: P. Schreier (Editor), Analysis of Volatiles. W. de Gruyter & Co., Berlin, pp. 77 – 91.

Laing, D.G. and Panhuber, H., 1979. Application of anatomical and psychophysical methods to studies of odour interactions. In: D.G. Land and H.E. Nursten (Editors), Progress in Flavour Research. Applied Science Publishers Ltd., London, pp. 27 – 46.

Latrasse, A., Rigaud, J. and Sarris, J., 1982. Aroma of the blackcurrant berry (*Ribes nigrum*, L.): main odor and secondary notes. *Sci. Aliment.*, 2: 145 – 162.

Lehmushovi, A. and Säkö, J., 1975. Domestication of the cowberry (*Vaccinium vitis idaea* L.) in Finland. *Ann. Agric. Fenn.*, 14: 227 – 230.

Lewis, M.J. and Williams, A.A., 1980. Potential artefacts from using porous polymers for collection of aroma components. *J. Sci. Food Agric.*, 31: 1017 – 1026.

Likens, S.T. and Nickerson, G.B., 1964. Detection of certain hop oil constituents in brewing products. *Proc. Am. Soc. Brew. Chem.*, 5 – 13.

Maarse, H. and Belz, R. (Editors), 1981. Handbuch der Aromaforschung, Isolation, Separation and Identification of Volatile Compounds in Aroma Research. Akademie-Verlag, Berlin.

Maarse, H. and Kepner, R.E., 1970. Changes in composition of volatile terpenes in Douglas fir needles during maturation. *J. Agric. Food Chem.*, 18: 1095 – 1101.

Mansell, R.L., Stockigt, J. and Zenk, M.H., 1973. Reduction of ferulic acid to coniferyl alcohol in a cell free system from a higher plant. *Z. Pflanzenphysiol.*, 68: 286 – 288.

Marriott, R.J., 1986. Biogenesis of blackcurrant (*Ribes nigrum*) aroma. In: T. Parliment and R. Croteau. (Editors), Biogeneration of Aromas. ACS Symp. Ser. 317, Am. Chem. Soc., Washington.

McFadden, W.H., Teranishi, R., Corse, J., Black, D.R. and Mon, T.R., 1965. Volatiles from strawberries. II Combined mass spectrometry and gas chromatography on complex mixtures. *J. Chromatogr.*, 18: 10–19.

Mulders, E.J., 1973. The odour of white bread. *Z. Lebensm.-Unters.-Forsch.*, 151: 310–317.

Murray, K.E., 1977. Concentration of headspace, airborne and aqueous volatiles on Chromosorb 105 for examination by gas chromatography and gas chromatography–mass spectrometry. *J. Chromatogr.*, 135: 49–60.

Mussinan, C.J. and Walradt, J.P., 1975. Organic acid from fresh California strawberries. *J. Agric. Food Chem.*, 23: 482–484.

Neish, A.C., 1964. Major pathways of biosynthesis of phenols. In: J.B. Harborne (Editor), Biochemistry of Phenolic Compounds. Academic Press, London, New York, pp. 295–359.

Nickerson, G.B. and Likens, S.T., 1966. Gas chromatographic evidence for the occurrence of hop oil components in beer. *J. Chromatogr.*, 21: 1–5.

Nishimura, O., Masuda, H. and Mihara, S., 1987. Hydroxy nitriles in blackcurrant buds absolute (*Ribes nigrum* L.). *J. Agric. Food Chem.*, 35: 338–340.

Novak, J., Golias, J. and Drozd, J., 1981. Determination of hydrophilic volatiles in gas–aqueous liquid systems by Grob's closed-loop strip/trap method and standard-addition calibration. *J. Chromatogr.*, 204: 421–428.

Nursten, H.E., 1982. Flavour chemistry. *Food*, 14–21.

Nursten, H.E. and Williams, A.A., 1967. Fruit aromas: survey of components identified. *Chem. Ind.*, 486–497.

Nursten, H.E. and Williams, A.A., 1969a. Volatile constituents of the blackcurrant, *Ribes nigrum* L. I. *J. Sci. Food Agric.*, 20: 91–98.

Nursten, H.E. and Williams, A.A., 1969b. Volatile constituents of the blackcurrant, *Ribes nigrum* L. II. *J. Sci. Food Agric.*, 20: 613.

Parliment, T.H. and Kolor, M.G., 1975. Identification of the major volatile components of blueberry. *J. Food Sci.*, 40: 762–763.

Parliment, T.H. and Scarpellino, R., 1977. Organoleptic techniques in chromatographic food flavor analysis. *J. Agric. Food Chem.*, 25: 97–99.

Patte, F., Etcheto, M. and Laffort, P., 1975. Selected and standardized values of suprathreshold odor intensities for 110 substances. *Chem. Senses Flavor*, 1: 283–305.

Pickenhagen, W., Velluz, A., Passerat, J.-P. and Ohloff, G., 1981. Estimation of 2,5-dimethyl-4-hydroxy-3(2H)-furanone (FURANEOL) in cultivated and wild strawberries, pineapples and mangoes. *J. Sci. Food Agric.*, 32: 1132–1134.

Potts, R., 1973. 4-Hydroxylation of cinnamic acid by sorghum microsomes. Ph.D. Theses University of California, p. 77.

Pyysalo, T., 1976a. Studies on the volatile compounds of some berries in genus *Rubus*, especially cloudberry (*Rubus chamaemorus* L.) and hybrids between raspberry (*Rubus idaeus* L.) and arctic bramble (*Rubus arcticus* L.). Acad. Diss., Espoo, Technical University of Helsinki, Finland.

Pyysalo, T., 1976b. Identification of volatile compounds in hybrids between raspberry (*Rubus idaeus* L.) and arctic bramble (*Rubus arcticus* L.) *Z. Lebensm.-Unters.-Forsch.*, 162: 263–372.

Pyysalo, T. and Honkanen, E., 1977. The influence of heat on the aroma of cloudberries (*Rubus chamaemorus* L.). *Z. Lebensm.-Unters.-Forsch.*, 163: 25–30.

Pyysalo, T., Suihko, M. and Honkanen, E., 1977a. Odour thresholds of the major volatiles in cloudberry (*Rubus chamaemorus*, L.) and arctic bramble (*Rubus arcticus*, L.). *Lebensm.-Wiss. Technol.*, 10: 36–40.

Pyysalo, T., Torkkeli, H. and Honkanen, E., 1977b. The thermal decarboxylation of some substituted cinnamic acids. *Lebensm.-Wiss. Technol.*, 10: 145–147.

Pyysalo, T., Honkanen, E. and Hirvi, T., 1979. Volatiles of wild strawberries, *Fragaria vesca* L., compared to those of cultivated berries, *Fragaria* × *ananassa* Cv. Senga sengana. *J. Agric. Food Chem.*, 27: 19–22.

Qvist, I.H., von Sydow, E.C.F. and Åkesson, C.A., 1976. Unconventional proteins as aroma precursors: Instrumental and sensory analysis of the volatile compounds in a canned meat product containing soy or rapeseed protein. *Lebensm.-Wiss.-Technol.*, 9: 311–320.

Rapp, A., 1982. Analysis of grapes, wines and brandies. *Chromatogr. Sci.*, 15: 579–621.

Ravid, U., Putievsky, E. and Snir, R., 1983. The volatile compounds of oleoresins and the essential oils of *Foeniculum vulgare* in Israel. *J. Nat. Prod.*, 46(6): 848 – 851.

Re, L., Maurer, B. and Ohloff, G., 1973. Ein einfacher Zuzang zu 4-Hydroxy-2,5-dimethyl-3(2*H*)furanon (Furaneol), einem Aromabestandteil von Ananas und Erdbeere. *Helv. Chim. Acta*, 56: 1882 – 1894.

Resvoll, T.R., 1925. *Rubus chamaemorus* L. Die geographische Verbreitung der Pflanze und ihre Verbreitungsmittel. *Ver. Off. Geobot. Inst. Rübel*, 3: 224 – 241.

Resvoll, T.R., 1928. *Rubus chamaemorus* L. A morphological biological study. *Nyt. Mag. Naturvidensk.*, 67: 55 – 129.

Rigaud, J., Etievant, P., Henry, R. and Latrasse, A., 1986. Le Methoxy-4-methyl-2-butanethiol-2 un constituant majeur del'arome du bourgeon de cassis (*Ribes nigrum*), L.). *Sci Aliment.*, 6: 213 – 220.

Rix, C.E., Lloyd, R.A. and Miller, C.W., 1977. Headspace analysis of tobacco with Tenax traps. *Tob. Sci.*, 93: 32 – 35.

Römer, G. and Renner, E., 1974. Einfache Methoden zur Isolierung und Anreicherung von Aromastoffen aus Lebensmitteln. *Z. Lebensm.-Unters.-Forsch.*, 156: 329 – 335.

Roselius, W., Vitzthum, O. and Hubert, P., 1982. Method of extracting coffee oil containing aroma constituents from roasted coffee. U.S. Pat. 4328255.

Rothe, M., 1975. Aroma values – a useful concept? Proc. Int. Symp. Aroma Research, Akademie-Verlag, Zeist. Pudoc, Wageningen, pp. 111 – 119.

Rothe, M., 1976. Aroma values – a useful concept? *Nahrung*, 20: 259 – 266.

Rousi, A., 1963. Hybridization between *Vaccinium uliginosum* and cultivated blueberry. *Ann. Agric. Fenn.*, 2: 12 – 18.

Rousi, A., 1965. Mesivadelman jalostuksen nykyinen vaihe Puutarhatutkimuslaitoksessa. *Puutarha*, 68: 36 – 38.

Rousi, A., 1966. The use of North-European *Vaccinium* species in blueberry breeding. *Acta Agric. Scand., Suppl.*, 16: 50 – 54.

Rousi, A., 1971. The genus *Hippophae* L. A taxonomic study. *Ann. Bot. Fenn.*, 8: 177.

Russel, D.W., 1971. Metabolism of aromatic compounds in higher plants. X Properties of the cinnamic acid 4-hydroxylase of pea seedlings and some aspect of its metabolic and developmental control. *J. Biol. Chem.*, 246: 3870 – 3878.

Sakodynskii, K., Panina, L. and Klinskaya, N., 1974. A study of some properties of Tenax, a porous polymer sorbent. *Chromatographia*, 7: 339 – 344.

Salo, P., 1975. Use of odour thresholds in sensorial testing and comparisons with instrumental analysis. In: H. Maarse and P.J. Groenen (Editors), Aroma Research. Centre for Agric, Publishing and Documentation, Wageningen, pp. 121 – 130.

Salo, P., Nykänen, L. and Suomalainen, H., 1972. Odor thresholds and relative intensities of volatile aroma components in an artificial beverage imitating whisky. *J. Food Sci.*, 37: 394 – 398.

Scanlan, R.A., Bills, D.D. and Libbey, L.M., 1970. Blackberry flavor components of commercial essence. *J. Agric. Food Chem.*, 18: 744.

Schaefer, J., 1981. Comparison of adsorbents in head space sampling. In: P. Schreier (Editor), Flavour 81. W. de Gruyter & Co., Berlin, New York, pp. 301 – 313.

Schinz, H. and Seidel, C.F., 1957. Untersuchungen über Aromastoffe. Über das Himbeeraroma. *Helv. Chim. Acta*, 40: 1839 – 1859.

Schreier, P., 1980. Quantitative composition of volatile constituents in cultivated strawberries, *Fragaria ananassa* cv. Senga Sengana, Senga Litessa and Senga Gourmella. *J. Sci. Food Agric.*, 31: 487 – 494.

Schreier, P., 1984. Chromatographic Studies of Biogenesis of Plant Volatiles. Hüthig-Verlag, Heidelberg.

Schreier, P. and Drawert, F., 1981. Über die Aromastoffzusammensetzung von Erdbeerweinen. I. Erdbeerwein. *Chem., Mikrobiol., Technol. Lebensm.*, 7: 23 – 27.

Schreier, P., Drawert, F., Kerenyi, F. and Junker, A., 1976. Gaschromatographisch – massenspektrometrische Untersuchung flüchtiger Inhaltstoffe des Weines. VI. Aromastoffe in Tokajer Trockenbeerenauslese (Aszu)-Weinen a) Neutralstoffe. *Z. Lebensm.-Unters.-Forsch.*, 161: 249 – 258.

Schreier, P., Drawert, F. and Junker, A., 1977. Über die quantitative Zusammensetzung natürlicher und technologisch veränderter pflanzlicher Aromen III. Veränderungen und Neubildung von Aromastoffen bei der Herstellung von Säften aus roten Johannisbeeren. *Lebensm.-Wiss. Technol.*, 10: 337 – 340.

Schreier, P., Drawert, F. and Schmid, M., 1978. Changes in the composition of neutral volatile components during the production of apple brandy. *J. Sci. Food Agric.*, 29: 728 – 736.

Schreier, P., Drawert, F. and Winkler, F., 1979. Composition of neutral volatile constituents in grape brandies. *J. Agric. Food Chem.*, 27: 365 – 372.

Schultz, T.H., Flath, R.A., Black, D.R., Guadagni, D.G., Schultz, W.G. and Teranishi, R., 1967. Volatiles from Delicious apple essence-extraction methods. *J. Food Sci.*, 32: 279 – 283.

Schultz, T.H., Flath, R.A., Mon, T.R., Eggling, S.B. and Teranishi, R., 1977. Isolation of volatile components from a model system. *J. Agric. Food Chem.*, 25: 446 – 449.

Schultz, W.G. and Randall, J.M. 1970. Liquid carbon dioxide for selective aroma extraction. *Food Technol.*, 24: 1282 – 1286.

Schwob, J. and Dupaigne, P., 1965. Acquisitions recentes dans le domaine des aromes de fruits et legumes. *Ann. Nutr. Aliment.*, 19: A 481.

Scott, D. and Lawrence, F., 1975. Strawberries. In: J. Janick and J. Moore (Editors), Advances in Fruit Breeding. Purdue University Press, West Lafayette, Indiana, pp. 71 – 77.

Shen, J.A., Montgomery, M.W. and Libbey, L.M., 1980. Subjective and objective evaluation of strawberry pomace essence. *J. Food Sci.*, 45: 41 – 46.

Singleton, J.A. and Pattee, H.E., 1980. A preconcentration and subsequent gas liquid chromatographic analysis method for trace volatiles. *J. Am. Oil Chem. Soc.*, 57: 405 – 408.

Skrede, G., 1980. Strawberry varieties for industrial jam production. *J. Sci. Food Agric.*, 31: 670 – 676.

Stahl, E. and Gerard, D., 1982. Hochdruck-Extraktion von Naturstoffen mit überkritischen und verflüssigten Gasen 9. Mitteilung: Verfahren zur schonenden Gewinnung etherischer Öle. *Parfuem., Kosmet.*, 63(3): 117 – 122, 124 – 125.

Stahl, E. and Schütz, E., 1978. Extraktion der Kamillen Blüten mit überkritischen Gasen. *Arch. Pharm.*, 331: 992 – 1001.

Stahl, E., Schütz, E. and Schilz, W., 1978. Kamillenextrakte und ihre Herstellung. DE Pat. 2709033.

Staudt, G., Drawert, F. and Tressl, R., 1975. Gaschromatographisch – massenspektrometrische Differenzierung der Aromastoffe von Erdbeerarten. II *Fragaria nilgerrensis. Z. Pflanzenzuecht.*, 75: 36 – 42.

Stolz, L.P., Kemp, T.R., Smith, W.O., Smith, W.I. and Chaplin C.E., 1970. 1,2-Dihydro-1,1,6-trimethylnaphthalene from strawberry oil. *Phytochemistry*, 9: 1157 – 1158.

van Straten, S. (Editor), 1981. Volatile compounds in food. Division for Nutrition and Food Research, TNO, The Netherlands, suppl. 7.

van Straten, S. (Editor), 1987. Volatile compounds in food. Division for Nutrition and Food Research, TNO, The Netherlands, suppl. 9.

Sugisawa, H., 1981. Sample preparation: Isolation and concentration. In: R. Teranishi, R. Flath and H. Sugisawa (Editors), Flavor Research, Recent Advances. Marcel Dekker, Inc., New York, pp. 11 – 51.

Sugisawa, H. and Hirose, T., 1981. Microanalysis of volatile compounds in biological materials in small quantities. In: P. Schreier (Editor), Flavour 81. W. de Gruyter & Co., Berlin, New York, pp. 287 – 299.

Sundt, E., 1970. The development and future of flavor research. *Proc. Scand. Symp. Aroma Research, Naeringsmiddelindustrien*, 23: 5 – 13.

Sundt, E. and Winter, M., 1960. Untersuchungen über Aromastoffe. III Die Isolierung von Geraniol aus Himbeeren. *Helv. Chim. Acta*, 43: 1120 – 1121.

Sundt, E. and Winter, M., 1962. Analyse de l'arome volatils des framboises. II Les alcools. *Helv. Chim. Acta*, 45: 2212 – 2218.

Sydor, R. and Pietrzyk, J.D., 1978. Comparison of porous copolymers and related absorbents for the stripping of low molecular weight compounds from a flowing air stream. *Anal. Chem.*, 50: 1842 – 1847.

von Sydow, E. and Anjou, K., 1969. The aroma of bilberries (*Vaccinium myrtillus* L.). I. Identification of volatile compounds. *Lebensm.-Wiss. Technol.*, 2: 78 – 81.

von Sydow, E. and Karlsson, G., 1971a. The aroma of blackcurrants, IV. The influence of heat measured by instrumental methods. *Lebensm.-Wiss. Technol.*, 4: 54 – 58.

von Sydow, E. and Karlsson, G., 1971b. The aroma of blackcurrants. V. The influence of heat measured by odour quality assessment techniques. *Lebensm.-Wiss. Technol.*, 4: 152 – 157.

von Sydow, E., Andersson, J., Anjou, K., Karlsson, G., Land, D.G. and Griffiths, N., 1970. The aroma of bilberries (*Vaccinium myrtillus* L.) II. Evaluation of the press juice by sensory methods and by gas

192

chromatography and mass spectrometry. *Lebensm.-Wiss. Technol.*, 3: 11 – 17.

Taylor, K., 1971. *Rubus chamaemorus*, L. *J. Ecol.*, 59: 293 – 306.

Teranishi, R., Corse, J.W., McFadden, W.H., Black, D.R. and Morgan, A.I., 1963. Volatiles from strawberries. I. Mass spectral identification of the more volatile components. *J. Food Sci.*, 24: 478 – 483.

Teranishi, R., Murphy, E.L. and Mon, T.R., 1977. Steam distillation – solvent extraction recovery of volatiles from fats and oils. *J. Agric. Food Chem.*, 25: 464 – 466.

Teranishi, R., Flath, R.A. and Sugisawa, H. (Editors), 1981. Flavor Research, Recent Advances. Marcel Dekker Inc., New York.

Trajkovski, V., 1978. Annelie – den första 'smulgubben' från Balsgård. *Tidskr. Frukt och Bärodl*, 20: 18 – 24.

Tressl, R. and Drawert, F., 1971. Über die Biogenese von Aromastoffen in Pflanzen und Früchten XIII Einbau von $8-^{14}C$-Caprylsäure in Bananen- und Erdbeeraromastoffe. *Z. Naturforsch.*, B26: 774 – 779.

Tressl, R. and Drawert, F., 1973. Biogenesis of banana volatiles. *J. Agric. Food Chem.*, 21: 560 – 565.

Tressl, R., Drawert, F. and Heimann, W., 1969. Gaschromatographisch – massenspektrometrische Bestimmung von Erdbeer-Aromastoffen. *Z. Naturforsch.*, B24: 1201 – 1202.

Tuorila-Ollikainen, H., Alanko, T. and Hirvi, T., 1983. Facial representation of sensory profiling data. *Lebensm.-Wiss.-Technol.*, 16: 376 – 377.

Vaarama, A., 1948. Cytogenetic studies on two *Rubus arcticus* hybrids. *J. Sci. Agric. Soc. Finland*, 20: 67 – 79.

Versino, B., Knöppel, H., de Groot, M., Peil, A., Poelman, J., Schauenburg, H., Vissers, H. and Geiss, F., 1976. Organic micropollutants in air and water. Sampling, gas chromatographic – mass spectrometric analysis and computer identification. *J. Chromatogr.*, 122: 373 – 388.

Wang, P.L., Du, C.T. and Francis, F.J., 1978. Isolation and characterization of polyphenolic compounds in cranberries. *J. Food Sci.*, 43: 1402 – 1404.

Weurman, C., 1969. Isolation and concentration of volatiles in food odor research. *J. Agric. Food Chem.*, 17: 370 – 384.

Willhalm, B., Palluy, E. and Winter, M., 1966. Recherches sur aromes. Sur l'arome des fraises fraiches. Identification des acides volatils et de quelques autres composes. *Helv. Chim. Acta*, 49: 65 – 67.

Williams, A.A., 1978a. Interpretation of the sensory significance of chemical data in flavour research. Part. I. Methods based on evaluating sensory properties of separated fractions and individual compounds. *Int. Flavours Food Addit.*, 9: 80 – 85.

Williams, A.A., 1978b. Interpretation of the sensory significance of chemical data in flavour research. Part 2. Statistical methods. *Int. Flavours Food Addit.*, 9: 131 – 133.

Williams, A.A.., 1978c. Interpretation of the sensory significance of chemical data in flavour research. Part. 3. Sensory analysis. *Int. Flavours Food Addit.*, 9: 171 – 175.

Williams, A.A. and Ismail, H.M.M., 1981. The volatile flavour components of plums and their sensory evaluation. In: J. Solms and R.L. Hall (Editors), Criteria of Food Acceptance. Forster Verlag Ag, Switzerland, pp. 333 – 354.

Williams, A.A. and Tucknott, O.G., 1972. Volatile components of scotch whisky. *J. Sci. Food Agric.*, 23: 1 – 7.

Winter, M. and Enggist, P., 1971. Recherches sur les aromes. Sur l'arome de framboise IV. *Helv. Chim. Acta*, 54: 1891 – 1898.

Winter, M. and Sundt, E., 1962. Analyse de l'arome des framboises I. Les constituants carbonyles volatils. *Helv. Chim. Acta*, 45: 2195 – 2211.

Winter, M. and Willhalm, B., 1964. Recherches sur les aromes. Sur l'arome des fraises. Analyse des composes carbonyles, esters et alcools volatils. *Helv. Chim. Acta*, 47: 1215 – 1227.

Winter, M., Palluy, E., Hinder, M. and Willhalm, B., 1962. Procede d'isolement des constituants volatils de la fraise et de la framboise. *Helv. Chim. Acta*, 45: 2186 – 2218.

Withycombe, D.A., Mookherjee, B.D. and Hruza, A., 1978. Isolation of trace volatile constituents of hydrolyzed vegetable protein via porous polymer entrainment. In: G. Charalambous (Editor), Analysis of Food and Beverages, Headspace Techniques. Academic Press Inc., New York, pp. 81 – 94.

Wüst, R., Pfeiffer, H. and van der Mei, H., 1982. Verfahren zur Herstellung von Pflanzenextraktion mit verbesserten sensorischen Eigenschaften. DE Pat. 3115157.

Yamashita, I., Nemoto, Y. and Yoshikawa, S., 1975. Formation of volatile esters in strawberries. *J. Agric. Biol. Chem.*, 39: 2303 – 2307.

Yamashita, I., Nemoto, Y. and Yoshikawa, S., 1976. Formation of volatile alcohols and esters from aldehydes in strawberries. *Phytochemistry*, 15: 1633 – 1637.

Yamashita, I., Iino, K., Nemoto, Y. and Yoshikawa, S., 1977. Studies on flavor development in strawberries. 4. Biosynthesis of volatile alcohols and esters from aldehydes during ripening. *J. Agric. Food Chem.*, 25: 1165 – 1168.

Chapter V

THE FLAVOUR OF TROPICAL FRUITS (BANANA, MELON, PINEAPPLE)

KARL-HEINZ ENGEL, JÜRGEN HEIDLAS and ROLAND TRESSL
Technische Universität Berlin, Institute of Biotechnology, Seestr. 13, D-1000 Berlin 65, (West Germany)

1. INTRODUCTION

Flavour and aroma are properties of food which strongly influence its acceptance by consumers. This essential impact of sensory qualities is especially true for (sub)tropical fruits. Not so much the nutritional value but their unique and exotic flavours are major reasons for the popularity and consumption of these foods. This chapter covers three major tropical fruits: pineapple, banana and melon. Due to sophisticated and well-organized transportation systems, these fruits are offered world-wide to consumers at high quality standards.

Overall flavour impressions are determined by odour and taste. This review will concentrate on volatile compounds, i.e. those essential to odour qualities of a food. The investigation of volatile constituents of tropical fruits has been the object of intensive research activities. Comprehensive data on volatiles identified in banana, pineapple and melon are available (Maarse and Visscher, 1987). Therefore, this contribution will not be limited to a compilation of qualitative data. Each section will present an overview of major publications and a description of the characteristic features of the volatile composition of the fruit. The major goal of the chapter, however, is to cover two aspects, reflecting recent (and future) trends and developments in flavour and aroma research:

(a) the contribution of individual (trace) constituents to the characteristic flavour, and

(b) the investigation of pathways involved in the biogenesis of 'natural' flavour and aroma compounds.

2. BANANA

2.1 Volatile Composition

If one follows the publications on banana volatiles starting from the first reports in 1905 (Rothenbach and Eberlein) until the most recent paper (Macku and Jennings, 1987), the tremendous development of analytical techniques and its importance to the identification of volatile (trace) constituents becomes obvious.

The first investigations (Rothenbach and Eberlein, 1905; Kleber, 1912; Loesecke, 1950; Hultin and Procter, 1961) were limited to classical techniques of organic chemistry. The separation of volatiles by gas chromatography (Issenberg and Wick, 1963; McCarthy et al., 1963, 1964; Wick, 1965) was a decisive step towards more detailed studies. The use of capillary columns, the pre-separation of total aroma extracts on the basis of functional groups using liquid – solid chromatography on silica gel (Murray et al., 1968; Palmer, 1973), characterization of compounds by chemical micro-methods in combination with the comparison of retention data on several GC-columns, and finally the application of gas chromatography – mass spectrometry (Issenberg, 1969; Wick et al., 1969; Tressl et al., 1969, 1970a,b; Mattei, 1973a; Berger et al., 1986a) made possible the identification of a broad spectrum of banana volatiles. Detailed reviews of these investigations (Dupaigne, 1970, 1975; Palmer, 1970) and comprehensive compilations of the compounds identified (Nursten, 1970; Forsyth, 1980; Maarse and Visscher, 1987) are available.

Some typical banana volatiles are presented in Fig. V.1. Esters are qualitatively and quantitatively the dominating group of banana volatiles. More

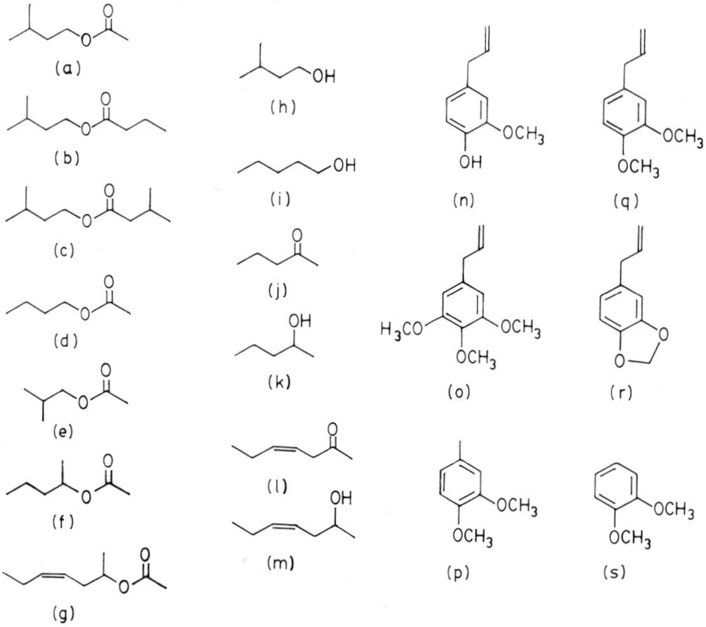

Fig. V.1. Typical banana volatiles. (a) 3-Methylbutyl acetate, (b) 3-methylbutyl butanoate, (c) 3-methylbutyl 3-methylbutanoate, (d) butyl acetate, (e) 2-methylpropyl acetate, (f) pent-2-yl acetate, (g) (Z)-hept-4-en-2-yl acetate, (h) 3-methylbutan-1-ol, (i) pentan-1-ol, (j) pentan-2-one, (k) pentan-2-ol, (l) (Z)-hept-4-en-2-one, (m) (Z)-hept-4-en-2-ol, (n) eugenol, (o) elemicin, (p) 3,4-dimethoxytoluene, (q) O-methyleugenol, (r) safrol, (s) 1,2-dimethoxybenzene.

than 100 esters have been identified, the typical 'banana esters', however, consist of only a few alkyl and acyl moieties: major alkyl moieties are from 3-methylbutan-1-ol, 2-methylpropan-1-ol, pentan-2-ol, heptan-2-ol and (Z)-hept-4-en-2-ol; they are mainly esterified with acetic, butanoic and 3-methylbutanoic acid (Tressl et al., 1970b).

If the fruit enzymes are inhibited before the preparation of an aroma extract, e.g. by addition of methanol (Tressl et al., 1970a), the ratio of esters, alcohols and carbonyl compounds is approximately 95 : 4 : 1. A significantly changed distribution of volatiles is obtained if isolation procedures without enzyme inhibition are applied (Hultin and Procter, 1961; Issenberg and Wick, 1963). Due to hydrolytic activities of carboxylesterases, the amounts of esters decrease; enzyme-catalyzed oxidative splitting of unsaturated fatty acids leads to the formation of additional aldehydes and alcohols (Tressl and Drawert, 1973).

Other typical banana constituents are odd-carbon-numbered aliphatic methyl-ketones, the corresponding secondary alcohols and their esters. A class of compounds not commonly found in fruits are phenolether derivatives (Fig. V.1).

2.2 Sensory Aspects

In their early experiments, McCarthy et al. (1963) related gas chromatographic patterns of banana volatiles to the flavour profiles of the fruits, and reported amyl (pentyl) esters of acetic, propanoic and butanoic acids as responsible for the characteristic 'banana-like' aroma. By using the more sophisticated capillary GC-sniffing technique (Nitz et al., 1984), the essential rôles of 3-methylbutyl acetate, butanoate and 3-methylbutanoate in banana odour have been confirmed (Berger et al., 1986a). These major constituents of natural banana extracts are also used as key compounds in commercially available constituted flavours (Fenaroli, 1975; Ziegler, 1982). The sensory quality of freeze-dried bananas could be improved by storage of banana slices in an atmosphere containing 3-methylbutan-1-ol prior to the dehydration process. Using the natural pathways the precursor is incorporated and a dry product containing amounts of 3-methylbutyl esters comparable to the fresh fruit is obtained (Berger et al., 1986b).

The distinctive 'fruity, estery' notes of bananas have been attributed to butyl, isobutyl (2-methylpropyl) and 2-pentyl (1-methylbutyl) acetates and butanoates, and hexyl acetate, respectively (McCarthy et al., 1963; Berger et al., 1986a). Yabumoto et al. (1975) showed that the chromatographic peak data of the major esters were highly related to the sensory impression described as 'artificial banana odour'; natural banana odour, however, was not. Green, woody or musty odours have been assigned to alcohols and carbonyl compounds (McCarthy et al., 1963; Palmer, 1973).

The unusual phenol derivatives eugenol, 5-methoxyeugenol, eugenol-methylether, and elemicin contribute to the full-bodied, mellow aroma of the ripe bananas (Nursten, 1970). Even the trace constituent 2-methoxy-4-methylphenol has been detected by GC-sniffing (Berger et al., 1986a). Eugenol was used in synthetic banana essences long before its presence in banana was proved.

The full natural flavour of banana still awaits elucidation. A first step towards this goal is the identification of sensorially active trace components by capillary GC-sniffing technique (Nitz *et al.,* 1984; Berger *et al.,* 1986a). Sensory properties of a series of newly identified minor constituents are listed in Table V.I. They are described as useful compounds to impart characteristic 'fatty-green' notes to banana aromas.

Despite numerous descriptions of odour properties of banana volatiles, objective data concerning the importance of individual constituents to banana flavour are still lacking. The application of the concept of odour units (Rothe and Thomas, 1963; Guadagni *et al.,* 1966), which involves a quantitative analysis of volatiles and a systematic measuring of their odour detection thresholds, should help to add valuable information in this field.

2.3 Biogenetic Aspects

Compared with other fruits, such as apple or pears, bananas are typical 'climacteric' fruits. The flavour and aroma compounds are not produced during growth, nor are they present at the time of harvest. In the preclimacteric stage, the 'ripening hormone' ethylene induces biochemical, physical and chemical changes resulting, for example, in increased synthesis of proteins and enzyme activities. The ripening process is initiated by the climacteric rise in respiration. The metabolism changes to mainly catabolic pathways; the amounts of high molecular weight structures, such as polysaccharides and lipids, decrease, and one consequence of this change in metabolism in the postclimacteric stage is the production of volatile compounds.

The volatiles produced by postclimacteric bananas show maximum concentrations 10 days after the climacteric rise in the respiration, then they either plateau or decrease again (Drawert *et al.,* 1972). Various internal and external

TABLE V.I

Recently identified aroma-active banana trace constituents (Berger *et al.,* 1986a)

Component	Average concentration (ppb)	Odour description on GC-sniffing
(E)-Hept-4-en-2-one	150	Fatty, sweet (banana-like)
(Z)-Hept-4-en-2-one	200	Green-fatty (banana-like)
(Z)-Oct-4-en-1-ol	300	Fruity-oily
(Z)-Oct-5-en-1-ol	50	Weak oily
(Z)-Hex-4-enyl acetate	200	Green, strong
(Z)-Oct-4-enyl acetate	250	Intensively green
(Z)-Oct-5-enyl acetate	50	

factors influence this formation of volatiles. Cyclic phenomena in the behaviour of major aroma constituents depending on the physiological stage of the fruit have been observed by investigation of slices from climacteric and postclimacteric bananas (Drawert and Künanz, 1975). Mattei (1973b) demonstrated that the production of volatiles is an exponential function of temperature in the range of $5-25°C$. Higher temperature leads to decreased amounts of volatiles and formation of fermentation products.

There are several discrepancies in the reports on concentrations of individual components during ripening of bananas. Tressl and Jennings (1972) monitored acetate and butanoate esters over a period of 12 days. They reported a cyclic production of these volatiles; the cycle of total acetate esters was out-of-phase with that observed for the butanoates. In a recent re-investigation of this phenomenon, Macku and Jennings (1987) demonstrated that the amounts of individual volatiles increased continuously to the onset of peel browning. There was a linear correlation between the ratio (total acetate esters/total butanoate esters) and the time of fruit ripening.

Macku and Jennings (1987) rationalized these contradicting results by possible errors in the sampling techniques or variations in the isolation conditions. This example demonstrates the critical rôle of the method chosen for the isolation of volatiles from live plant tissue.

In one of the first detailed reviews on the flavour and biochemistry of volatile banana components, Wick et al. (1966) pointed out that insight into the mechanism of the production of volatiles and the identity of their precursors is one of the ultimate goals of basic flavour research. In fact, banana became one of the first fruits where experimental data on the biogenesis of volatiles were obtained.

Addition of isotopically-labelled precursors and the application of suitable analytical techniques, such as radio gas chromatography (Tressl et al., 1972, 1973), revealed that the biosynthesis of various banana volatiles can be explained by a few known metabolic pathways (Tressl et al., 1970d):

(a) Conversion of amino acids

Obvious structural similarities and results obtained from studies of higher alcohol formation during fermentation (Guymon, 1965), indicated that the branched-chain esters and alcohols and the phenolethers in banana may be derived from the amino acids leucine, valine and phenylalanine. The amounts of these amino acids increase during the ripening of bananas after the climacteric rise in respiration. Myers et al. (1969, 1970), Palmer (1970), and Tressl et al. (1970e) demonstrated the transformation of $(U-^{14}C)$-leucine to 3-methylbutan-1-ol, 3-methylbutyl esters, 3-methylbutanoates and 2-keto-4-methylpentanoates by banana tissue slices. Analogously, $(U-^{14}C)$-L-valine was converted to 2-methylpropan-1-ol, 2-methylpropyl acetate, 2-methylpropanoic acid and 2-keto-3-methylbutanoic acid. Labelling experiments with $(1-^{14}C)$-phenylalanine and $(1-^{14}C)$-caffeic acid (3,4-dihydroxycinnamic acid) demonstrated that biogenetic sequences of the phenylpropanoid metabolism are involved in the biogenesis of phenolic ethers in banana (Tressl and Drawert, 1973).

(b) Fatty acid metabolism

Addition of ^{14}C-labelled fatty acid precursors to banana tissue slices demonstrated their conversion to esters, ketones and alcohols. The spectrum of metabolization products depends on the chain lengths of the acids. Postclimacteric banana slices incorporate (U-^{14}C)-acetate and (U-^{14}C)-butanoate only into the corresponding acetate and butanoate esters (Tressl *et al.*, 1970f). The rôle of acyl-coenzyme A as acylating agent in ester formation has been demonstrated by significant increases in concentrations of acetates and 2-methylpropyl and 3-methylbutyl butanoates, respectively, after the addition of acetyl-coenzyme A and butyryl-coenzyme A to banana slices (Gilliver and Nursten, 1976). The metabolization of longer chain acids, such as hexanoic, octanoic and decanoic acids, involves additional reduction to the primary alcohols and esterification of these alcohol moieties. Depending on the ripening stage, chain shortening of the fatty acid by α- and β-oxidation, and conversion to the

TABLE V.II

Metabolites obtained after addition of (8-^{14}C)-octanoic acid to banana tissue

	Radioactivity in aroma extract of banana disks (%)	
	Climacteric fruits[a] 69	Postclimacteric fruits[a] 31
Distribution of radioactivity among the volatile components (%)		
Pentan-2-one	0.12	0.15
Pentan-2-ol[b]		
Heptan-2-one	0.02	
(Z)-Hept-4-en-2-ol	0.9	1.25
Octan-1-ol	16.4	14.7
Butanoic acid	0.06	
Hexanoic acid	2.1	0.9
Heptanoic acid	0.6	1.1
Octanoic acid	65.0	43.6
Methyl octanoate		2.1
Ethyl octanoate/unidentified acid	2.1	3.4
Octyl acetate	1.6	4.5
2-Methylpropyl octanoate		0.2
Octyl butanoate/butyl octanoate	0.5	5.8
3-Methylbutyl octanoate/unidentified acid	0.4	5.5
Pentyl octanoate	0.2	0.3
Octyl hexanoate/hexyl octanoate	3.3	6.5
Octyl octanoate	6.7	10.0

[a] 50 g banana disks in 0.4 M saccharose solution; precursor, (8 − [14]C)-octanoic acid (25 μCi); incubation time, 5 h.
[b] Detected only in saccharose solution (1.6% of aroma activity).

corresponding odd-carbon-numbered methylketones and secondary alcohols is observed (Tressl and Drawert, 1971, 1973). Metabolites obtained after addition of (8-^{14}C)-octanoic acid to banana tissue are listed in Table V.II. Capillary gas chromatographic investigation of diastereoisomeric derivatives of (R)-($+$)-1-phenethyl isocyanate revealed that (Z)-hept-4-en-2-ol and the corresponding acetate are mainly present as (S)-enantiomers (Tressl and Albrecht, 1986).

(c) Enzymic oxidative splitting of unsaturated fatty acids

The rôle of unsaturated fatty acids as precursors of volatile compounds was confirmed by oxidative splitting of (U-^{14}C)-linoleic acid by homogenates of green bananas (Tressl and Drawert, 1973). The concentrations of C_{16}- and C_{18}-fatty acids increase sharply before the climacteric rise in respiration, but decrease in the stage of maximum formation of aroma compounds (Tressl *et al.*, 1970c).

The spectrum of aldehydes isolated from banana homogenates depends also on the ripening stage of the fruits. (E)-Non-2-enal, (E, Z)-nona-2,6-dienal, and 9-oxononanoic acid, the carbonyl compounds produced by green bananas, are identical to those obtained from cucumbers. Bananas, treated with ethylene and stored for 4 days at 15°C, produce hexanal, (E)-hex-2-enal and (E)-12-oxo-dodec-10-enoic acid. The same compounds are also produced by climacteric and postclimacteric bananas. Ethylene treatment and storage of only 2 days at 15°C leads to a mixture of C_6- and C_9-aldehydes and C_9- and C_{12}-oxoacids in almost equal amounts.

3. MELON

3.1   Volatile Composition

The generally used term 'melon' is commonly related to muskmelon (*Cucumis melo* L.) and watermelon (*Citrullus vulgaris*), prominent members of the botanical family *Cucurbitaceae,* which enjoy world-wide popularity as dessert fruits (Pratt, 1971).

Several groups of researchers have investigated the volatile constituents of different cultivars of muskmelon (Kemp *et al.,* 1971, 1972a,b, 1973, 1974a; Yabumoto *et al.,* 1974, 1977, 1978; Yabumoto and Jennings 1977; Buttery *et al.,* 1982; Hashinaga *et al.,* 1984) and watermelon (Katayama and Kameko, 1969; Kemp *et al.,* 1974a; Kemp, 1975; Yajima *et al.,* 1985).

Compared with other fruits, the flavour of melons is due to a complex but well-defined mixture of different classes of compounds. The data compiled by Maarse and Visscher (1987) offer an overview of 168 melon volatiles, comprising hydrocarbons, alcohols, aldehydes, ketones, esters, sulphur-containing compounds, and furans.

One group of compounds stands out of this diverse mixture and is a characteristic feature of the volatile composition of melon fruits: saturated and unsaturated aliphatic aldehydes and alcohols with skeletons of nine carbon atoms (Fig. V.2).

Another important class of melon volatiles is the esters. Yabumoto and Jennings (1977) and Yabumoto *et al.* (1974, 1978) identified a broad spectrum of esters as major constituents of various muskmelon cultivars. Buttery *et al.* (1982) reported esters of unsaturated C_9-alcohols (Fig. V.3) as typical constituents of honeydew melons.

In the course of their investigations of watermelons, Yajima *et al.* (1985) identified 4-oxononanal, 2-hydroxy-5-pentyltetrahydrofuran and several alkyl- and alkenyl-substituted 2-hydroxy-5-pentyltetrahydrofurans for the first time as naturally occurring flavour components (Fig. V.4). It remains uncertain whether these compounds are produced enzymically or whether they are artifacts formed by chemical reactions in the course of the isolation of volatiles. 4-Oxononanal has been postulated as a possible precursor of 2-pentylfuran formed by autoxidation of linoleic acid (Chang *et al.*, 1966); chemically, the

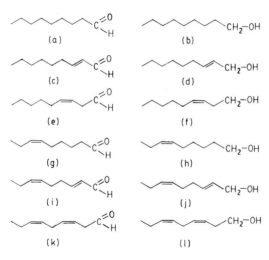

Fig. V.2. Typical melon volatiles. (a) Nonanal, (b) nonan-1-ol, (c) (*E*)-non-2-enal, (d) (*E*)-non-2-en-1-ol, (e) (*Z*)-non-3-enal, (f) (*Z*)-non-3-en-1-ol, (g) (*Z*)-non-6-enal, (h) (*Z*)-non-6-en-1-ol, (i) (*E*,*Z*)-nona-2,6-dienal, (j) (*E*,*Z*)-nona-2,6-dien-1-ol, (k) (*Z*,*Z*)-nona-3,6-dienal, (l) (*Z*,*Z*)-nona-3,6-dien-1-ol.

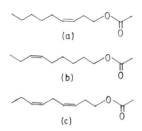

Fig. V.3. Esters of unsaturated C_9-alcohols isolated from honeydew melons (Buttery *et al.*, (1982)).

5-pentyltetrahydrofuran derivatives can be considered as acetals of 4-hydroxy-nonanal.

There are striking differences between the spectra of volatiles isolated from muskmelon and watermelon by different research groups. Yabumoto and Jennings (1977) and Yabumoto et al. (1974, 1978) isolated primarily esters, and they detected none of the C_9-components described by Kemp et al. (1972a,b, 1973). They rationalized these differences by the different techniques employed to obtain the samples. Yabumoto and Jennings (1977) used headspace adsorption on Porapak, direct sampling of cavity gases by means of a septum-sealed glass probe connected to the melon, and steam distillation – continuous extraction of small pieces of the fruit tissue. On the other hand, Kemp et al. (1972a,b, 1973) homogenized the fruits before vacuum steam distillation, and used frozen material for their investigations. Fleming et al. (1968) observed that the concentrations of C_9-compounds in cucumber increased rapidly upon cutting or blending the fruit. Kemp et al. (1973) studied the effect of storage conditions on volatiles isolated from muskmelons. Essences obtained from fruits stored refrigerated for 2 weeks before distillation did not show marked differences from those obtained from freshly harvested melons. However, after 2 weeks of freezing, the concentrations of C_9-components increased significantly, e.g. (E)-non-2-enal approximately 20-fold.

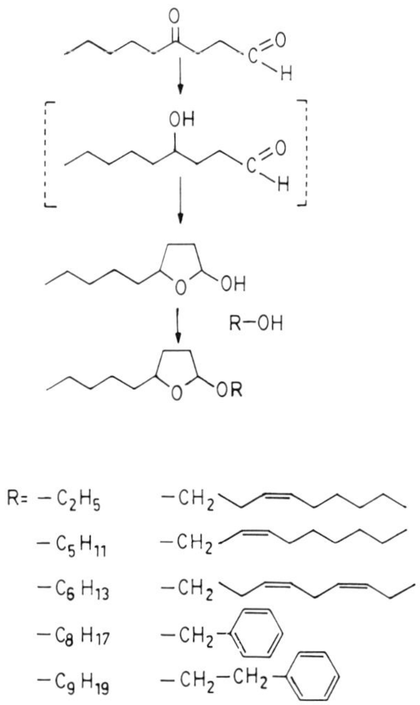

Fig. V.4. Volatile constituents isolated from watermelons (Yajima et al., 1985).

3.2 Sensory Aspects

Many of the melon volatiles with C_9-skeletons are characterized by high aroma activities. In Table V.III odour detection thresholds (in water) of some C_9-aldehydes, C_9-alcohols, and their esters are listed. On one hand, some of these C_9-aldehydes and alcohols are known as off-flavour components of certain hydrogenated vegetable oils (Keppler *et al.,* 1965; Yasuda *et al.,* 1975); on the other hand, they contribute significantly to desired aroma impressions in foods, especially members of the family *Cucurbitaceae.*

Significant importance to the aromas of muskmelon and watermelon has been attributed to (Z)-non-6-enal and (Z, Z)-nona-3,6-dien-1-ol, respectively (Kemp *et al.,* 1972b, 1974a). An aqueous solution containing 0.9 ppb (Z)-non-6-enal was described as having a strong melon-like flavour. The flavour threshold of this compound, determined according to the method of Patton and Josephson (1957), was found to be 0.02 ppb in water (Kemp *et al.,* 1972b). Using the squeeze-bottle method described by Guadagni and Buttery (1978), an odour threshold in water of 0.005 ppb has been determined (Buttery *et al.,* 1982). In mature muskmelon fruits, amounts of approximately 0.9 ppb (Z)-non-6-enal are present; this indicates a strong impact of this compound on the aroma of these fruits (Kemp *et al.,* 1972b).

The (E)-isomer has been shown to be responsible for the off-flavour of foam-spray-dried milks (Parks *et al.,* 1969). Both (Z)- and (E)-non-6-enal are off-flavour components formed during hardening of linseed and soybean oil (Keppler *et al.,* 1965). The stereoisomeric (E)-non-2-enal and the unsaturated homolog

TABLE V.III

Odour detection thresholds (ppb) in water of some C_9-aldehydes, C_9-alcohols and their esters

Compound	Odour threshold (ppb)	Reference
Nonanal	1	Guadagni *et al.* (1963)
(E)-Non-2-enal	0.5 – 1	Forss *et al.* (1962)
	0.08	Guadagni *et al.* (1963) Buttery *et al.* (1971)
(Z)-Non-6-enal	0.02	Kemp *et al.* (1972b)
	0.005	Buttery *et al.* (1982)
(E,Z)-Nona-2,6-dienal	0.1	Forss *et al.* (1962)
	0.01	Buttery *et al.* (1969)
(E,E)-Nona-2,6-dienal	0.5 – 1	Forss *et al.* (1962)
(Z)-Non-6-en-1-ol	1	Buttery *et al.* (1982)
(E,Z)-Nona-3,6-dien-1-ol	3	Buttery *et al.* (1982)
(Z,Z)-Nona-3,6-dien-1-ol	10	Kemp *et al.* (1974a)
Nonanyl acetate	200	Buttery *et al.* (1982)
(E)-Non-3-enyl acetate	60	Buttery *et al.* (1982)
(Z)-Non-6-enyl acetate	2	Buttery *et al.* (1982)
(E,Z)-Nona-3,6-dienyl acetate	15	Buttery *et al.* (1982)

(*E, Z*)-nona-2,6-dienal are essential for the flavour of cucumbers (Forss *et al.,* 1962).

(*Z, Z*)-Nona-3,6-dien-1-ol has been characterized in both muskmelons and watermelons (Kemp *et al.,* 1974a). Its flavour has been described as reminiscent of watermelon or watermelon rind. A concentration of about 200 ppb in fresh watermelon and a flavour threshold in water of 10 ppb demonstrate the significance of this compound for watermelon aroma.

Both 4-oxononanal and 2-hydroxy-5-pentyltetrahydrofuran have pleasant, fruity and green odours, and they are thought to contribute to the fresh note in the aroma of watermelon (Yajima *et al.,* 1985).

It is noteworthy that the same spectrum of C_9-compounds isolated from muskmelon and watermelon has also been identified in cucumber (*Cucumis sativus* L.) (Forss *et al.,* 1962; Kemp *et al.,* 1974b; Hatanaka *et al.,* 1975; Sekiya *et al.,* 1977). Saturated and unsaturated C_9-alcohols and aldehydes are common and characteristic constituents of these three cucurbits. Quantitative differences in the amounts of positional and stereo-isomers may contribute to the final flavour and aroma qualities of these related foods.

The C_9-components, however, are to be considered as typical 'secondary' aroma constituents (Drawert, 1978). To determine their actual concentration in intact fruits and to evaluate their contribution to melon flavour, studies involving well-defined enzyme-inhibiting experimental conditions comparable to investigations of tomato flavour by Buttery *et al.* (1987) are necessary. There must be additional constituents essential to the flavour of intact melons.

Yabumoto and Jennings (1977) isolated large quantities of esters from cantaloupe melons. They concluded that these highly volatile compounds play a critical rôle in the integrated flavours of melons, and are necessary for the strong and characteristic fruity aroma. The importance of esters to the aroma

TABLE V.IV

Synthetic honeydew melon aroma mixture (Buttery *et al.,* 1982)

Compound	ppb
Ethanol	18 100
2-Methylpropyl acetate	180
Ethyl butanoate	180
Ethyl 2-methylbutanoate	220
Butyl acetate	250
3-Methylbutyl acetate	290
Ethyl hexanoate	70
Hexyl acetate	100
Nonyl acetate	20
(*Z*)-Non-6-enyl acetate	20
(*E*)-Non-3-enyl acetate	10
Ethyl acetate	360
Benzyl acetate	200

has also been emphasized by Buttery *et al.* (1982). Due to its odour detection threshold of 2 ppb, (*Z*)-non-6-enyl acetate, which has a pleasant honeydew melon-like aroma, is thought to be an important contributor to this aroma. Based upon their qualitative and quantitative results obtained from honeydew melons, Buttery *et al.* (1982) created a synthetic mixture of major aroma compounds (Table V.IV). Sensory panel studies indicated that this mixture, which mainly consists of esters, was not totally identical, but very similar, to natural honeydew juice as far as humans are concerned. It is interesting that insects (*Drosophila* spp.) showed an overwhelming preference for the natural honeydew melon aroma compared with the synthetic mixture.

Unsaturated C_9-compounds related to the pattern found in *Cucurbitaceae* also play important rôles as fruit fly attractants. (*E*)-Non-6-enyl acetate has been described as a strong attractant for the female melon fly (*Daucus cucurbitaceae*) by Jacobson *et al.* (1971). Methyl (*E*)-non-6-enoate and (*E*)-non-6-en-1-ol are sex pheromones for the Mediterranean fruit fly, *Ceratitus capitata* Wiedemann (Jacobson *et al.*, 1973).

3.3 Biogenetic Aspects

Commercially, melons are harvested prior to the climacteric peak at the green-ripe stage, and permitted to ripen during shipment to the markets. Several factors, such as ethylene, fruit maturity, and attachment to the parent plant, influence the production of volatiles by muskmelons during the ripening period. Ethylene treatment of preclimacteric muskmelons initiated an earlier production of volatile esters; the maximum level of accumulation, however, was not significantly changed compared with the untreated fruits (Bliss and Pratt, 1979).

By means of computer-generated three-dimensional displays, Yabumoto *et al.* (1974, 1978) followed the changes in individual melon constituents as a function of the ripening time. Depending on the exhibited patterns of production, the volatiles can be fitted into groups; one group (e.g. ethyl esters, acetaldehyde, ethanol) showed a continuously accelerating rate of production; other compounds (e.g. acetate esters) increased rapidly and then reached a plateau. Correlations were observed between the sensory qualities of muskmelon fruits and the production of ethylene and esters during ripening (Yamaguchi *et al.*, 1977).

The biosynthesis of some low molecular weight esters was investigated by means of injection of (U-^{14}C)-leucine and (U-^{14}C)-isoleucine into muskmelon fruit tissue (Yabumoto *et al.*, 1977). Isoleucine was converted into volatile compounds at a much faster rate than leucine. The branched-chain skeleton of isoleucine was incorporated in volatile esters at either the acid or alcohol moiety.

The formation of C_9-compounds by means of enzyme-catalyzed splitting of unsaturated fatty acids has been studied extensively in cucumber fruits. Grosch and Schwarz (1971) incubated cucumber homogenate with (U-^{14}C)-linoleic and linolenic acid. Lipoxygenase (Galliard and Phillips, 1976), a hydroperoxide lyase (Galliard *et al.,* 1976) and a (*Z*)-3:(*E*)-2-enal isomerase (Phillips *et al.,* 1979) are involved in the formation of volatiles from fatty acid precursors by cucumber tissue. A hydroperoxide lyase has also been identified in watermelon seedlings (Vick and Zimmerman, 1976).

Based upon structural similarities in the positions of the double bonds, Kemp *et al.* (1974a) proposed oleic, linoleic, isolinoleic and linolenic acids as precursors of the C_9-compounds identified in melon essences.

The chemical formation of non-6-enal from isolinoleic acid, formed during hydrogenation of linolenic acid-containing oils, has been demonstrated (Keppler *et al.,* 1967). However, the mechanism of the enzyme-catalyzed formation of these compounds in melon fruit tissue still awaits full elucidation.

4. PINEAPPLE

4.1 Volatile Composition

Pineapple flavour has been the subject of extensive studies. Earlier work (Haagen-Smit *et al.,* 1945a,b; Gawler, 1962; Connell, 1964; Rodin *et al.,* 1965, 1966; Silverstein *et al.,* 1966; Howard and Hoffmann, 1967; Creveling *et al.,* 1968; Flath and Forrey, 1970; Näf-Müller and Willhalm, 1971) has been comprehensively reviewed by Dupaigne (1976) and Flath (1980). In their compilation covering data up to 1981, Maarse and Visscher (1987) listed a total of 116 pineapple volatiles, among them 1 hydrocarbon, 18 alcohols, 4 aldehydes, 7 ketones, 2 acids, 64 esters, 7 lactones, 4 sulphur-containing compounds, 1 acetal, 2 phenols and 5 furans. Some typical constituents are listed in Fig. V.5.

Esters constitute a major part of pineapple volatiles. They make up about 81% of the headspace samples obtained from intact ripe pineapples; they are also major compounds found in pineapple pulp samples (Takeoka *et al.,* 1988). Methyl esters are quantitatively dominating, while esters of higher alcohols are lacking. The presence of some unique sulphur-containing esters was already established in early investigations (Haagen-Smit *et al.,* 1945b; Connell, 1964; Rodin *et al.,* 1966).

A characteristic property of the spectrum of pineapple volatiles is the presence of numerous hydroxy acid esters and the corresponding acetoxy acid esters. Biogenetically related are unsaturated esters with double bonds in the 3- and 4-positions, respectively, and γ- and δ-lactones.

A major compound among the more polar pineapple constituents is 2,5-dimethyl-4-hydroxy-(2*H*)-furan-3-one.

4.2 Sensory Aspects

Despite detailed knowledge of the qualitative composition of pineapple volatiles, solid information about the importance of individual constituents to the aroma and flavour of pineapple has long been lacking.

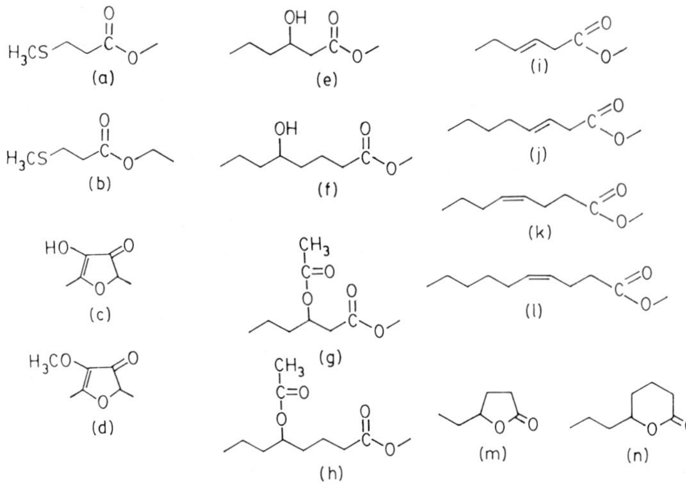

Fig. V.5. Typical pineapple volatiles. (a) Methyl 3-(methylthio)propanoate, (b) ethyl 3-(methylthio)propanoate, (c) 2,5-dimethyl-4-hydroxy-(2H)-furan-3-one, (d) 2,5-dimethyl-4-methoxy-(2H)-furan-3-one, (e) methyl 3-hydroxyhexanoate, (f) methyl 5-hydroxyoctanoate, (g) methyl 3-acetoxyhexanoate, (h) methyl 5-acetoxyoctanoate, (i) methyl (E)-hex-3-enoate, (j) methyl (E)-oct-3-enoate, (k) methyl (Z)-oct-4-enoate, (l) methyl (Z)-dec-4-enoate, (m) γ-hexalactone, (n) δ-octalactone.

No doubt an essential pineapple flavour component is 2,5-dimethyl-4-hydroxy-(2H)-furan-3-one. Isolation and identification of this furanone, also called 'furaneol', was first reported by Rodin et al. (1965). Pittet et al. (1970) determined its odour and taste thresholds in water as $0.1 - 0.2$ ppm and 0.03 ppm, respectively. Its odour changes from strongly caramel-like at high concentrations to pleasantly fruity, reminiscent of strawberry and pineapple, near its threshold concentration (Re et al., 1973). Pickenhagen et al. (1981) reported amounts of 7.4 ppm furaneol in pineapples from the Ivory Coast. Storage of pineapples for 10 days at 22°C led to a decrease of furaneol of more than 50%. The biogenetically related methyl ether, 4-methoxy-2,5-dimethyl-(2H)-furan-3-one is more stable; this compound has a higher threshold and is described as having a more sherry-like aroma (Hunter et al., 1974).

Rodin et al. (1965) were thought to have isolated the 'character-impact' component (Jennings and Sevenants, 1964) of pineapple flavour. The complex aroma impression of pineapple, however, can only be explained by a contribution of various compounds. Therefore, recent research activities have focused on the identification of sensorially important trace constituents and the investigation of odour properties and significance of various constituents.

In the course of their investigation of pineapple flavour, Berger et al. (1983) pre-separated pineapple aroma extracts by means of fractionation on silica gel. A non-polar fraction obtained by elution with pentane exhibited a balsamic, fruity odour which was believed to be important to the overall flavour of the fruit. Capillary GC – MS investigation revealed the presence in this fraction of sesquiterpene hydrocarbons with bi- or tri-cyclic skeletons. α-Copaene, β-ylangene, α-patchoulene, γ-gurjunene, germacrene D, α-muurolene, and δ-cadinene were identified for the first time as pineapple constituents. Of all these sesquiterpene hydrocarbons, only α-patchoulene seems to contribute to the fruity-spicy odour of this fraction. The major impact on the aroma character, however, has been attributed to some trace constituents. Figure V.6 presents the structures of four unsaturated hydrocarbons which were identified in this non-polar fraction. Application of a GC-sniffing technique (Christoph and Drawert, 1985) showed that these compounds possess unique odour properties. They are summarized in Table V.V.

(E,Z)-Undeca-1,3,5-triene and (E,Z,Z)-undeca-1,3,5,8-tetraene possess the lowest odour detection thresholds in air reported for hydrocarbons (Gemert and Nettenbreijer, 1977). Despite their low concentrations, these compounds may contribute to the overall impression of pineapple. However, it has to be noted that their concentrations decrease below the sensory detection thresholds within several hours in homogenates prepared from pineapple without inhibition of en-

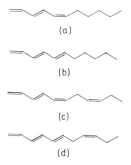

(a)

(b)

(c)

(d)

Fig. V.6. Undecatrienes and undecatetraenes identified as pineapple volatiles (Berger *et al.,* 1985). (a) (*E,Z*)-Undeca-1,3,5-triene, (b) (*E,E*)-undeca-1,3,5-triene, (c) (*E,Z,Z*)-undeca-1,3,5,8-tetraene, (d) (*E,E,Z*)-undeca-1,3,5,8-tetraene.

TABLE V.V

Occurrence and sensory properties of undecatrienes and undecatetraenes in pineapple (Berger *et al.,* 1985)

Compound	Average concentration (μg/kg)	Odour detection threshold in air (ng/l)	Odour description (sniffing-GC)
(*E,Z*)-Undeca-1,3,5-triene[a]	1	0.01 – 0.02	Balsamic, spicy, pinewood
(*E,E*)-Undeca-1,3,5-triene	< 0.5	7500 – 10 000	Musty
(*E,Z,Z*)-Undeca-1,3,5,8-tetraene	1	0.02 – 0.04	Resembling[a], more fruity
(*E,E,Z*)-Undeca-1,3,5,8-tetraene	< 0.5	200 – 300	Sweet, fruity

zyme activities. The instability of these constituents may be responsible for the pronounced losses of the aroma during processing of pineapple fruits, and the difficulties in creating an aroma mixture showing the aroma character of fresh pineapples.

A class of compounds which are of special importance for pineapple aroma are esters. An ester which has long been used as an aroma-active constituent of artificial pineapple essences is prop-2-enyl hexanoate (Fenaroli, 1975). Nitz and Drawert (1982) identified the compound as naturally occurring in pineapple; however, due to its low concentration it does not contribute to pineapple flavour in the fruit. A comparison of spatial models revealed similarities in size and shape of (E,Z)-undeca-1,3,5-triene and prop-2-enyl hexanoate. A series of naturally occurring and synthetic pineapple flavour compounds possesses common structural features which may be essential for their odour qualities (Berger and Kollmannsberger, 1985).

Takeoka et al. (1989) determined the odour contribution of various volatiles from crown, pulp and intact pineapple fruit by correlating quantitative and sensory data. The naturally present amounts of volatiles were determined by means of standard-controlled dynamic head-space sampling (Buttery et al., 1987); odour thresholds data of pineapple constituents were obtained by applying the method developed by Guadagni et al. (1963). Odour units, defined as the ratio

TABLE V.VI

Odour thresholds and odour units for selected constituents in blended Hawaiian pineapples (pulp) (Takeoka et al., 1989)

Constituent	Odour threshold in water (ppb)	Odour unit	Odour description (Fenaroli, 1975)
Methyl 2-methylbutanoate	0.25	8316	Pungent, fruity
Ethyl 2-methylbutanoate	0.3	220	Apple (Flath et al., 1967)
Ethyl acetate	5	101	Ether-like, reminiscent of pineapple
Ethyl hexanoate	1	99	Powerful fruity with pineapple-banana note
Ethyl butanoate	1	92	Fruity with pineapple undernote
Ethyl 2-methylpropanoate	0.1	60	Apple-like
Methyl hexanoate	70	49	Ether-like, reminiscent of pineapple
Methyl butanoate	60	34	Apple-like
Methyl heptanoate	4	12	Fruity; orchis-like odour; currant-like flavour
Methyl (Z)-dec-4-enoate	3	4	
Ethyl pentanoate	1.5	4	Fruity, suggestive of apple
3-Methylbutyl acetate	2	4	
Ethyl 3-methylthiopropanoate	7	4	
Methyl 3-methylthiopropanoate	180	3	Onion-like at high concentrations; pineapple flavour at high dilutions
Ethyl propanoate	10	2	Reminiscent of rum and pineapple
Methyl 3-acetoxyhexanoate	190	1	

of the concentration of the compound to its odour threshold (Guadagni *et al.*, 1966), of some pineapple volatiles are listed in Tabel V.VI. They are a measurement of the relative contribution of a constituent to the total odour. These data demonstrate the importance of esters to the fruity note of pineapple flavour.

Further systematic studies of the sensory properties of pineapple constituents, especially the more polar compounds such as furaneol and γ- and δ-lactones, are needed.

4.3 Biogenetic Aspects

Sequences of fatty acid metabolism play an important rôle in the biogenesis of aroma-active esters in pineapple. Exogenous precursors are metabolized by pineapple tissue. Incubation of pineapple segments with saturated medium-chain fatty acids and alcohol precursors, respectively, leads to significant increases of the corresponding esters (Berger *et al.*, 1984).

The metabolization of unsaturated fatty acids by pineapple tissue is more complex. Addition of D_5-deuterated sorbic acid (hexa-2,4-dienoic acid) revealed that (a) an acyl-CoA-oxidoreductase catalyzing the formation of hex-4-enoates, (b) an (E,E)-2, 4-dienoyl-CoA-reductase leading to (E)-hex-3-enoates, and (c) an enoyl-CoA-3(E),2(E)-isomerase, finally leading to butanoates after thioclastic cleavage, are involved (Fig. V.7). Berger and Kollmannsberger (1985) postulated that a combination of these enzyme-catalyzed reaction steps starting from the key intermediate (Z,Z)-dodeca-3,6-dienoyl-CoA may lead to a series of typical esters and lactones in pineapple. In analogy to the results obtained by Boland and Mertes (1985), the biogenesis of (E,Z)-undeca-1,3,5-triene may also start from this intermediate, involving a lipoxygenase-type hydrogen ab-

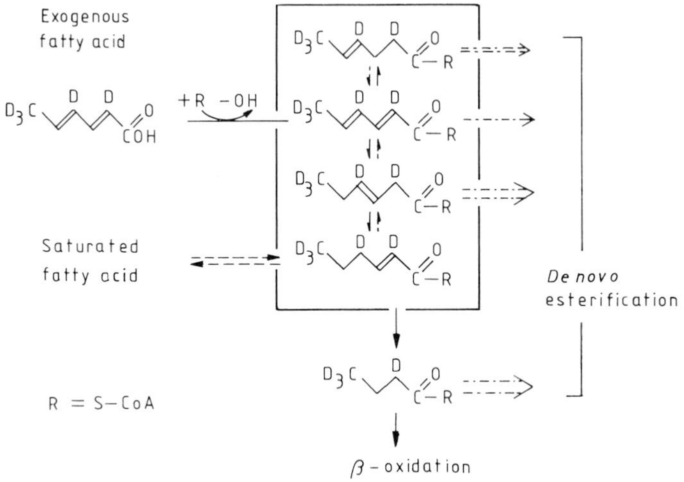

Fig. V.7. Formation of labelled carboxylic esters from D_5-sorbic acid in pineapple tissue (from Berger and Kollmannsberger, 1985).

straction from the penta-1,4-dienyl system and subsequent oxidation and fragmentation to the unsaturated hydrocarbon and carbon dioxide.

Numerous chiral compounds, such as 3- and 5-hydroxyesters, 3-, 4- and 5-acetoxyesters, and γ- and δ-lactones are major pineapple volatiles (Silverstein *et al.*, 1966; Creveling *et al.*, 1968; Näf-Müller and Willhalm, 1971; Ohta *et al.*, 1987). Neither the naturally occurring configurations nor the pathways involved in the biosynthesis of these constituents have been investigated in the past. Recent developments of analytical micro methods have now made possible the investigation of chiral components at trace levels. Pineapple is an ideal natural system to apply these methods to, and hence gain more insight into the biogenesis of this class of compound.

Capillary gas chromatographic separation of diastereoisomeric derivatives of (S)-(+)-α-methoxy-α-trifluoromethylphenylacetic acid (MTPA) and (R)-(+)-1-phenethyl isocyanate (PEIC) revealed that the chiral pineapple constituents are not present in optically pure form, but as mixtures of enantiomers (Tressl *et al.*, 1985, 1988). Investigation of pineapples subjected to postharvest ripening at room temperature over a period of 5 days, showed that the concentrations of chiral components increased significantly; the enantiomeric compositions, however, remained virtually unchanged (Fig. V.8). This constancy of enantiomeric ratios at different physiological stages is important, because it is one of the premises to use the investigation of chiral constituents for detection of adulteration in fruit products. Addition of chemically synthesized precursors to pineapple tissue was used to trace some of the biogenetic pathways involved in the biogenesis of chiral volatiles. The metabolization of keto acids and keto esters comprises esterification, reduction to hydroxy compounds, formation of acetoxy esters, and cyclization to the corresponding lactones (Fig. V.9). Metab-

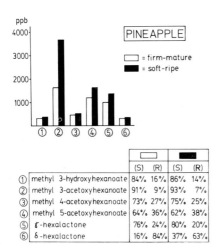

			(S)	(R)	(S)	(R)
①	methyl 3-hydroxyhexanoate		84%	16%	86%	14%
②	methyl 3-acetoxyhexanoate		91%	9%	93%	7%
③	methyl 4-acetoxyhexanoate		73%	27%	75%	25%
④	methyl 5-acetoxyhexanoate		64%	36%	62%	38%
⑤	γ-hexalactone			76%	24%	80%	20%
⑥	δ-hexalactone			16%	84%	37%	63%

Fig. V.8. Concentrations and enantiomeric compositions of chiral pineapple volatiles at different stages of ripeness.

olization rates and distribution of formed products strongly depend on the structures of the precursors (Engel *et al.,* 1989).

The enantiomeric compositions of most of the chiral metabolites obtained in precursor experiments are different from those determined for the naturally occurring compounds. δ-Octalactone obtained after the addition of 5-oxo-octanoic acid to pineapple tissue is almost optically pure (92% (*S*)); on the other hand, δ-octalactone is naturally present in pineapple fruit as the racemic mixture. Addition of deuterium-labelled compounds showed that, in addition to the reduction of the 5-oxoprecursor, at least two other pathways, chain elongation of 3-hydroxyhexanoate and hydration of (Z)-oct-4-enoic acid, may be involved in the biogenesis of δ-octalactone (Fig. V.10). These results demonstrate the complexity of the biogenesis of chiral compounds in natural systems (Engel, 1988).

There are still many unanswered questions concerning the biogenetic pathways leading to pineapple volatiles. In addition to routes involving fatty acids and amino acids, the metabolism of carbohydrates may play an important rôle. The chemical formation of furaneol during non-enzymic browning reactions of sugars has long been established (Mills *et al.,* 1969; Shaw *et al.,* 1969). Enzyme-catalyzed model reactions have been investigated (Wong *et al.,* 1983), but the natural biogenesis of this compound and its methyl ether in pineapple remains unclear.

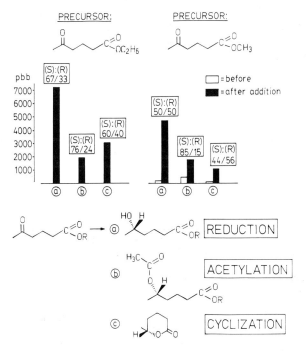

Fig. V.9. Concentrations and enantiomeric compositions of metabolites isolated after addition of chemically synthesized precursors to pineapple tissue.

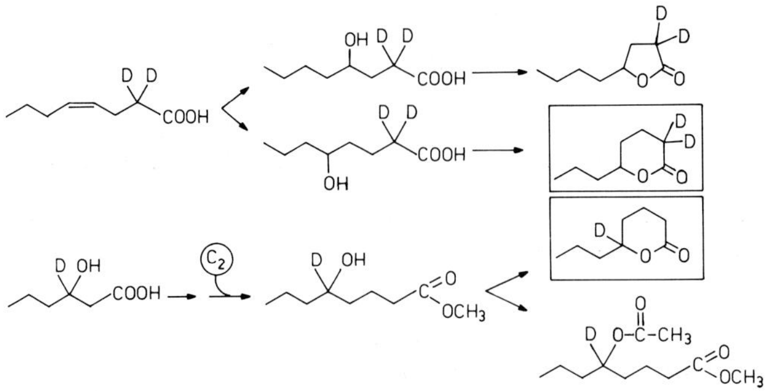

214

Fig. V.10. Pathways involved in the biosynthesis of δ-octalactone in pineapple.

5. SUMMARY

Due to the complexity of their aroma compositions, tropical fruits will continue to be 'profitable lodes to be mined' by modern analytical techniques. In our opinion, the two aspects emphasized in this chapter, (a) determination of objective sensory data, and (b) investigation of biogenetic routes, will be of major interest in future research activities. These data are a necessary basis for the present world-wide renaissance of 'natural' flavour and aroma. Fundamental knowledge about the chemical and enzymic reactions in live plant tissue is needed to improve or produce natural fruit flavour.

REFERENCES

Berger, R.G. and Kollmannsberger, H., 1985. Pineapple flavours, biosynthetic roots and sensorial properties. In: R.G. Berger, S. Nitz and P. Schreier (Editors), Topics in Flavour Research. Eichhorn, Marzling, pp. 305 – 320.

Berger, R.G., Drawert, F. and Nitz, S., 1983. Sesquiterpene hydrocarbons in pineapple fruit. *J. Agric. Food Chem.,* 31: 1237 – 1239.

Berger, R.G., Drawert, F. and Kollmannsberger, H., 1984. Precursorabhängige Aromaakkumulation im Fruchtgewebe der Ananas. *Dtsch. Lebensm.-Rundsch.,* 80: 299 – 304.

Berger, R.G., Drawert, F., Kollmannsberger, H., Nitz, S. and Schraufstetter, B., 1985. Novel volatiles in pineapple fruit and their sensory properties. *J. Agric. Food Chem.,* 33: 232 – 235.

Berger, R.G., Drawert, F. and Kollmannsberger, H., 1986a. Geruchsaktive Spurenkomponenten des Bananenaromas. *Chem. Mikrobiol. Technol. Lebensm.,* 10: 120 – 124.

Berger, R.G., Drawert, F. and Kollmannsberger, H., 1986b. Biotechnologische Erzeugung von Aromastoffen II. PA-Lagerung zur Kompensation von Aromaverlusten bei der Gefriertrocknung von Bananenscheibchen. *Z. Lebensm. Unters. Forsch.,* 183: 169 – 171.

Bliss, M.L. and Pratt, H.K., 1979. Effect of ethylene, maturity, and attachment to the parent plant on production of volatile compounds by muskmelon. *J. Am. Soc. Hortic. Sci.,* 104: 273 – 277.

Boland, W. and Mertes, K., 1985. Biosynthesis of algae pheromenes. A model study with the Composite Senecio Isatideus. *Eur. J. Biochem.,* 147: 83 – 91.

Buttery, R.G., Seifert, R.M., Guadagni, D.G. and Ling, L.C., 1969. Characterization of some volatile

215

constituents of bell peppers. *J. Agric. Food Chem.*, 17: 1322 – 1327.

Buttery, R.G., Seifert, R.M., Guadagni, D.G. and Ling, L.C., 1971. Characterization of additional volatile components of tomato. *J. Agric. Food Chem.*, 19: 524 – 529.

Buttery, R.G., Seifert, R.M., Ling, L.C., Soderstrom, E.L., Ogawa, J.W. and Turnbaugh, J.G., 1982. Additional aroma components of honeydew melon. *J. Agric. Food Chem.*, 30: 1208 – 1211.

Buttery, R.G, Teranishi, R. and Ling, L.C., 1987. Fresh tomato volatiles: A quantitative study. *J. Agric. Food Chem.*, 35: 540 – 544.

Chang, S.S., Smouse, T.H., Krishnamurthy, R.G., Mookherjee, B.D. and Reddy, R.B., 1966. Isolation and identification of 2-pentylfuran as contributing to the reversion flavour of soybean oil. *Chem. Ind.*, 1926 – 1927.

Christoph, N. and Drawert, F., 1985. Olfactory thresholds of odor stimuli determined by gas chromatographic sniffing technique; structure – activity relationships. In: R.G. Berger, S. Nitz and P. Schreier (Editors), Topics in Flavour Research. Eichhorn, Marzling: pp. 59 – 77.

Connell, D.W., 1964. Volatile flavouring constituents of the pineapple. *Aust. J. Chem.*, 17: 130 – 140.

Creveling, R.K., Silverstein, R.M. and Jennings, W.G., 1968. Volatile components of pineapple. *J. Food Sci.*, 33: 284 – 287.

Drawert, F., 1978. Primär- und Sekundäraromastoffe und deren Analytik. *IFU, Ber. Wiss. Tech. Komm. XV,* Juris, Bern, p. 1.

Drawert, F. and Künanz, H.-J., 1975. Verhalten der Hauptkomponenten in Gewebescheibchen aus der Fruchtpulpe von Bananen in Abhängigkeit vom physiologischen Zustand der Frucht. *Chem. Mikrobiol. Technol. Lebensm.*, 3: 185 – 192.

Drawert, F., Heimann, W., Emberger, R. and Tressl, R., 1972. Bedeutung des Klimakteriums für die Entwicklung der Aromastoffe von Apfeln und Bananen. *Chem. Mikrobiol. Technol. Lebensm.*, 1: 201 – 205.

Dupaigne, P., 1970. L'Arome de la Banane. *Fruits*, 25: 281 – 291.

Dupaigne, P., 1975. L'Arome de la Banane. *Fruits*, 30: 783 – 789.

Dupaigne, P., 1976. L'Arome de l'Ananas. *Fruits*, 31: 631 – 639.

Engel, K.-H., 1988. Investigation of chiral flavor compounds by chromatographic micro-methods. In: P. Schreier (Editor), Bioflavour '87. Walter de Gruyter, Berlin, New York, pp. 75 – 88.

Engel, K.-H., Heidlas, J., Albrecht, W. and Tressl, R., 1989. Biosynthesis of chiral flavor and aroma compounds in plants and microorganisms. In: R. Teranishi, R.G. Buttery and F. Shahidi (Editors), Flavor Chemistry, Trends and Developments. American Chemical Society, Washington, DC, pp. 8 – 22.

Fenaroli, G., 1975. In: E.T. Furia and N. Bellanca (Editors), Fenaroli's Handbook of Flavor Ingredients. 2nd ed., CRC Press, Cleveland.

Flath, R.A., 1980. Pineapple. In: S. Nagy and P. Shaw (Editors), Tropical and Subtropical Fruits, Composition, Properties and Uses. AVI, Westport, pp. 157 – 183.

Flath, R.A. and Forrey, R.R., 1970. Volatile components of smooth cayenne pineapple. *J. Agric. Food Chem.*, 18: 306 – 309.

Flath, R.A., Black, D.R., Guadagni, D.G., McFadden, W.H. and Schultz, T.H., 1967. Identification and organoleptic evaluation of compounds in Delicious Apple essence. *J. Agric. Food Chem.*, 15: 29 – 35.

Fleming, H.P., Cobb, W.Y., Etchells, J.L. and Bell, T.A., 1968. The formation of carbonyl compounds in cucumbers. *J. Food Sci.*, 33: 572 – 576.

Forss, D.A., Dunstone, E.A., Ramshaw, E.H. and Stark, W., 1962. The flavor of cucumbers. *J. Food Sci.*, 27: 90 – 93.

Forsyth, W.G.C., 1980. Banana and plantain. In: S. Nagy and P. Shaw (Editors), Tropical and Subtropical Fruits, Composition, Properties and Uses. AVI, Westport, pp. 259 – 278.

Galliard, T. and Phillips, D.R., 1976. The enzymic cleavage of linoleic acid to C_9 carbonyl fragments in the extracts of cucumber (*Cucumis sativus)* fruit and the possible role of lipoxygenase. *Biochim. Biophys. Acta,* 431: 278 – 287.

Galliard, T. Phillips, D.R. and Reynolds, J., 1976. The formation of *cis*-3-nonenal, *trans*-2-nonenal and hexanal from the linoleic acid hydroperoxide isomers by a hydroperoxide cleavage enzyme system in cucumber (*Cucumis sativus*) fruits. *Biochim. Biophys. Acta,* 441: 181 – 192.

Gawler, J.H., 1962. Constituents of canned Malayan pineapple juice. *J. Sci. Food Agric.*, 13: 57 – 61.

216

Gemert, L.J. van and Nettenbreijer, A.H., 1977. Compilation of Odour Threshold Values in Air and Water. TNO, Zeist, The Netherlands.

Gilliver, P.J. and Nursten, H.E., 1976. The source of acyl moiety in the biosynthesis of volatile banana esters. *J. Sci. Food Agric.,* 27: 152–158.

Grosch, W. and Schwarz, J.M., 1971. Linoleic and linolenic acid as precursors of the cucumber flavor. *Lipids,* 6: 351–352.

Guadagni, D.G. and Buttery, R.G., 1978. Odor threshold of 2, 3, 6-trichloroanisole in water. *J. Food Sci.,* 43: 1346–1347.

Guadagni, D.G., Buttery, R.G. and Okana, S., 1963. Odour thresholds of some organic compounds associated with food flavours. *J. Sci. Food Agric.,* 14: 761–765.

Guadagni, D.G. and Buttery, R.G. and Harris, J., 1966. Odour intensities of hop oil components. *J. Sci. Food Agric.,* 17: 142–144.

Guymon, J.F., 1965. Mutant strains of *Saccharomyces cerevisiae* applied to studies of higher alcohol formation during fermentation. *Dev. Ind. Microbiol.,* 7: 88–93.

Haagen-Smit, A.J., Kirchner, J.G., Prater, A.N. and Deasy, C.L., 1945a. Chemical studies of pineapple I. The volatile flavor and odor constituents of pineapple. *J. Am. Chem. Soc.,* 67: 1646–1650.

Haagen-Smit, A.J., Kirchner, J.G., Deasy, C.L. and Prater, A.N., 1945b. Chemical studies of pineapple II. Isolation and identification of a sulfur-containing ester in pineapple. *J. Am. Chem. Soc.,* 67: 1651–1652.

Hashinaga, F., Koga, T. and Ishida, K., 1984. Some changes in the chemical constituents occurring in accordance with the ripening of 'Prince' melon fruits and comparison of the fermented fruits with the normal ones. *Bull. Fac. Agric. Kagoshima Univ.,* 34: 29–37.

Hatanaka, A., Kajiwara, T. and Harada, T., 1975. Biosynthetic pathway of cucumber alcohol: *trans*-2, *cis*-6 nonadienol via *cis*-3, *cis*-6 nonadienal. *Phytochemistry,* 14: 2589–2592.

Howard, G.E. and Hoffmann, A., 1967. A study of the volatile flavouring constituents of canned Malayan pineapple. *J. Sci. Food Agric.,* 18: 106–110.

Hultin, H.O. and Procter, B.E., 1961. Changes in some volatile constituents of the banana during ripening, storage, and processing. *Food Technol.,* 15: 440–443.

Hunter, G.L.K., Bucek, W.A. and Radford, T., 1974. Volatile components from canned alphonso mango. *J. Food Sci.,* 39: 900–903.

Issenberg, P., 1969. Mass spectra for flavor research. *Food Technol.,* 23: 103–110.

Issenberg, P. and Wick, E.L., 1963. Volatile components of bananas. *J. Agric. Food Chem.,* 11: 3–8.

Jacobson, M., Keiser, I. Chambers, D.L., Miyashita, D.H. and Harding, C., 1971. Synthetic nonenyl acetates as attractants for female melon flies. *J. Med. Chem.,* 14: 236–239.

Jacobson, M., Ohinata, K., Chambers, D.L., Jones, W.A. and Fujimoto, M.S., 1973. Insect sex attractants. 13. Isolation, identification, and synthesis of sex pheromones of the male mediterranean fruit fly. *J. Med. Chem.,* 16: 248–251.

Jennings, W.G. and Sevenants, M.R., 1964. Volatile esters of Barlett Pear. *J. Food Sci.,* 29: 158–163.

Katayama, O. and Kameko, K., 1969. *Nippon Shokuhin Kogyo Gakkai-Shi,* 16: 474.

Kemp, T.R., 1975. Identification of some volatile compounds from *Citrullus vulgaris. Phytochemistry,* 14: 2637–2638.

Kemp, T.R., Knavel, D.E. and Stoltz, L.P., 1971. Characterization of some volatile components of muskmelon fruit. *Phytochemistry,* 19: 1925–1928.

Kemp, T.R., Stoltz, L.P. and Knavel, D.E., 1972a. Volatile components of muskmelon fruit. *J. Agric. Food Chem.,* 20: 196–198.

Kemp, T.R., Knavel, D.E. and Stoltz, L.P., (1972b) *cis*-6-Nonenal: A flavor component of muskmelon fruit. *Phytochemistry,* 11: 3321–3322.

Kemp, T.R., Knavel, D.E. and Stoltz, L.P. 1973. Volatile *Cucumis melo* components: Identification of additional compounds and effects of storage conditions. *Phytochemistry,* 12: 2921–2924.

Kemp, T.R., Knavel, D.E. and Stoltz, L.P., 1974a. 3, 6-Nonadien-1-ol from *Citrullus vulgaris* and *Cucumis melo. Phytochemistry,* 13: 1167–1170.

Kemp, T.R., Knavel, D.E. and Stoltz, L.P., 1974b. Identification of some volatile compounds from cucumber. *J. Agric. Food Chem.,* 22: 717–718.

Keppler, J.G., Schols, J.A., Feenstra, W.H. and Meijboom, P.W., 1965. Components of the hardening flavor present in hardened linseed oil and soybean oil. *J. Am. Oil Chem. Soc.,* 42: 246–249.

Keppler, J.G., Horikx, M.M., Meijboom, P.W. and Feenstra, W.H., 1967. Iso-linoleic acids responsible for the formation of the hardening flavor. *J. Am. Oil Chem. Soc.,* 44: 543–544.

Kleber, C., 1912. The occurrence of amyl acetate in bananas. *Am. Perfum.,* 7: 235.

Loesecke von, H.W., 1950. Bananas. 2nd edn, Interscience Publishers, Inc., New York.

Maarse, H. and Visscher, C.A., 1987. Volatile Compounds in Food – Qualitative Data. Supplement 4; TNO- CIVO, Zeist, The Netherlands.

Macku, C. and Jennings, W.G., 1987. Production of volatiles by ripening bananas. *J. Agric. Food Chem.,* 35: 845–848.

Mattei, A., 1973a. Analyse de l'emission volatile de la banane. *Fruits,* 28: 231–238.

Mattei, A., 1973b. Variations in the emission of volatiles from the banana, *Musa cavendishii,* in the course of ripening and as a function of temperature. *Physiol. Veg.,* 11: 721–738.

McCarthy, A.I., Palmer, J.K., Shaw, C.P. and Anderson E.E., 1963. Correlation of gas chromatographic data with flavor profiles of fresh banana fruit. *J. Food Sci.,* 28: 379–384.

McCarthy, A.I., Wyman, H. and Palmer, J.K., 1964. Gas chromatographic identification of banana fruit volatiles. *J. Gas Chromatogr.,* 2: 121–124.

Mills, F.D., Baker, B.G. and Hodge, J.E., 1969. Amadori compounds as nonvolatile flavor precursors in produced food. *J. Agric. Food Chem.,* 17: 723–727.

Murray, K.E., Palmer, J.K., Whitfield, F.B., Kennett, B.H. and Stanley, G., 1968. The volatile alcohols of ripe bananas. *J. Food Sci.,* 33: 632–634.

Myers, M.J., Issenberg, P. and Wick, E.L., 1969. Vapor analysis of the production by banana fruit of certain volatile constituents. *J. Food Sci.,* 34: 504–509.

Myers, M.J., Issenberg, P. and Wick, E.L., 1970. L-leucine as a precursor of isoamyl alcohol and isoamyl acetate volatile aroma constituents of banana fruit discs. *Phytochemistry,* 9: 1693–1700.

Näf-Müller, R. and Willhalm B., 1971. Uber die flüchtigen Anteile der Ananas. *Helv. Chim. Acta,* 34: 1880–1890.

Nitz, S. and Drawert, F., 1982. Vorkommen von Allylhexanoat in Ananasfrüchten. *Chem. Mikrobiol. Technol. Lebensm.,* 7: 148.

Nitz, S., Berger, R.G., Leupold, G. and Drawert, F., 1984. Detektion geruchtsaktiver Verbindungen durch Kombination einer SCOT-Glaskapillarsäule mit einem variablen Ausgangsteiler. *Chem. Mikrobiol. Technol. Lebensm.,* 8: 121–125.

Nursten, H.E., 1970. Volatile compounds: The aroma of fruits. In: A.C. Hulme (Editor), The Biochemistry of Fruits and Their Products, Vol. 2. Academic Press, New York, London, pp. 239–268.

Ohta, H., Kinjo, S. and Osajima, Y., 1987. Glass capillary gas chromatographic analysis of volatile components of canned Philippine pineapple juice. *J. Chromatogr.,* 409: 409–412.

Palmer, J.K., 1970. The banana. In: A.C. Hulme (Editor), The Biochemistry of Fruits and Their Products, Vol. 2. Academic Press, New York, London, pp. 65–105.

Palmer, J.K., 1973. Separation of components of aroma concentrates on the basis of functional group and aroma quality. *J. Agric. Food Chem.,* 21: 923–925.

Parks, O.W., Wong, N.P., Allen, C.A. and Schwartz, D.P., 1969. 6-*trans*-Nonenal: An off-flavor component of foam-spray-dried milks. *J. Dairy Sci.,* 52: 953–956.

Patton, S. and Josephson, D.V., 1957. A method for determining significance of volatile flavor compounds in foods. *J. Food Res.,* 22: 316–318.

Phillips, D.R., Matthew, J.A., Reynold, Y. and Fenwick, G.R., 1979. Partial purification and properties of a *cis*-3-*trans*-2-enal isomerase from cucumber fruit. *Phytochemistry,* 18: 401–404.

Pickenhagen, W., Velluz, A., Passerat, J.-P. and Ohloff, G., 1981. Estimation of 2, 5-dimethyl-4-hydroxy-3(2*H*)-furanone (furaneol) in cultivated and wild strawberries, pineapples and mangoes. *J. Sci. Food Agric.,* 32: 1132–1134.

Pittet, A.O., Rittersbacher, P. and Muralidhara, R., 1970. Flavor properties of compounds related to maltol and isomaltol. *J. Agric. Food Chem.,* 18: 929–933.

Pratt, H.K., 1971. Melons. In: A.C. Hulme (Editor) The Biochemistry of Fruits and Their Products, Vol. 2. Academic Press, New York, London, pp. 207–232.

Re, L., Maurer, B. and Ohloff, G., 1973. Ein einfacher Zugang zu 4-Hydroxy-2, 5-dimethyl-3(2*H*)-

furanon (Furaneol), einem Aromastandteil von Ananas und Erdbeere. *Helv. Chim. Acta.,* 56: 1882 – 1894.

Rodin, J.O., Himel, C.M., Silverstein, R.M., Leeper, R.W. and Gortner, W.A., 1965. Volatile flavor and aroma components of pineapple. I. Isolation and tentative identification of 2, 5-dimethyl-4-hydroxy-3(2*H*)-furanone. *J. Food Sci.,* 30: 280 – 285.

Rodin, J.O., Coulson, D.M. and Silverstein, R.M., 1966. Volatile flavor and aroma components of pineapple III. The sulfur-containing components. *J. Food Sci.,* 31: 721 – 725.

Rothe, M. and Thomas, B., 1963. Aromastoffe des Brotes. *Z. Lebensm.-Unters.-Forsch.,* 119: 302 – 310.

Rothenbach, F. and Eberlein, L., 1905. Uber das Vorkommen von Estern in den Früchten der Bananen. *Dtsch. Essigind.,* 9: 81.

Sekya, R., Kajiwara, T. and Hatanaka, A., 1977. *trans*-2-*cis*-6-Nonadienal and *trans*-2-Nonenal in cucumber fruits. *Phytochemistry,* 16: 1043 – 1044.

Shaw, P.E., Tatum, J.H. and Berry, R.E., 1969. Base-catalyzed sucrose degradation studies. *J. Agric. Food Chem.,* 17: 907 – 908.

Silverstein, R.M., Rodin, J.O. and Himel, C.M., 1966. Volatile flavor and aroma components of pineapple II. Isolation and identification of chavicol and γ-caprolactone. *J. Food Sci.,* 31: 668 – 672.

Takeoka, G., Buttery, R.G., Flath, R.A., Stern, D.J., Teranishi, R., Wheeler, E.L. and Wieczorek, R.L., 1989. Volatile constituents of pineapple (*Ananas comosus* L. Merr.). In: R. Teranishi, R.G. Buttery and F. Shahidi (Editors), Flavor Chemistry, Trends and Developments. American Chemical Society, Washington, DC, pp. 223 – 237.

Tressl, R. and Albrecht, W., 1986. Biogeneration of aroma compounds through acyl pathways. In: T.H. Parliament and R. Croteau (Editors), Biogeneration of Aromas. ACS, Washington, D.C, pp. 114 – 133.

Tressl, R. and Drawert, F., 1971. Einbau von 8-^{14}C-Caprylsaüre in Bananen- und Erdbeeraromastoffe. *Z. Naturforsch.,* 26b: 774 – 779.

Tressl, R. and Drawert, F., 1973. Biogenesis of banana volatiles. *J. Agric. Food Chem.,* 21: 560 – 565.

Tressl, R. and Jennings, W.G., 1972. Production of volatile compounds in the ripening banana. *J. Agric. Food Chem.,* 20: 189 – 192.

Tressl, R. Drawert, F., Heimann, W. and Emberger, R., 1969. Gaschromatographische Bestandsaufnahme von Bananen-Aromastoffen. *Z. Naturforsch.,* 24b: 781 – 783.

Tressl, R., Drawert, F. and Heimann, W., 1970a. Anreicherung, Trennung und Identifizierung von Bananenaromastoffen. *Z. Lebensm.-Unters.-Forsch.,* 142: 249 – 263.

Tressl, R., Drawert, F., Heimann, W. and Emberger, R., 1970b. Ester, Alkohole, Carbonylverbindungen und Phenoläther des Bananenaromas. *Z. Lebensm.-Unters.-Forsch.* 142: 313 – 321.

Tressl, R., Drawert, F. and Heimann, W., 1970c. Fettsaüren des Bananenaromas: Verhalten von C$_{16}$- und C$_{18}$-Fettsaüren während der Reifung und ihre Verteilung in den Lipoiden. *Z. Lebensm.-Unters.-Forsch.,* 142: 393 – 397.

Tressl, R., Drawert, F., Heimann, W. and Emberger, R., 1970d. Zur Biogenese der in Bananen gefundenen Aromastoffe. *Z. Lebensm.-Unters.-Forsch.,* 142: 4 – 12.

Tressl, R., Emberger, R., Drawert, F. and Heimann, W., 1970e. Einbau von ^{14}C-Leucin und -Valin in Bananenaromastoffe. *Z. Naturforsch.,* 25b: 704 – 707.

Tressl, R., Emberger, R., Drawert, F. and Heimann, W., 1970f. Einbau von ^{14}C-Acetat und ^{14}C-Butyrat in Bananenaromastoffe. *Z. Naturforsch.,* 25b: 893 – 894.

Tressl, R., Emberger, R., Drawert, F. and Prenzel, U., 1972. Anwendung der Reaktions-Radio-Gas-Chromatographie auf Aromaprobleme. *Chromatographia,* 5: 319 – 323.

Tressl, R., Drawert, F. and Prenzel, U., 1973. Anwendung der Reaktions-Radio-Gas-Chromatographie zur Untersuchung der Biogenese von Aromastoffen. *Chromatographia,* 6: 7 – 13.

Tressl, R., Engel, K.-H., Albrecht, W. and Bille-Abdullah, H., 1985. Analysis of chiral aroma components in trace amounts. In: D.D. Bills and C.J. Mussinan, Characterization and Measurement of Flavor Compounds. ACS, Washington, DC, pp. 61 – 78.

Tressl, R., Heidlas, J., Albrecht, W. and Engel, K.-H., 1988. Biogenesis of chiral hydroxyacid esters. In: P. Schreier (Editor), Bioflavour '87. Walter de Gruyter, Berlin, New York, pp. 221 – 236.

Vick, B.A. and Zimmerman, D.C., 1976. Lipoxygenase and hydroperoxide lyase in germinating watermelon seedlings. *Plant Physiol.,* 57: 780.

Wick, E.L., 1965. Chemical and sensory aspects of the identification of odor constituents in foods. *Food Technol.,* 19: 145 – 151.

Wick, E.L., McCarthy, A.I., Myers, M., Murray, E., Nursten, H.E. and Issenberg, P., 1966. Flavor and biochemistry of volatile banana components. In: R.E. Gould (Editor), Flavor Chemistry. Advances in Chemistry, Series 56. Am. Chem. Soc., Washington, DC, pp. 241 – 260.

Wick, E.L., Yamanishi, T., Kobayashi, A., Valenzuela, S. and Issenberg, P., 1969. Volatile constituents of banana. *J. Agric. Food Chem.,* 17: 751 – 759.

Wong, C., Mazenod, F.P. and Whitesides, G.M., 1983. Chemical and enzymatic synthesis of 6-desoxyhexoses. Conversion to 2, 5-dimethyl-4-hydroxy-2, 3-dihydrofuran-3-one (furaneol) and analogues. *J. Org. Chem.,* 48: 3493 – 4497.

Yabumoto, K. and Jennings, W.G., 1977. Volatile constituents of cantaloupe, *Cucumis melo,* and their biogenesis. *J. Food Sci.,* 42: 32 – 37.

Yabumoto, K., Yamaguchi, M., Hughes, D. and Jennings, W.G., 1974. Changes in the volatile constituents of ripening cantaloupes. Proc. IV. Int. Congr. Food Sci. and Technol. Selegraf, Valencia, Spain, Vol. I, pp. 93 – 98.

Yabumoto, K., Jennings, W.G. and Pangborn, R.M., 1975. Evaluation of lactose as a transfer carrier for volatile flavor constituents. *J. Food Sci.,* 40: 105 – 108.

Yabumoto, K., Yamaguchi, M. and Jennings, W.G., 1977. Biosynthesis of some volatile constituents of muskmelon, *Cucumis melo. Chem. Mikrobiol. Technol. Lebensm.,* 5: 53 – 56.

Yabumoto, K., Yamaguchi, M. and Jennings, W.G., 1978. Production of volatile compounds by muskmelon, *Cucumis melo. Food Chem.,* 3: 7 – 16.

Yajima, I., Sakakibara, H., Ide, J., Yanai, T. and Hayashi, K., 1985. Volatile flavor components of watermelon (*Citrullus vulgaris*). *Agric. Biol. Chem.,* 49: 3145 – 3150.

Yamaguchi, M., Hughes, D.L., Yabumoto, K. and Jennings, W.G., 1977. Quality of cantaloupe muskmelons: Variability and attributes. *Sci. Hort.,* 6: 59 – 70.

Yasuda, K., Peterson, R.J. and Chang, S.S., 1975. Identification of volatile flavor compounds developed during the storage of deodorized hydrogenated soybean oil. *J. Am. Oil Chem. Soc.,* 52: 307 – 311.

Ziegler, E., 1982. Die natürlichen und künstlichen Aromen. Hüthig Verlag, Heidelberg.

Chapter VI

'MINOR' TROPICAL FRUITS – MANGO, PAPAYA, PASSION FRUIT, AND GUAVA

TAKAYUKI SHIBAMOTO and CHUNG-SHIH TANG

Department of Environmental Toxicology, University of California, Davis, CA 95616 (U.S.A.),
Department of Agricultural Biochemistry, University of Hawaii, Honolulu, HI 96822 (U.S.A.)

1. INTRODUCTION

Tropical fruits have attracted people with their exotic appearance and unique flavours since antiquity. Several hundred species of edible tropical fruits are known to grow over the world, but only a few, such as bananas, citrus, and pineapple have been commonly consumed outside of their growing locale. In addition to these major tropical fruits, certain minor ones, such as mango, papaya, passion fruit, and guava, are also important in some regions. For example, world production of mango reached nearly 15 million tons in 1986. This amount is about one-half of banana production, but twice that of pineapple production. Despite large production or economic importance, the flavours of these 'minor' fruits have received less attention than those of the major fruits. The flavour characteristics and volatile components of these fruits should be studied to assess their quality; although current world markets are limited, the potential for growth is there.

In this chapter, the volatile compounds reported in mango, papaya, passion fruit, and guava are summarized and tabulated in five groups: hydrocarbons, esters, aldehydes and ketones, alcohols and acids, and miscellaneous compounds. Throughout the tables, systematic chemical nomenclature has been used as appropriate, and some names used in some papers have been changed for consistency. Odour qualities of components are from MacLeod and Snyder (1985) and MacLeod and de Troconis (1982a) for mango, MacLeod and Pieris (1983) for papaya, and MacLeod and de Troconis (1982b) for guava. Additional information on odour qualities was obtained from Arctander (1969). Some terminology used for odour qualities in the tables has been changed to more commonly used terms for readers who are not familiar with certain technical words.

2. MANGO

Mango (*Mangifera indica* L.) is an evergreen tree that originated in the Indo-

TABLE VI.I

Volatile hydrocarbons found in mango

Compound	Cultivar	Source	Form	Odour quality	References
Cyclopentane	Alphonso	India	Canned puree		a
Cyclohexane	Tommy Atkins	Florida	Fresh fruit	Fruity, sweet	b
Methylcyclohexane	Tommy Atkins	Florida	Fresh fruit	Fruity, sweet	b
		Venezuela	Fresh fruit		c
A dimethylcyclohexane	Tommy Atkins	Florida	Fresh fruit	Fruity, sweet	b
		Venezuela	Fresh fruit		c
Ethylcyclohexane	Tommy Atkins	Florida	Fresh fruit	Fruity, sweet	b
Octane		Venezuela	Fresh fruit		c
Tetradecane	Alphonso/Baladi	Egypt	Fruit pulp		d
Hexadecane	Alphonso/Baladi	Egypt	Fruit pulp		d
Octadecane	Alphonso/Baladi	Egypt	Fruit pulp		d
Eicosane	Alphonso/Baladi	Egypt	Fruit pulp		d
Benzene	Alphonso/Baladi	Egypt	Fruit pulp		d
Toluene	Tommy Atkins	Florida	Fresh fruit	Caramel, ethereal	b
	Keitt	Florida	Fruit[1,2,4,5]		b
		Venezuela	Fresh fruit		c
	Alphonso/Baladi	Egypt	Fruit pulp		d
p-Xylene	Tommy Atkins	Florida	Fresh fruit	Cold meat fat-like	b
m-Xylene	Tommy Atkins	Florida	Fresh fruit		b
Dimethylstyrene		Venezuela	Fresh fruit	Mango-like	c
Methylpropenylbenzene		Venezuela	Fresh fruit	Sickly, cereal-, hay-like	c
α-Pinene	Tommy Atkins	Florida	Fresh fruit	Floral, warm, resinous,	b
	Keitt	Florida	Fruit[1,2,3,4,5]	pine, cedar wood-like	b
		Venezuela	Fresh fruit		c
	Alphonso/Baladi	Egypt	Fruit pulp		d
	Alphonso	Philippines	Fresh fruit		e
β-Pinene	Alphonso/Baladi	Egypt	Fruit pulp	Wood-, pine-like	d
	Tommy Atkins	Florida	Fresh fruit		b
	Alphonso	Philippines	Canned puree		e
Myrcene	Alphonso	India	Canned puree	Fresh-green, grass-like	a
	Tommy Atkins	Florida	Fresh fruit		b
	Alphonso/Baladi	Egypt	Fruit pulp	Warm, balsamic	d
	Alphonso	Philippines	Canned puree		e

Compound	Cultivar	Origin	Product	Odor	
Limonene	Alphonso	India	Canned puree	Fresh, sweet, lemon-like	a
	Tommy Atkins	Florida	Fresh fruit		b
		Venezuela	Fresh fruit		c
	Alphonso/Baladi	Egypt	Fruit pulp		d
	Alphonso	Philippines	Canned puree		e
	Alphonso	Philippines	Fresh fruit		e
(Z)-Ocimene	Alphonso	India	Canned puree	Warm, herbaceous	a
		Venezuela	Fresh fruit		c
	Alphonso/Baladi	Egypt	Fruit pulp		d
	Alphonso	Philippines	Canned puree		e
(E)-Ocimene	Alphonso	India	Canned puree	Warm, herbaceous	a
	Alphonso/Baladi	Egypt	Fruit pulp		d
(Z)-Alloocimene	Alphonso	India	Canned puree	Fresh-gassy	a
(E)-Alloocimene	Alphonso	India	Canned puree	Fresh-gassy	a
α-Terpinolene	Alphonso/Baladi	Egypt	Fruit pulp	Floral, sweet, pine-like	d
	Tommy Atkins	Florida	Fruit[1,2,3,4,5]		b
	Keitt	Florida	Fresh fruit		b
	Alphonso	Philippines	Fresh fruit		e
γ-Terpinene	Keitt	Florida	Fruit[2]	Refreshing, citrus-like	b
		Venezuela	Fresh fruit		c
Sabinene	Tommy Atkins	Florida	Fresh fruit		b
Car-3-ene	Tommy Atkins	Florida	Fresh fruit	Floral, mango leaf-like	b
	Keitt	Florida	Fruit[1,2,3,4,5]		b
		Venezuela	Fresh fruit		c
	Alphonso	Philippines	Fresh fruit		e
α-Phellandrene	Tommy Atkins	Florida	Fresh fruit	Estery, pepper-like	b
	Keitt	Florida	Fruit[1,3,4,5]		b
		Venezuela	Fresh fruit		c
α-Fenchene	Keitt	Florida	Fruit[1,2,5]	Fruity	b
Camphene	Tommy Atkins	Florida	Fresh fruit	Mild-oily, camphor-like	b
β-Phellandrene	Tommy Atkins	Florida	Fresh fruit	Fatty, oily	b
		Venezuela	Fresh fruit		c
α-Copaene	Tommy Atkins	Florida	Fresh fruit	Earthy, mango-like	b
α-Gurjunene	Alphonso/Baladi	Egypt	Fruit pulp		d
	Baladi	Egypt	Fruit pulp		d

(continued)

TABLE VI.1 (*continued*)

Compound	Cultivar	Source	Form	Odour quality	References
β-Caryophyllene	Alphonso	India	Canned puree	Sweet, floral, dry wood-, clove leaf oil-like	a
	Tommy Atkins	Florida	Fresh fruit		b
	Keitt	Florida	Fruit[1,2,3,4,5]		b
		Venezuela	Fresh fruit		c
	Alphonso/Baladi	Egypt	Fruit pulp		d
α-Humulene	Alphonso	India	Canned puree	Fresh, green, floral	a
	Tommy Atkins	Florida	Fresh fruit		b
	Keitt	Florida	Fruit[1,2,3,4,5]		b
		Venezuela	Fresh fruit		c
	Alphonso/Baladi	Egypt	Fruit pulp		d
α-Selinene	Baladi	Egypt	Fruit pulp	Pepper-like	d
β-Selinene		Venezuela	Fresh fruit	Floral, wood-, almond-like	c
Germacrene D	Baladi	Egypt	Fruit pulp		d
Bicyclogermacrene	Baladi	Egypt	Fruit pulp		d
γ-Cadinene	Baladi	Egypt	Fruit pulp	Mild, dry wood-like	d
δ-Cadinene	Baladi	Egypt	Fruit pulp	Mild, dry wood-like	d
p-Cymene	Tommy Atkins	Florida	Fresh fruit	Herbaceous, mint-like	b
	Keitt	Florida	Fruit[1,2,3,4]		b
		Venezuela	Fresh fruit		c
	Alphonso	Philippines	Fresh fruit		e
	Alphonso	Philippines	Canned puree		e

[1] The fully ripe fruit.
[2] The ripe fruit stored in the deep-freeze compartment of a refrigerator at −15°C.
[3] The ripe fruit sliced and wrapped in polythene film and stored for 13 days at 10°C.
[4] The fruit kept at ambient temperature for 12 days after fully ripe.
[5] The fruit about 9 days before fully ripe.
[a] Hunter *et al.* (1974).
[b] MacLeod and Snyder (1985).
[c] MacLeod and de Troconis (1982a).
[d] Engel and Tressl (1983a).
[e] Yamaguchi *et al.* (1983).

Malaysian region. It has been cultivated in India for over 4000 years. It gradually spread into other parts of Southeast Asia such as Malaysia, Burma, and the Philippines. Mango trees were introduced to South America by the Portuguese, and were found growing in Florida in the mid-19th century (Wolfe, 1962). Over 50 countries cultivate mango today. Mango can be classified as one of the most important commercial fruits in the world, and especially so in India, Pakistan, and Brazil.

Only small numbers of products with artificial mango flavours have appeared on the market compared with those of banana or pineapple. This is mainly due to the lack of sufficient information on mango flavours.

2.1 Flavour Characteristics

Green mango fruits are astringent, sour, and even bitter, but they ripen rapidly and develop a unique and pleasant flavour. Ashraf et al. (1981) reported the role of pectinesterase in mango ripening, and suggested that some nonvolatile components broke down into flavour chemicals during the ripening stage. Gholap and Bandyopadhyay (1980) suggested that biogenesis of fatty acids was involved in the development of the characteristic aroma in the ripe fruits. Gholap and Bandyopadhyay (1976), who described the aroma profiles of some mango cultivars, reported that essences from Alphonso and Langra mangoes had green mango-like, estery, burnt sugar-like, and soily notes. Langra mango had, in addition, camphor-like, peach-like, and woody notes. Alphonso had unique almond-like and coconut-like notes. MacLeod and de Troconis (1982a) assessed the individual components of mango essence and found that car-3-ene gave an aroma of mango leaves and dimethylstyrene was the only other component providing a mango aroma. There is, however, no single chemical known to have a characteristic whole mango flavour. MacLeod and Snyder (1985), who studied the relative percentage abundances of some important volatile constituents of various mango cultivars, indicated that there could not be any typical aroma component listing nor typical aroma formulation for mango.

2.2 Hydrocarbons

Hydrocarbons found in mangoes are shown in Table VI.I. Monoterpene hydrocarbons, some of which play an important rôle in fruit flavours, are the major group of volatiles for all mango cultivars. MacLeod and de Troconis (1982a) found car-3-ene as the major component and the odour-contributing compound of Venezuelan mangoes. Myrcene, which has a fresh, green grass-like odour, was found to be the major constituent of two mango varieties, Alphonso and Baladi, from Egypt (Engel and Tressl, 1983a). Limonene, which is a main constituent of citrus oil, was also found in large quantities in Baladi but much less in Alphonso (Engel and Tressl, 1983a). α-Terpinolene, which has a floral, sweet, and pine-like aroma, comprised 37% of the total GC peak area of Alphonso mango fruit extract from the Philippines (Yamaguchi et al., 1983).

TABLE VI.II

Volatile esters found in mango

Compound	Cultivar	Source	Form	Odour quality	References
Methyl butanoate	Baladi	Egypt	Fruit pulp	Apple, banana, pineapple-like	d
Methyl pyruvate	Alphonso	India	Canned puree		a
Ethyl acetate	Alphonso/Baladi	Egypt	Fruit pulp	Ethereal, fruity, banana-like	d
	Alphonso	India	Canned puree		a
Ethyl 2-methylpropanoate	Baladi	Egypt	Fruit pulp		d
Ethyl butanoate	Alphonso/Baladi	Egypt	Fruit pulp	Ethereal, fruity, banana-like	d
An ethyl butenoate	Keitt	Florida	Fruit[1,2]	Estery, acidic	b
Ethyl 3-methylbutanoate	Alphonso/Baladi	Egypt	Fruit pulp	Slightly nutty, coconut-like	d
Ethyl 3-hydroxybutanoate	Alphonso	Philippines	Canned puree		e
	Alphonso/Baladi	Egypt	Fruit pulp		d
Ethyl hexanoate	Baladi	Egypt	Fruit pulp	Fruity, floral, wine-like	d
Ethyl 3-hydroxyhexanoate	Baladi	Egypt	Fruit pulp	Fresh, fruity, slightly green	d
Ethyl octanoate	Baladi	Egypt	Fruit pulp		d
	Keitt	Florida	Fruit[3]		b
Ethyl decanoate	Keitt	Florida	Fruit[3]	Floral, sweet, oily nut-like	b
	Baladi	Egypt	Fruit pulp		d
Ethyl undecanoate	Alphonso	India	Canned puree	Fatty, fruity	a
Ethyl dodecanoate	Alphonso/Baladi	Egypt	Fruit pulp	Slightly mango-like	d
	Keitt	Florida	Fruit[3]		b
Ethyl tetradecanoate	Alphonso/Baladi	Egypt	Fruit pulp	Oily, ethereal, violet-like	d
Ethyl hexadecanoate	Alphonso/Baladi	Egypt	Fruit pulp	Faint-waxy, mild-sweet	d
Butyl acetate	Alphonso/Baladi	Egypt	Fruit pulp	Ethereal, fruity, pear-like	d
Butyl butanoate	Alphonso	India	Canned puree	Fresh, sweet, fruity	a
	Baladi	Egypt	Fruit pulp		d
Butyl 3-hydroxybutanoate	Baladi	Egypt	Fruit pulp		d
2-Methylpropyl 3-hydroxy-butanoate	Baladi	Egypt	Fruit pulp		d
Butyl hexanoate	Baladi	Egypt	Fruit pulp	Heavy-vinous, fruity	d
2-Methylpropyl acetate	Alphonso/Baladi	Egypt	Fruit pulp	Banana-, pear-like	d
2-Methylpropyl butanoate	Alphonso	India	Canned puree	Ethereal, fruity	a
3-Methylbutyl acetate	Alphonso/Baladi	Egypt	Fruit pulp	Pungent, pear-like	d
	Alphonso	Egypt	Fruit pulp	Fruity, pear-, banana-like	d
3-Methylbutyl butanoate	Baladi	Egypt	Fruit pulp	Sweet, apricot-, banana-like	d

227

Compound	Variety	Origin	Source	Odour	Ref.
Hexyl acetate	Alphonso	Egypt	Fruit pulp	Sweet, berry-, pear-like	d
(Z)-Hex-3-enyl acetate	Alphonso/Baladi	Egypt	Fruit pulp	Green, sharp-fruity	d
(E)-Hex-3-enyl acetate	Alphonso	Egypt	Fruit pulp		d
(E)-Hex-3-eny acetate	Alphonso	Egypt	Fruit pulp	Fresh, fruity, banana-like	d
(Z)-Hex-3-enyl propanoate	Alphonso	Egypt	Fruit pulp	Green vegetable-like	d
(Z)-Hex-3-enyl butanoate	Alphonso	Philippines	Canned puree	Green, banana-, wine-like	e
(Z)-Hex-3-enyl 3-hydroxy-butanoate	Alphonso	Egypt	Fruit pulp		d
2-Phenethyl acetate	Alphonso	Egypt	Fruit pulp	Fruity, rose-, honey-like	d
A 2-phenethyl butenoate	Alphonso	Egypt	Fruit pulp	Dry, sweet, musty, rose-like	d
A (Z)-hex-3-enyl butanoate	Alphonso	Egypt	Fruit pulp		d
(Z)-Hex-3-enyl (E)-but-2-enoate	Alphonso	Egypt	Fruit pulp		d
A (Z)-hex-3-enyl pentenoate	Alphonso	Egypt	Fruit pulp		d
(Z)-Hex-3-enyl (E)-hex-2-enoate	Alphonso	Egypt	Fruit pulp		d
4-Butanolide	Alphonso/Baladi	Egypt	Fruit pulp	Sweet aromatic, caramel	d
	Alphonso	India	Canned puree		a
		Philippines	Canned puree		e
1-Methyl-4-butanolide	Alphonso	India	Canned puree		a
4-Pentanolide	Alphonso/Baladi	Egypt	Fruit pulp	Warm, hay-, tobacco-like	d
4-Hexanolide	Alphonso/Baladi	Egypt	Fruit pulp	Herbaceous, coumarin-like	d
	Alphonso	India	Canned puree		a
4-Heptanolide	Alphonso	India	Canned puree	Caramel, malt-, nut-like	a
4-Octanolide	Alphonso/Baladi	Egypt	Fruit pulp	Coconut-, tonka bean-like	d
	Alphonso	India	Canned puree		a
5-Octanolide	Alphonso/Baladi	Egypt	Fruit pulp		d
	Alphonso	India	Canned puree		a
4-Nonanolide	Alphonso/Baladi	Egypt	Fruit pulp	Creamy, coconut-like	d
	Alphonso	India	Canned puree		a
4-Decanolide	Alphonso/Baladi	Egypt	Fruit pulp	Creamy, fruity, peach-like	d
	Alphonso	India	Canned puree		a
5-Decanolide	Alphonso/Baladi	Egypt	Fruit pulp	Sweet, creamy, nut-like	a

1, 2, 3 See Table VI.1.
a Hunter et al. (1974).
b MacLeod and Snyder (1985).
d Engel and Tressl (1983a).
e Yamaguchi et al. (1983).

TABLE VI.III

Volatile aldehydes and ketones found in mango

Compound	Cultivar	Source	Form	Odour quality	References
Acetaldehyde	Alphonso	Philippines	Fresh fruit/canned puree	Pungent, ethereal, nauseating	e
Butanal	Alphonso	Philippines	Canned puree	Pungent, irritating	e
Hexanal	Alphonso	Egypt	Fruit pulp	Fatty, green grass-like	d
	Keitt	Florida	Fruit[1]		b
(E)-Hex-2-enal	Alphonso	Egypt	Fruit pulp	Green, fruity, vegetable-like	d
	Alphonso	Philippines	Fresh fruit		e
(E)-Non-2-enal	Baladi	Egypt	Fruit pulp	Penetrating, fatty, orris-like	d
(E, Z)-Nona-2,6-dienal	Alphonso/Baladi	Egypt	Fruit pulp	Powerful green vegetable-like	d
Furfural (2-furaldehyde)	Keitt	Florida	Fruit[1, 2, 4, 5]	Weak nutty, cold meat-, gravy-, coconut-like	b
	Alphonso	India	Canned puree		a
		Venezuela	Fresh fruit		c
	Alphonso/Baladi	Egypt	Fruit pulp		d
5-Methylfurfural	Alphonso	Philippines	Canned puree		e
	Alphonso	India	Canned puree	Warm, sweet, caramel	a
		Venezuela	Fresh fruit		c
Benzaldehyde	Alphonso	Egypt	Fruit pulp	Wood bark-, green twigs-like	d
		Venezuela	Fresh fruit		c
Phenylacetaldehyde	Alphonso	Egypt	Fruit pulp	Pungent green, floral, sweet	d
		Venezuela	Fresh fruit	Sweet, floral, wallflowers-like	c
	Alphonso	Philippines	Canned puree		e
β-Cyclocitral	Alphonso	Egypt	Fruit pulp		d
Viridifeoral	Baladi	Egypt	Fruit pulp		d
Acetone	Alphonso	Philippines	Fresh fruit/canned puree	Light ethereal, nauseating	e
Pentan-3-one	Baladi	Egypt	Fruit pulp	Ethereal, fruity	d
Heptan-2-one	Alphonso	Egypt	Fruit pulp	Spicy, cinnamon bark-like	d
Tridecan-2-one	Alphonso	Egypt	Fruit pulp		d
2-Acetylfuran	Alphonso	Philippines	Canned puree	Green, resinous, woody	e
	Alphonso	India	Canned puree		a
		Venezuela	Fresh fruit		c
2,5-Dimethyl-2H-furan-3-one	Alphonso	Philippines	Canned puree		e
	Alphonso	India	Canned puree		a

Compound	Cultivar	Country	Product	Odor	Ref.
2,5-Dimethyl-4-methoxy-2H-furan-3-one	Alphonso	Philippines	Canned puree		e
	Alphonso	Philippines	Fresh fruit		e
	Alphonso	India	Canned puree		a
	Alphonso/Baladi	Egypt	Fruit pulp		d
Acetophenone		Venezuela	Fresh fruit		c
4-Methylacetophenone	Alphonso	Philippines	Canned puree	Floral, bitter almond-like	e
β-Ionone	Alphonso	India	Canned puree	Warm, woody, violet-like	a
	Alphonso	Philippines	Canned puree		e
Damascenone	Baladi	Egypt	Fruit pulp		d
Isophorone (3,5,5-trimethyl-cyclohex-2-enone)	Alphonso	Philippines	Canned puree	Sweet camphor-like	e

1, 2, 3, 4, 5 See Table VI.I.

[a] Hunter et al. (1974).

[b] MacLeod and Snyder (1985).

[c] MacLeod and de Troconis (1982a).

[d] Engel and Tressl (1983a).

[e] Yamaguchi et al. (1983).

TABLE VI.IV

Volatile alcohols and acids found in mango

Compound	Cultivar	Source	Form	Odour quality	References
Ethanol	Keitt	Florida	Fruit[3, 4, 5]	Sickly, ethereal	b
2-Methylpropan-1-ol	Keitt	Florida	Fruit[3, 4]	Pungent	b
Butan-1-ol	Alphonso/Baladi	Egypt	Fruit pulp	Sweat, sickly	d
	Alphonso/Baladi	Egypt	Fruit pulp		d
	Alphonso	India	Canned puree		a
	Keitt	Florida	Fruit[1, 2, 3, 4]		b
Butan-2-ol	Alphonso/Baladi	Egypt	Fruit pulp	Oily, vinous	d
3-Methylbutan-1-ol	Keitt	Florida	Fruit[3]	Sweet, valeric	b
	Alphonso/Baladi	Egypt	Fruit pulp		d
	Alphonso	India	Canned puree		a
	Alphonso	Philippines	Canned puree		e
3-Methylbut-2-en-1-ol	Alphonso/Baladi	Egypt	Fruit pulp	Green, oily, gassy	d
2-Methylbut-3-en-1-ol	Alphonso/Baladi	Egypt	Fruit pulp		d
	Alphonso	Philippines	Fresh fruit/canned puree		e
2-Methylbut-3-en-2-ol	Alphonso/Baladi	Egypt	Fruit pulp	Oily, herbaceous, fungus-like	d
Pentan-1-ol	Alphonso/Baladi	Egypt	Fruit pulp	Harsh, fusel oil-like	d
Pent-3-en-2-ol	Alphonso	Philippines	Canned puree		e
Pentan-2-ol	Alphonso/Baladi	Egypt	Fruit pulp	Ethereal, wine-like	d
Hexan-1-ol	Alphonso/Baladi	Egypt	Fruit pulp	Fatty, fruity, wine-like	d
	Alphonso	Philippines	Canned puree		e
(Z)-Hex-3-en-1-ol	Alphonso/Baladi	Egypt	Fruit[1, 3]	Intensely green, gassy	d
	Keitt	Florida	Fruit[1, 3]	Fruity, green grass-like	b
	Alphonso	Philippines	Canned puree		e
(E)-Hex-3-en-1-ol	Alphonso/Baladi	Egypt	Fruit pulp	Intensely green, foliage-like	d
	Alphonso	Philippines	Canned puree		e
(E)-Hex-2-en-1-ol	Alphonso/Baladi	Egypt	Fruit pulp	Fruity green, caramel	d
Hexadecan-1-ol	Alphonso/Baladi	Egypt	Fruit pulp	Faint, sweet, oily	d
2-Phenylethanol	Alphonso	Egypt	Fruit pulp	Mild, warm, rose-, honey-like	d
	Alphonso	India	Canned puree		a
	Alphonso	Philippines	Canned puree		e
Furfuryl alcohol	Alphonso	Philippines	Canned puree	Warm oily, burning	e

Compound	Variety	Origin	Product	Odour	Ref.
α-Terpineol	Alphonso/Baladi	Egypt	Fruit pulp	Floral, lilac-like	d
	Alphonso	India	Canned puree		a
	Alphonso	Philippines	Canned puree		e
β-Terpineol	Baladi	Egypt	Fruit pulp	Pungent, woody, earthy	d
Linalool	Alphonso	Egypt	Fruit pulp	Refreshing, floral, woody	d
	Alphonso	India	Canned puree		a
	Alphonso	Philippines	Fresh fruit/canned puree		e
Terpinen-4-ol	Alphonso	India	Canned puree		a
(Z)-Carveol	Baladi	Egypt	Fruit pulp		d
(E)-Carveol	Baladi	Egypt	Fruit pulp		d
Nerol	Alphonso	Philippines	Canned puree	Sweet, refreshing, rose-like	e
p-Cymen-8-ol	Alphonso	Philippines	Canned puree		e
Globulol	Baladi	Egypt	Fruit pulp		d
Cubenol	Baladi	Egypt	Fruit pulp		d
α-Cadinol	Baladi	Egypt	Fruit pulp		d
δ-Cadinol	Baladi	Egypt	Fruit pulp		d
(Z)-Linalool oxide	Alphonso	India	Canned puree	Sweet woody	a
	Alphonso	Philippines	Canned puree	Floral	e
(E)-Linalool oxide	Alphonso	India	Canned puree		a
	Alphonso	Philippines	Canned puree		e
Selin-11-en-4-ol	Baladi	Egypt	Fruit pulp		d
4-Vinylphenol	Alphonso	Egypt	Fruit pulp		d
Acetic acid	Alphonso	India	Canned puree	Pungent, stinging-sour	a
	Alphonso	Philippines	Fresh fruit/canned puree		e
Butanoic acid	Alphonso	Philippines	Canned puree	Sour, rancid butter-like	e
2-Methylbutanoic acid	Alphonso	Philippines	Canned puree	Pungent, cheese-like	e
Pentanoic acid	Alphonso	Philippines	Canned puree	Animal perspiration-like	e
Hexanoic acid	Alphonso	Philippines	Canned puree	Fatty, rancid, sweat-like	e
Octanoic acid	Alphonso	Philippines	Canned puree	Burning, sour cheese-like	e

1, 2, 3, 4, 5 See Table VI.I.
[a] Hunter et al. (1974).
[b] MacLeod and Snyder (1985).
[d] Engel and Tressl (1983a).
[e] Yamaguchi et al. (1983).

Some sesquiterpenes were identified in relatively large amounts in some mangoes. For example, β-selinene made up 8.7% of the mango essence obtained from Venezuelan mango (MacLeod and de Troconis, 1982a). The floral note of mango may come from these terpenes.

2.3 Esters

Table VI.II shows the volatile esters found in mango. Esters may be the most odorous group of compounds found in natural products, and the low molecular weight esters have been used widely as ingredients of artificial fruit flavours (Arctander, 1969). Engel and Tressl (1983a) identified the complete series of ethyl esters of the even-numbered fatty acids from C_2 to C_{16} in the Baladi mango from Egypt. They also reported (Z)-hex-3-enyl esters, which have an intensely green note, in the same mango. The esters of unsaturated acids seem to be characteristic constituents of the Alphonso mango, because they were also found in that variety from the Philippines (Yamaguchi et al., 1983). Ethyl butanoate, which has a fruity, banana-like, odour, was reported as the major ester in Baladi mango, but was found in only trace levels or not at all in other varieties (Engel and Tressl, 1983a). The series of γ-lactones, which have a somewhat creamy flavour, was reported in Alphonso and Baladi mangoes (Hunter et al., 1974; Engel and Tressl, 1983a).

2.4 Aldehydes and Ketones

Table VI.III shows the volatile aldehydes and ketones found in mangoes. Certain aldehydes and ketones also play important rôles in flavours of other fruits. However, only a few aldehydes have been reported in mangoes. (E)-Hex-2-enal, which has a green vegetable-like odour, was found in Alphonso mango, but not in any other cultivars (Engel and Tressl, 1983a; Yamaguchi et al., 1983). Furfural (2-furaldehyde), which is a well known sugar degradation product, was found in almost all varieties. It was found in especially large amount in processed fruits, such as canned puree. For example, 18% of the canned puree extract of mango from the Philippines was furfural (Yamaguchi et al., 1983). β-Ionone, which is known to be naturally present in fruits, was reported only in canned puree (Hunter et al., 1974; Yamaguchi et al., 1983). Some furan derivatives related to sugar degradation products have also been found in processed mangoes. They are 2,5-dimethyl-2H-furan-3-one and 2,5-dimethyl-4-methoxy-2H-furan-3-one (Hunter et al., 1974; Yamaguchi et al., 1983), which are closely related to 2,5-dimethyl-4-hydroxy-2H-furan-3-one, which had been identified in pineapple and strawberry, and in the products from browning reactions.

2.5 Alcohols and Acids

Table VI.IV shows the volatile alcohols and acids found in mangoes. Ethanol has been reported only in the Keitt cultivar, but it is interesting that extracts obtained from the ripe mango stored for 13 days at 10°C consisted of 90% ethanol (MacLeod and Snyder, 1985). MacLeod and Snyder (1985) suggested that, in addition to ethanol, 2-methylpropan-1-ol and 3-methylbutan-1-ol could reasonably be considered as indicators that the fruit was 'going off'. Unsaturated C_6 alcohols, which have powerful green notes, have been found in Alphonso, Baladi, and Keitt cultivars (Engel and Tressl, 1983a; MacLeod and Snyder, 1985). They may contribute a fresh-green note to immature mangoes. Monoterpene alcohols, such as α-terpineol, linalool and nerol, have been reported in various mangoes, but the quantities found were much less than those of aliphatic or olefinic alcohols. These monoterpene alcohols are widely used in artificial flavours and fragrances because of their characteristic floral note (Arctander, 1969). Butanoic acid, which has a sour and rancid butter-like odour, accounted for 22% of the total GC peak area of an extract obtained from canned puree (Yamaguchi et al., 1983). Free organic acids have not been reported in large quantities from fruit extracts, except for the one reported by Yamaguchi et al. (1983). The presence of these organic acids in mangoes is, however, obvious. Some sour taste also indicates the presence of organic acids.

2.6 Miscellaneous Compounds

Table VI.V shows the miscellaneous compounds found in mango. Chlorinated hydrocarbons and dimethylformamide may be solvent contaminants. Epoxide derivatives of humulene and caryophyllene were reported in the mangoes from Egypt (Engel and Tressl, 1983a), but their odour qualities are unknown. Acetoin (3-hydroxybutanone) may contribute a creamy flavour to canned puree, because it was found in relatively large amounts. It accounted for 9% of the total GC peak area of the canned puree extract, and was the third largest constituent after butanoic acid (22%) and furfural (18%) (Yamaguchi et al., 1983).

3. PAPAYA

Papaya (*Carica papaya* L., or *Carica pubescens*) is a short-lived perennial tropical plant that originated in Central America. By the mid-19th century, papaya had spread over the tropical region, including Florida, India, Sri Lanka, Malaysia, Indonesia, Africa, and Hawaii. Papaya is one of the most important commercial fruits produced in Hawaii, and is air-freighted to the U.S. mainland and other parts of the world. The world production of papaya reached 2.74 million tons in 1986. This amount is still about one-fifth of that of mango production, but about equal to that of avocado and strawberry. World production is increasing gradually.

TABLE VI.V

Volatile miscellaneous compounds found in mango

Compound	Cultivar	Source	Form	Odour quality	References
Acetol (hydroxypropanone)	Alphonso	Philippines	Canned puree	Sweet, caramel	e
Acetoin (3-hydroxybutanone)	Alphonso	India	Canned puree	Creamy, fatty, butter-like	a
	Alphonso/Baladi	Egypt	Fruit pulp		d
	Alphonso	Philippines	Fresh fruit/canned puree		e
Humulene epoxide I	Alphonso/Baladi	Egypt	Fruit pulp		d
Humulene epoxide II	Alphonso/Baladi	Egypt	Fruit pulp		d
Caryophyllene epoxide	Alphonso/Baladi	Egypt	Fruit pulp		d
Trichloroethene		Venezuela	Fresh fruit		c
A trichloropropane	Alphonso	Philippines	Fresh fruit		e
Dimethylformamide	Alphonso	Philippines	Canned puree		e
Dimethyleneglycol monoethyl ether	Alphonso	Philippines	Canned puree		e
N-Methylpyrrolidone	Alphonso	Philippines	Canned puree	Pungent, nauseating	e
Pyridine	Alphonso	Philippines	Canned puree	Pungent, penetrating, fishy	e
(Methylthio)benzaldehyde	Alphonso	Egypt	Fruit pulp		d
Benzothiazole	Alphonso	Egypt	Fruit pulp		d

[a] Hunter et al. (1974).
[c] MacLeod and de Troconis (1982a).
[d] Engel and Tressl (1983a).
[e] Yamaguchi et al. (1983).

3.1 Flavour Characteristics

Fresh papaya fruit has a strong green, metallic note, which is generally due to linalool content. Idstein *et al.* (1985) described one fraction of papaya essence as a heavy, adherent floral-fruity odour, which was most likely to correspond to the original fruit odour. As with most fruits, the colour and aroma of papaya fruits develop parallel to the progress of ripening. A typical papaya aroma emanates from the fruit when it ripens halfway to yield a yellow colour. This development of aroma is in line with the changes in concentration of the various constituents, which are produced from the biogenesis of high molecular weight components such as carbohydrate, proteins, and lipids. MacLeod and Pieris (1983) reported that methyl benzoate was the only volatile constituent having a characteristic papaya aroma among the volatile constituents identified in papaya. They suggested that methyl butanoate is mainly, if not exclusively, responsible for the characteristic sweaty note of some papaya fruit.

3.2 Hydrocarbons

Table VI.VI shows the volatile hydrocarbons found in papayas. Very few monoterpenes, which can provide fruity aromas, have been reported. The quantities found were very low. A rather diverse group of hydrocarbons, both aliphatic and aromatic, has, however, been reported in papayas. The true sources of these hydrocarbons are not clear, even though they have been reported in many other fruits. It is possible that the fruits have been exposed to many kinds of petroleum solvents during growth and transportation (Chan *et al.*, 1973). Some hydrocarbons can be contaminants of an extracting solvent used to isolate volatiles from a fruit. Any hydrocarbons that could be significant to papaya flavour have not been reported.

3.3 Esters

Table VI.VII shows the volatile esters found in papaya. Papaya may be one of the most ester-rich fruits; more than 130 esters, including 66 methyl and 34 ethyl esters, have been reported. Flath and Forrey (1977) identified 32 esters in the Solo cultivar of papaya. They noted that the presence of a rather diverse group of esters in papaya is unusual in studies of fruit volatiles composition. Idstein and Schreier (1985) identified 103 esters in the mountain papaya. MacLeod and Pieris (1983) found only 15 esters, but they comprised 53% w/w of the extract from Sri Lankan papaya. On the other hand, Flath and Forrey (1977) identified 32 esters that comprised less than 1% of the volatiles isolated from Hawaiian papaya. This relatively low ester content in Hawaiian papaya is due to the presence of terpenoids in large amounts. MacLeod and Pieris (1983) reported that there are major differences between the main groups of volatile components of papaya from Hawaii and from Sri Lanka, which would be expected to provide quite different flavour characteristics to the fruits.

TABLE VI.VI

Volatile hydrocarbons found in papaya

Compound	Species	Source	Form	Odour quality	References
Hexane	C. papaya	Hawaii	Fresh fruit	Ethereal, kerosene-like	a
Methylcyclohexane	C. papaya	Sri Lanka	Fresh fruit	Fruity, sweet	b
Propylcyclohexane	C. papaya	Hawaii	Fresh fruit		a
Isopropylcyclohexane	C. papaya	Hawaii	Fresh fruit		a
Butylcyclohexane	C. papaya	Hawaii	Fresh fruit		a
Heptane	C. papaya	Hawaii	Fresh fruit		a
Decane	C. papaya	Hawaii	Fresh fruit		a
Tetradecane	C. pubescens	Chile	Fruit pulp	Pungent, choking, dry tarry	c
Toluene	C. papaya	Hawaii	Fresh fruit	Ethereal	a
	C. papaya	Sri Lanka	Fresh fruit		b
Ethylbenzene	C. papaya	Hawaii	Fresh fruit		a
	C. pubescens	Chile	Fruit pulp		c
Styrene	C. papaya	Sri Lanka	Fresh fruit		b
o-Xylene	C. papaya	Hawaii	Fresh fruit	Oily, fatty, pungent	a
	C. papaya	Sri Lanka	Fresh fruit		b
	C. pubescens	Chile	Fruit pulp		c
	C. papaya	Hawaii	Fresh fruit		e
p-Xylene	C. pubescens	Chile	Fruit pulp	Gassy, kerosene-like	c
	C. papaya	Hawaii	Fresh fruit		a
	C. pubescens	Chile	Fruit pulp		c
	C. papaya	Hawaii	Fresh fruit		e
1,2,3-Trimethylbenzene	C. pubescens	Chile	Fruit pulp	Aromatic, herbaceous	c
	C. papaya	Hawaii	Fresh fruit	Aromatic, herbaceous	a
1,3,5-Trimethylbenzene	C. papaya	Hawaii	Fresh fruit	Aromatic, herbaceous	a
1,2,4-Trimethylbenzene	C. papaya	Hawaii	Fresh fruit	Aromatic, herbaceous	a
Propylbenzene	C. papaya	Hawaii	Fresh fruit		a
Isopropylbenzene	C. papaya	Hawaii	Fresh fruit	Aromatic, herbaceous	a
Isopropenylbenzene	C. pubescens	Chile	Fruit pulp		c
1,2,4,5-Tetramethylbenzene	C. pubescens	Chile	Fruit pulp		c
m-Ethyltoluene	C. papaya	Hawaii	Fresh fruit		a
p-Ethyltoluene	C. papaya	Hawaii	Fresh fruit		a

Compound	Species	Origin	Part	Odor description	Ref
o-Ethyltoluene	*C. papaya*	Hawaii	Fresh fruit		a
Butylbenzene	*C. papaya*	Hawaii	Fresh fruit		a
2-Methylnaphthalene	*C. pubescens*	Chile	Fruit pulp		c
p-Cymene	*C. papaya*	Hawaii	Fresh fruit	Gassy, kerosene-like	a
β-Pinene	*C. papaya*	Hawaii	Fresh fruit	Dry woody, resinous, pine-like	a
α-Ocimene	*C. papaya*	Hawaii	Fresh fruit		a
(Z)-β-Ocimene	*C. papaya*	Sri Lanka	Fresh fruit	Warm, herbaceous, mint-like	b
Limonene	*C. papaya*	Hawaii	Fresh fruit	Sweet citrusy, lemon-like	a
Car-3-ene	*C. pubescens*	Chile	Fruit pulp	Sweet, diffusive, penetrating	c
β-Farnesene	*C. pubescens*	Chile	Fruit pulp	Mild, sweet, warm	c

[a] Flath and Forrey (1977).
[b] MacLeod and Pieris (1983).
[c] Idstein et al. (1985).

TABLE VI.VII

Volatile esters found in papaya

Compound	Species	Source	Form	Odour quality	References
Methyl propanoate	C. papaya	Hawaii	Fresh fruit	Ethereal, rum-like	a
Methyl methylpropanoate	C. papaya	Hawaii	Fresh fruit	Sweet apricot-, peach-like	a
Methyl butanoate	C. papaya	Hawaii	Fresh fruit	Apple, banana, pineapple-like	a
	C. pubescens	Sri Lanka	Fresh fruit		b
	C. pubescens	Colombia	Fresh fruit		d
	C. pubescens	Chile	Fruit pulp		c
Methyl but-2-enoate	C. papaya	Hawaii	Fresh fruit	Sharp green, fruity	a
	C. pubescens	Chile	Fruit pulp		c
Methyl 3-hydroxybutanoate	C. pubescens	Chile	Fruit pulp		c
Methyl 2-methylbutanoate	C. pubescens	Chile	Fruit pulp	Pungent, ethereal, fruity	c
Methyl 3-methylbutanoate	C. pubescens	Chile	Fruit pulp	Sweet, ethereal, apple-like	c
2-Methylpropyl acetate	C. papaya	Hawaii	Fresh fruit		a
Methyl pentanoate	C. papaya	Hawaii	Fresh fruit	Sweet, ethereal, apple-like	a
	C. pubescens	Chile	Fruit pulp		c
Methyl (E)-hex-2-enoate	C. pubescens	Chile	Fruit pulp	Green, musty, earthy, sweet	c
Methyl (Z)-hex-3-enoate	C. pubescens	Chile	Fruit pulp		c
Methyl 3-hydroxyhexanoate	C. pubescens	Chile	Fruit pulp	Ethereal, fruity, wine-like	c
Methyl hexanoate	C. papaya	Hawaii	Fresh fruit	Pineapple-, apricot-like	a
	C. papaya	Sri Lanka	Fresh fruit		b
	C. pubescens	Colombia	Fresh fruit		d
	C. pubescens	Chile	Fruit pulp		c
Methyl heptanoate	C. papaya	Hawaii	Fresh fruit	Fruity, green, waxy, berry-like	a
	C. pubescens	Colombia	Fresh fruit		d
	C. pubescens	Chile	Fruit pulp		c
Methyl octanoate	C. pubescens	Colombia	Fresh fruit		d
	C. pubescens	Chile	Fruit pulp		c
Methyl 3-hydroxyoctanoate	C. pubescens	Chile	Fruit pulp		c
Methyl (E)-oct-3-enoate	C. pubescens	Chile	Fruit pulp		c
Methyl (E)-oct-2-enoate	C. pubescens	Chile	Fruit pulp		c
Methyl nonanoate	C. pubescens	Chile	Fruit pulp	Fruity, green, foliage-like	c
	C. pubescens	Chile	Fruit pulp	Fruity, nut-, coconut-like	c
Methyl decanoate	C. papaya	Hawaii	Fresh fruit	Oily, slightly fruity, wine-like	a
	C. papaya	Sri Lanka	Fresh fruit		b
	C. pubescens	Chile	Fruit pulp		c

(continued)

Compound	Species	Origin	Tissue	Odour description	
Methyl dodecanoate	C. papaya	Sri Lanka	Fresh fruit	Oily, fatty, floral, wine-like	b
	C. pubescens	Chile	Fruit pulp		c
Methyl undecanoate	C. pubescens	Chile	Fruit pulp	Sweet, oily, wine-, brandy-like	c
Methyl tridecanoate	C. pubescens	Chile	Fruit pulp		c
Methyl tetradecanoate	C. papaya	Sri Lanka	Fresh fruit	Weakly oily, orris-like	b
Methyl pentadecanoate	C. pubescens	Chile	Fruit pulp		c
Methyl hexadecanoate	C. pubescens	Chile	Fruit pulp		c
Methyl benzoate	C. papaya	Hawaii	Fresh fruit	Pungent, heavy-sweet, floral	a
	C. papaya	Sri Lanka	Fresh fruit		b
	C. pubescens	Chile	Fruit pulp		c
Methyl furoate	C. papaya	Sri Lanka	Fresh fruit	Wine-, berry-like	b
	C. pubescens	Chile	Fruit pulp		c
Methyl 2-methylbenzoate	C. pubescens	Chile	Fruit pulp		c
Methyl 2-hydroxybenzoate	C. pubescens	Chile	Fruit pulp		c
Methyl phenylacetate	C. pubescens	Chile	Fruit pulp	Diffusive, honey, jasmin-like	c
Methyl 4-methoxybenzoate	C. pubescens	Chile	Fruit pulp	Sweet, herbaceous, lilac-like	c
Methyl nicotinate (methyl pyridine-3-carboxylate)	C. pubescens	Chile	Fruit pulp	Nauseating, sweet-herbaceous	c
Ethyl acetate	C. papaya	Sri Lanka	Fresh fruit	Mild, tobacco-like	b
	C. pubescens	Colombia	Fresh fruit	Ethereal, fruity, banana-like	d
	C. papaya	Hawaii	Fresh fruit	Brandy-like	a
	C. pubescens	Chile	Fruit pulp		c
Ethyl propanoate	C. pubescens	Colombia	Fresh fruit	Fruity, ethereal, rum-like	d
	C. pubescens	Chile	Fruit pulp		c
Ethyl 3-mercaptopropanoate	C. pubescens	Chile	Fruit pulp		c
Ethyl methylpropanoate	C. papaya	Hawaii	Fresh fruit	Sweet, ethereal, banana-like	a
Ethyl butanoate	C. papaya	Hawaii	Fresh fruit	Ethereal, fruity, banana-like	a
	C. pubescens	Colombia	Fresh fruit		d
	C. pubescens	Chile	Fruit pulp		c
Ethyl (E)-but-2-enoate	C. pubescens	Colombia	Fresh fruit	Sour, caramel, fruity	d
	C. papaya	Hawaii	Fresh fruit		a
	C. pubescens	Chile	Fruit pulp		c
Ethyl 2-methylbutanoate	C. papaya	Hawaii	Fresh fruit	Green, fruity, apple peel-like	a
	C. pubescens	Colombia	Fresh fruit		d
	C. pubescens	Chile	Fruit pulp		c
Ethyl 3-methylbutanoate	C. papaya	Hawaii	Fresh fruit		a
	C. pubescens	Chile	Fruit pulp	Slightly nutty, coconut-like	c
Ethyl 3-hydroxybutanoate	C. pubescens	Colombia	Fresh fruit		d
	C. pubescens	Chile	Fruit pulp		c

TABLE VI.VII (*continued*)

Compound	Species	Source	Form	Odour quality	References
Ethyl 3-acetoxybutanoate	*C. pubescens*	Chile	Fruit pulp	Ethereal, fruity, apple-like	c
Ethyl pentanoate	*C. pubescens*	Chile	Fruit pulp	Fruity, floral, wine-like	c
Ethyl hexanoate	*C. pubescens*	Colombia	Fresh fruit		d
Ethyl (*E*)-hex-2-enoate	*C. pubescens*	Chile	Fruit pulp		c
Ethyl (*Z*)-hex-3-enoate	*C. pubescens*	Chile	Fruit pulp		c
Ethyl 3-hydroxyhexanoate	*C. pubescens*	Chile	Fruit pulp		c
Ethyl heptanoate	*C. pubescens*	Colombia	Fresh fruit	Fruity, wine-, brandy-like	d
Ethyl octanoate	*C. pubescens*	Colombia	Fresh fruit	Apricot-, banana-, wine-like	d
	C. pubescens	Chile	Fruit pulp		c
Ethyl oct-2-enoate	*C. pubescens*	Colombia	Fresh fruit		d
Ethyl 3-hydroxyoctanoate	*C. pubescens*	Chile	Fruit pulp		c
Ethyl decanoate	*C. pubescens*	Chile	Fruit pulp	Sweet, nut-, brandy-like	c
Ethyl dodecanoate	*C. pubescens*	Chile	Fruit pulp	Faint-fruity, flower petal-like	c
Ethyl tetradecanoate	*C. pubescens*	Chile	Fruit pulp	Ethereal, orris-, violet-like	c
Ethyl hexadecanoate	*C. pubescens*	Chile	Fruit pulp	Faint-waxy, creamy	c
Ethyl benzoate	*C. papaya*	Hawaii	Fresh fruit	Blackcurrant-like	a
	C. pubescens	Colombia	Fresh fruit		d
	C. pubescens	Chile	Fruit pulp		c
Ethyl furoate	*C. pubescens*	Chile	Fruit pulp	Warm, fruity, plum-like	c
Ethyl 2-methylbenzoate	*C. pubescens*	Chile	Fruit pulp		c
Ethyl 2-hydroxybenzoate	*C. pubescens*	Chile	Fruit pulp		c
Ethyl benzylacetate	*C. pubescens*	Chile	Fruit pulp	Powerful-sweet, honey-like	c
Ethyl nicotinate (ethyl pyridine-3-carboxylate)	*C. pubescens*	Chile	Fruit pulp		c
Ethyl salicylate	*C. pubescens*	Colombia	Fresh fruit	Heavy, sweet, floral	d
Propyl acetate	*C. pubescens*	Chile	Fruit pulp	Ethereal, pear-, raspberry-like	c
Propyl propanoate	*C. papaya*	Hawaii	Fresh fruit	Apple-, pineapple-like	a
Propyl butanoate	*C. papaya*	Hawaii	Fresh fruit	Sweet, fruity, pineapple-like	a
2-Methylpropyl 2-methyl-butanoate	*C. pubescens*	Colombia	Fresh fruit	Heavy-fruity, pineapple-like	d
	C. papaya	Hawaii	Fresh fruit		a
2-Methylpropyl benzoate	*C. papaya*	Hawaii	Fresh fruit		a
Butyl acetate	*C. pubescens*	Colombia	Fresh fruit	Ethereal, fruity, pear-like	d
	C. pubescens	Chile	Fruit pulp		c

(continued)

Compound	Species	Location	Tissue	Odour description	Ref.
3-Methylbutyl acetate	C. papaya	Hawaii	Fresh fruit	Fresh, pear, banana-like	a
Butyl propanoate	C. pubescens	Colombia	Fresh fruit	Sweet, fruity, rum-like	d
Butyl methylpropanoate	C. pubescens	Chile	Fruit pulp		c
Butyl butanoate	C. pubescens	Chile	Fruit pulp		c
	C. papaya	Hawaii	Fresh fruit	Fresh, sweet, fruity	a
Butyl (E)-but-2-enoate	C. pubescens	Chile	Fruit pulp		c
	C. pubescens	Colombia	Fresh fruit		d
3-Methylbutyl 2-methyl-butanoate	C. papaya	Hawaii	Fresh fruit		a
Butyl 3-hydroxybutanoate	C. pubescens	Colombia	Fresh fruit		d
	C. pubescens	Chile	Fruit pulp		c
Butyl 2-methylbutanoate	C. pubescens	Chile	Fruit pulp		c
Butyl hexanoate	C. pubescens	Colombia	Fresh fruit	Heavy, vinous-fruity	d
	C. pubescens	Chile	Fruit pulp		c
Butyl (Z)-hex-3-enoate	C. pubescens	Chile	Fruit pulp		c
Butyl 3-hydroxyhexanoate	C. pubescens	Chile	Fruit pulp		c
Butyl octanoate	C. pubescens	Colombia	Fresh fruit	Fruity, green, oily, floral	d
	C. pubescens	Chile	Fruit pulp		c
	C. pubescens	Chile	Fruit pulp		c
Butyl 3-hydroxyoctanoate	C. pubescens	Colombia	Fresh fruit		d
Butyl oct-2-enoate	C. pubescens	Colombia	Fresh fruit		d
Butyl decanoate	C. pubescens	Chile	Fruit pulp	Heavy oily, banana-like	c
	C. pubescens	Chile	Fruit pulp		c
Butyl dodecanoate	C. pubescens	Chile	Fruit pulp	Mild-oily, virtually odourless	c
Butyl tetradecanoate	C. pubescens	Chile	Fruit pulp	Fatty, castor oil-, orris-like	c
Butyl benzoate	C. papaya	Hawaii	Fresh fruit	Mild floral, balsamic	a
	C. pubescens	Colombia	Fruit pulp		c
Butyl furoate	C. pubescens	Chile	Fruit pulp	Sweet, wine-, brandy-like	c
Butyl 2-hydroxybenzoate	C. pubescens	Chile	Fruit pulp	Sweet, rose-, honey-like	c
Butyl phenylacetate	C. pubescens	Chile	Fruit pulp	Mushroom-, tobacco-like	c
Butyl nicotinate (butyl pyridine-3-carboxylate)	C. pubescens	Chile	Fruit pulp		c
Butyl salicylate	C. pubescens	Colombia	Fresh fruit	Rough herbaceous	d
2-Methylbutyl acetate	C. pubescens	Chile	Fruit pulp		c
2-Methylbutyl butanoate	C. pubescens	Chile	Fruit pulp		c
3-Methylbutyl acetate	C. pubescens	Chile	Fruit pulp		c
	C. papaya	Hawaii	Fresh fruit		e

TABLE VI.VII (continued)

Compound	Species	Source	Form	Odour quality	References
3-Methylbutyl butanoate	C. pubescens	Chile	Fruit pulp		c
2-Methylbutyl decanoate	C. pubescens	Chile	Fruit pulp		c
2-Methylbutyl dodecanoate	C. pubescens	Chile	Fruit pulp		c
Pentyl acetate	C. pubescens	Chile	Fruit pulp		c
Pentyl butanoate	C. pubescens	Chile	Fruit pulp	Apricot-, pineapple-like	c
Pentyl (E)-but-2-enoate	C. pubescens	Chile	Fruit pulp		c
Pentyl 3-hydroxybutanoate	C. pubescens	Chile	Fruit pulp		c
Pentyl hexanoate	C. pubescens	Chile	Fruit pulp	Pungent, pineapple-like	c
Pentyl (Z)-hex-3-enoate	C. pubescens	Chile	Fruit pulp		c
Pentyl 3-hydroxyhexanoate	C. pubescens	Chile	Fruit pulp		c
Pentyl octanoate	C. pubescens	Chile	Fruit pulp		c
Hexyl acetate	C. pubescens	Colombia	Fresh fruit	Sweet, fruity, berry-, pear-like	d
	C. pubescens	Chile	Fruit pulp		c
Hexyl butanoate	C. pubescens	Chile	Fruit pulp	Powerful-fruity	c
Hexyl hexanoate	C. papaya	Hawaii	Fresh fruit	Fresh vegetable-like	e
	C. pubescens	Colombia	Fresh fruit	Oily herbaceous	d
Hexyl decanoate	C. pubescens	Chile	Fruit pulp		c
Octyl acetate	C. pubescens	Colombia	Fresh fruit	Waxy, floral, apple-like	d
(Z)-Hex-3-enyl pentanoate	C. papaya	Taiwan	Canned puree	Creamy, butter-, apple-like	e
Benzyl acetate	C. papaya	Taiwan	Canned puree	Floral, jasmin-, lily-like	e
	C. pubescens	Chile	Fruit pulp		c
Benzyl butanoate	C. papaya	Hawaii	Fresh fruit	Sweet, floral, jasmin-like	a
	C. papaya	Taiwan	Canned puree		e
	C. pubescens	Chile	Fruit pulp		c
2-Phenethyl acetate	C. pubescens	Chile	Fruit pulp	Sweet, rose-, honey-like	c
2-Phenethyl butanoate	C. pubescens	Chile	Fruit pulp	Warm, floral, rose-like	c
Linalyl acetate	C. pubescens	Chile	Fruit pulp	Sweet, bergamot-, pear-like	c
3-Methylhepta-2,6-dienyl acetate	C. pubescens	Chile	Fruit pulp		c
Benzyl but-2-enoate	C. papaya	Sri Lanka	Fresh fruit	Warm, herbaceous, mild spicy	b
γ-Hexalactone	C. papaya	Hawaii	Fresh fruit	Herbaceous, coumarin-like	a
	C. pubescens	Chile	Fruit pulp		c
	C. papaya	Hawaii	Fresh fruit		e
γ-Heptalactone	C. papaya	Hawaii	Fresh fruit	Malty, caramel, nut-like	a
	C. pubescens	Chile	Fruit pulp		c
		Hawaii	Fresh fruit		

Compound	Species	Origin	Part	Description	Ref
γ-Heptalactone	C. papaya	Hawaii	Fresh fruit	Malty, caramel, nut-like	a
δ-Heptalactone	C. pubescens	Chile	Fruit pulp		c
γ-Octalactone	C. pubescens	Chile	Fruit pulp		c
	C. papaya	Hawaii	Fresh fruit	Coconut-, tonka bean-like	a
δ-Octalactone	C. pubescens	Chile	Fruit pulp		c
	C. papaya	Sri Lanka	Fresh fruit		b
γ-Nonalactone	C. pubescens	Chile	Fruit pulp		c
	C. papaya	Hawaii	Fresh fruit	Creamy, coconut-like	a
δ-Nonalactone	C. pubescens	Chile	Fruit pulp		c
	C. pubescens	Chile	Fruit pulp	Fatty, oily, mild nut-like	c
γ-Decalactone	C. papaya	Hawaii	Fresh fruit	Creamy, fruity, peach-like	a
γ-Undecalactone	C. pubescens	Chile	Fruit pulp	Oily, fruity, peach-like	c

[a] Flath and Forrey (1977).
[b] MacLeod and Pieris (1983).
[c] Idstein et al. (1985).
[d] Morales and Duque (1987).
[e] Yamaguchi et al. (1983).

Although the composition of volatiles between the two cultivars differs greatly, they still have a strong, similar characteristic papaya aroma. Butyl acetate which has a fruity, pear-like odour, is the most abundant ester in mountain papaya from Chile (Idstein *et al.,* 1985). Methyl butanoate, which has an apple- or banana-like odour, is the major ester in papaya from Sri Lanka (MacLeod and Pieris, 1983). On the other hand, the ethyl butanoate content (GC peak area = 32.5%) was much larger than that of methyl butanoate (4.0%) in the mountain papaya from Colombia (Morales and Duque, 1987). Those low molecular weight esters play an important rôle in characteristic papaya flavour. Many lactones, which have a creamy, caramel and coconut-like flavour, have been found in various papayas (Table VI.VII). The somewhat creamy taste of papaya may be due to the presence of these lactones.

3.4 Aldehydes and Ketones

Volatile aldehydes and ketones found in papayas are shown in Table VI.VIII. Aldehyde content in papaya is rather low compared with other fruits. Acetaldehyde, which enhances the citrus note, was found only in Hawaiian papayas (Flath and Forrey, 1977). A series of methyl ketones has been reported, but their contribution to papaya flavour may be small. Butanedione, or diacetyl, which has an oily, butter-like odour, may give a creamy aroma to papaya, in addition to the lactones. The presence of certain aromatic aldehydes and ketones may be related to the unique component benzyl isothiocyanate, which will be discussed later.

3.5 Alcohols and Acids

Volatile alcohols and acids found in papayas are shown in Table VI.IX. Alcohols were the other group of important components observed in the papaya from Colombia, representing 30% of the total extract (Morales and Duque, 1987). Butan-1-ol, which has a sweet, fusel-like odour, comprised 25.7% of the extract of the mountain papaya from Colombia (Morales and Duque, 1987), but in the Solo variety from Hawaii it accounted for only 0.5% of the sample (Flath and Forrey, 1977). Linalool, which has a refreshing floral note, was the major component of Hawaiian papaya and comprised 68% of the extract. Linalool was reported in the other cultivars, but quantities were much less than that of Hawaiian papaya. For example, it was detected in only trace amounts (0.6%) in the Sri Lankan papaya (MacLeod and Pieris, 1983). The large content of linalool may be characteristic of the mountain papaya (*C. pubescens*). Acid content is very important because it gives well-balanced sour tastes to the fruits. Butanoic acid, which gives a sour, rancid butter-like taste, comprised 74% of the extract from canned papaya puree (Yamaguchi *et al.,* 1983). Yamaguchi *et al.* (1983) also found butanoic acid in large amounts (37%) in fresh fruit from Hawaii. Flath and Forrey (1977) found 16 acids as their methyl ester derivatives in Hawaiian papaya. Butanoic acid (73.8%) and hexanoic acid (19.0%) were the

TABLE VI.VIII

Volatile aldehydes and ketones found in papaya

Compound	Species	Source	Form	Odour quality	References
Acetaldehyde	C. papaya	Hawaii	Fresh fruit	Pungent, ethereal, nauseating	a
But-2-enal	C. pubescens	Chile	Fruit pulp		c
2-Methylbutanal	C. papaya	Sri Lanka	Fresh fruit	Choking, cocoa-, coffee-like	b
(E)-2-Methylbut-2-enal	C. pubescens	Chile	Fruit pulp		c
(E)-Pent-2-enal	C. pubescens	Chile	Fruit pulp		c
2-Methylpent-2-enal	C. papaya	Hawaii	Fresh fruit	Powerful gassy, green	a
Decanal	C. pubescens	Chile	Fruit pulp	Sweet, waxy, orange peel-like	c
Dodecanal	C. pubescens	Chile	Fruit pulp	Herbaceous, lily-, violet-like	c
Tetradecanal	C. pubescens	Chile	Fruit pulp	Faint fruity, citrus-like	c
Hexadecanal	C. pubescens	Chile	Fruit pulp	Mild floral	c
Benzaldehyde	C. papaya	Hawaii	Fresh fruit	Almond-, wood bark-like	a
	C. papaya	Sri Lanka	Fresh fruit		b
	C. pubescens	Chile	Fruit pulp		c
4-Methylbenzaldehyde	C. pubescens	Chile	Fruit pulp	Warm, mild floral, sweet, spicy	c
	C. pubescens	Colombia	Fresh fruit		d
Phenylacetaldehyde	C. papaya	Sri Lanka	Fresh fruit	Pungent, green, floral, sweet	b
	C. pubescens	Chile	Fruit pulp		c
Furfural (2-furaldehyde)	C. pubescens	Chile	Fruit pulp	Cold meat, gravy-, coconut-like	c
	C. papaya				c
Acetone	C. pubescens	Colombia	Fresh fruit	Light ethereal, nauseating	d
	C. papaya	Hawaii	Fresh fruit		e
	C. papaya	Hawaii	Canned puree		e
Butanone	C. papaya	Hawaii	Fresh fruit	Ethereal	a
Butenone	C. pubescens	Colombia	Fresh fruit	Sharp, pungent, irritating	d
Butanedione	C. pubescens	Chile	Fruit pulp		c
3-Acetoxybutanone	C. pubescens	Chile	Fruit pulp		c
Pentan-2-one	C. papaya	Hawaii	Fresh fruit	Sweet, ethereal, banana-like	a
Cyclohexanone	C. pubescens	Chile	Fruit pulp	Camphoraceous, mint-like	c
Hexan-2-one	C. pubescens	Chile	Fruit pulp	Fruity, cinnamon bark-like	c
Heptan-4-one	C. papaya	Hawaii	Fresh fruit	Fruity, cinnamon bark-like	a
	C. pubescens	Chile	Fruit pulp		c
	C. pubescens	Colombia	Fresh fruit		d

(continued)

TABLE VI.VIII (*continued*)

Compound	Species	Source	Form	Odour quality	References
Heptan-2-one	C. papaya	Hawaii	Fresh fruit	Ethereal, frtity, pungent	a
6-Methylhept-5-en-2-one	C. pubescens	Chile	Fruit pulp	Oily, green, banana-like	c
Octan-2-one	C. papaya	Hawaii	Fresh fruit	Floral, bitter, green	a
	C. pubescens	Chile	Fruit pulp		c
Oct-1-en-3-one	C. pubescens	Colombia	Fresh fruit		d
Nonan-2-one	C. pubescens	Chile	Fruit pulp	Fruity, floral, herbaceous	c
Decan-2-one	C. pubescens	Chile	Fruit pulp	Citrus, orange-like	c
Tridecan-2-one	C. pubescens	Chile	Fruit pulp	Warm, oily, nut-like	c
Tetradecan-2-one	C. pubescens	Chile	Fruit pulp		c
Acetophenone	C. pubescens	Chile	Fruit pulp	Sweet, floral, wallflower-like	c
3-Methylacetophenone	C. pubescens	Chile	Fruit pulp		c
2-Hydroxyacetophenone	C. pubescens	Chile	Fruit pulp	Sweet, heavy floral	c
Benzophenone	C. pubescens	Chile	Fruit pulp	Faint powdery, rose-like	c
Fenchone	C. pubescens	Chile	Fruit pulp		c
β-Damascenone	C. pubescens	Chile	Fruit pulp		c
β-Ionone	C. papaya	Hawaii	Fresh fruit	Warm, woody, violet-like	a
	C. pubescens	Chile	Fruit pulp		c
Dihydro-β-ionone	C. papaya	Hawaii	Canned puree		e
	C. papaya	Hawaii	Canned puree	Slightly violet-like	e

[a] Flath and Forrey (1977).
[b] MacLeod and Pieris (1983).
[c] Idstein *et al.* (1985).
[d] Morales and Duque (1987).
[e] Yamaguchi *et al.* (1983).

TABLE VI.IX

Volatile alcohols and acids found in papaya

Compound	Species	Source	Form	Odour quality	References
Methanol	C. papaya	Hawaii	Fresh fruit		a
Ethanol	C. papaya	Hawaii	Fresh fruit	Sickly, sweet, ethereal	a
	C. pubescens	Colombia	Fresh fruit		d
Propan-1-ol	C. papaya	Hawaii	Fresh fruit	Alcoholic, nauseating	a
Propan-2-ol	C. papaya	Hawaii	Fresh fruit	Alcoholic, ethereal	a
2-Methylpropan-1-ol	C. papaya	Hawaii	Fresh fruit	Pungent	a
Butan-1-ol	C. papaya	Hawaii	Fresh fruit	Sweet, sickly	a
	C. papaya	Hawaii	Fresh fruit		e
	C. pubescens	Chile	Fruit pulp		c
	C. pubescens	Taiwan	Canned puree		e
Butan-2-ol	C. papaya	Hawaii	Fresh fruit	Oily vinous	a
2-Methylbutan-1-ol	C. papaya	Hawaii	Fresh fruit	Sweet, valeric	a
2-Methylbutan-2-ol	C. papaya	Hawaii	Fresh fruit	Nauseating, camphor-like	a
3-Methylbutan-1-ol	C. papaya	Hawaii	Fresh fruit	Choking, cough provoking	a
	C. pubescens	Chile	Fruit pulp		c
Pentan-1-ol	C. papaya	Hawaii	Fresh fruit	Harsh, fusel oil-like	a
	C. pubescens	Chile	Fruit pulp		c
	C. pubescens	Taiwan	Canned puree		e
Pentan-2-ol	C. papaya	Hawaii	Fresh fruit	Ethereal, wine-like	a
	C. pubescens	Colombia	Fresh fruit		d
Pentan-3-ol	C. papaya	Hawaii	Fresh fruit		a
Pent-1-en-3-ol	C. papaya	Hawaii	Fresh fruit	Powerful gassy-green	a
Hexan-1-ol	C. papaya	Hawaii	Fresh fruit	Fatty, fruity, wine-like	a
	C. pubescens	Chile	Fruit pulp		c
	C. pubescens	Colombia	Fresh fruit		d
Hexan-2-ol	C. papaya	Hawaii	Fresh fruit		a
Hexan-3-ol	C. papaya	Hawaii	Fresh fruit		a
(E)-Hex-2-en-1-ol	C. papaya	Hawaii	Fresh fruit	Fruity green, caramel	a
(Z)-Hex-2-en-1-ol	C. papaya	Hawaii	Fresh fruit		a
Heptan-1-ol	C. papaya	Hawaii	Fresh fruit	Fresh, light green, nutty	a
Octan-1-ol	C. papaya	Hawaii	Fresh fruit	Powerful orange, rose-like	a
	C. pubescens	Chile	Fruit pulp		c

(continued)

TABLE VI.IX (*continued*)

Compound	Species	Source	Form	Odour quality	References
Oct-1-en-3-ol	C. papaya	Sri Lanka	Fresh fruit	Sweet, ethereal	b
	C. pubescens	Chile	Fruit pulp		c
(E)-2,6-Dimethyl-3,6-epoxy-oct-7-en-2-ol	C. papaya	Hawaii	Fresh fruit		a
(Z)-2,6-Dimethyl-3,6-epoxy-oct-7-en-2-ol	C. papaya	Hawaii	Fresh fruit		a
(E)-2,6-Dimethyl-3,6-epoxy-oct-7-en-3-ol	C. papaya	Hawaii	Fresh fruit		a
(Z)-2,6-Dimethyl-3,6-epoxy-oct-7-en-3-ol	C. papaya	Hawaii	Fresh fruit		a
	C. pubescens	Chile	Fruit pulp	Sweet, slightly fatty, oily	c
Benzyl alcohol	C. papaya	Hawaii	Fresh fruit	Faint sweet	a
	C. papaya	Sri Lanka	Fresh fruit		b
	C. pubescens	Chile	Fruit pulp		c
	C. pubescens	Colombia	Fresh fruit		d
	C. papaya	Taiwan	Canned puree		e
	C. papaya	Hawaii	Fresh fruit		e
2-Phenylethanol	C. papaya	Hawaii	Fresh fruit	Mild warm, rose-, honey-like	a
	C. papaya	Sri Lanka	Fresh fruit		b
	C. pubescens	Chile	Fruit pulp		c
	C. papaya	Taiwan	Canned puree		e
Eugenol	C. papaya	Taiwan	Canned puree	Warm, spicy, burning	e
Furfuryl alcohol	C. papaya	Hawaii	Fresh fruit	Floral, lilac-like	a
	C. pubescens	Chile	Fruit pulp		c
	C. papaya	Taiwan	Canned puree		e
Linalool	C. papaya	Sri Lanka	Fresh fruit		b
	C. pubescens	Chile	Fruit pulp		c
	C. papaya	Hawaii	Fresh fruit		e
	C. papaya	Taiwan	Canned puree		e
1,8-Cineol	C. pubescens	Chile	Fruit pulp	Fresh, camphor-like	c
	C. pubescens	Colombia	Fresh fruit		d
4-Terpineol	C. pubescens	Chile	Fruit pulp		c
Geraniol	C. papaya	Hawaii	Fresh fruit	Mild sweet, floral, rose-like	a
	C. papaya	Hawaii	Fresh fruit		e
	C. papaya	Taiwan	Canned puree		e

Compound	Species	Origin	Sample	Description	
Nerol	C. papaya	Taiwan	Canned puree	Sweet, refreshing, rose-like	e
Farnesol	C. pubescens	Chile	Fruit pulp	Mild sweet, oily	c
(Z)-Linalool oxide (furanoid)	C. papaya	Sri Lanka	Fresh fruit	Sweet, woody, floral, earthy	b
	C. pubescens	Chile	Fruit pulp		c
	C. papaya	Hawaii	Fresh fruit		e
(E)-Linalool oxide (furanoid)	C. papaya	Taiwan	Canned puree	Sweet, woody, floral, earthy	e
	C. papaya	Sri Lanka	Fresh fruit		b
	C. pubescens	Chile	Fruit pulp		c
	C. pubescens	Colombia	Fresh fruit		d
	C. papaya	Hawaii	Fresh fruit		e
	C. papaya	Taiwan	Canned puree		e
(Z)-Linalool oxide (pyranoid)	C. papaya	Hawaii	Fresh fruit	Sweet, woody, floral, earthy	e
	C. papaya	Sri Lanka	Fresh fruit		b
	C. papaya	Hawaii	Fresh fruit		e
(E)-Linalool oxide (pyranoid)	C. papaya	Taiwan	Canned puree	Sweet, woody, floral, earthy	e
	C. papaya	Hawaii	Fresh fruit		e
	C. papaya	Taiwan	Canned puree		e
Acetic acid	C. pubescens	Chile	Fruit pulp	Pungent, stingingly sour	c
Propanoic acid	C. papaya	Hawaii	Fresh fruit	Pungent, sour milk-like	a
Methylpropanoic acid	C. papaya	Hawaii	Fresh fruit	Powerful diffusive, sour	a
	C. papaya	Hawaii	Canned puree		e
Butanoic acid	C. papaya	Hawaii	Fresh fruit	Sour, rancid butter-like	a
	C. pubescens	Chile	Fruit pulp		c
	C. papaya	Hawaii	Fresh fruit		e
	C. papaya	Taiwan	Canned puree		e
Crotonic acid (but-2-enoic acid)	C. papaya	Hawaii	Fresh fruit		a
3-Methylbutanoic acid	C. papaya	Hawaii	Fresh fruit	Diffusive, cheese-like	a
	C. papaya	Taiwan	Canned puree		e
Pentanoic acid	C. papaya	Hawaii	Fresh fruit	Animal perspiration-like	a
	C. papaya	Hawaii	Fresh fruit		e
Hexanoic acid	C. papaya	Hawaii	Fresh fruit	Fatty, rancid, sweat-like	a
	C. papaya	Sri Lanka	Fresh fruit		b
	C. pubescens	Chile	Fruit pulp		c
	C. papaya	Taiwan	Canned puree		e
Hex-2-enoic acid	C. papaya	Hawaii	Fresh fruit	Sour, sweet, fatty	a
Heptanoic acid	C. papaya	Hawaii	Fresh fruit		a
	C. papaya	Taiwan	Canned puree		e

TABLE VI.IX (*continued*)

Compound	Species	Source	Form	Odour quality	References
Octanoic acid	C. *papaya*	Hawaii	Fresh fruit	Burning, sour cheese-like	a
	C. *papaya*	Sri Lanka	Fresh fruit		b
	C. *pubescens*	Chile	Fruit pulp		c
	C. *papaya*	Taiwan	Canned puree		e
	C. *papaya*	Hawaii	Fresh fruit		e
Nonanoic acid	C. *papaya*	Hawaii	Fresh fruit	Waxy, nut-, brandy-like	a
Decanoic acid	C. *papaya*	Hawaii	Fresh fruit	Sour, rancid fat-like	a
	C. *pubescens*	Chile	Fruit pulp		c
	C. *papaya*	Hawaii	Fresh fruit		e
	C. *papaya*	Taiwan	Canned puree		e
Dodecanoic acid	C. *papaya*	Hawaii	Fresh fruit	Fatty, waxy	a
Benzoic acid	C. *papaya*	Hawaii	Fresh fruit	Wine-like	a
Phenylacetic acid	C. *papaya*	Hawaii	Fresh fruit	Sweet honey-like	a
Salicylic acid	C. *papaya*	Hawaii	Fresh fruit		a

[a] Flath and Forrey (1977).
[b] MacLeod and Pieris (1983).
[c] Idstein et al. (1985).
[d] Morales and Duque (1987).
[e] Yamaguchi et al. (1983).

TABLE VI.X

Volatile miscellaneous compounds found in papaya

Compound	Species	Source	Form	Odour quality	References
Acetol (hydroxypropanone)	C. papaya	Taiwan	Canned puree	Sweet, caramel	e
Acetoin (3-hydroxybutanone)	C. pubescens	Chile	Fruit pulp	Creamy, fatty, butter-like	c
	C. papaya	Hawaii	Fresh fruit		e
	C. papaya	Taiwan	Canned puree		e
1-Ethoxy-1-methoxyethane	C. papaya	Hawaii	Fresh fruit		a
1,1-Diethoxyethane	C. papaya	Hawaii	Fresh fruit		a
1,1-Diethoxy-2-methylpropane	C. papaya	Hawaii	Fresh fruit		a
Pyridine	C. papaya	Sri Lanka	Fresh fruit	Pungent, penetrating, fishy	b
Dimethylformamide	C. papaya	Hawaii	Fresh fruit		e
	C. papaya	Taiwan	Canned puree		e
	C. papaya	Sri Lanka	Fresh fruit		b
Dichloromethane	C. papaya	Hawaii	Fresh fruit		a
Chloroform	C. papaya	Hawaii	Fresh fruit		a
1,2-Dibromoethane	C. papaya	Hawaii	Fresh fruit		a
Phenylacetonitrile	C. papaya	Hawaii	Fresh fruit	Herbaceous, green, floral	a
	C. papaya	Sri Lanka	Fresh fruit		b
	C. pubescens	Chile	Fruit pulp		c
	C. papaya	Taiwan	Canned puree		e
Methyl (methylthio)acetate	C. papaya	Sri Lanka	Fresh fruit		b
Benzothiazole	C. papaya	Hawaii	Fresh fruit		a
2-Methylthiophene	C. pubescens	Chile	Fruit pulp	Sulphurous	c
3-Methylthiophene	C. pubescens	Chile	Fruit pulp	Sulphurous	c
2,3-Dimethylthiophene	C. pubescens	Chile	Fruit pulp	Sulphurous	c
Benzyl isothiocyanate	C. papaya	Hawaii	Fresh fruit	Powerful cress-like, green, herbaceous, pungent	a
	C. papaya	Sri Lanka	Fresh fruit		b
	C. pubescens	Chile	Fruit pulp		c
	C. papaya	Hawaii	Fresh fruit		e
Methyl isothiocyanate	C. papaya	Hawaii	Fresh fruit		a
2-Phenethyl isothiocyanate	C. pubescens	Chile	Fruit pulp		c

[a] Flath and Forrey (1977).
[b] MacLeod and Pieris (1983).
[c] Idstein et al. (1985).
[d] Morales and Duque (1987).
[e] Yamaguchi et al. (1983).

major acids among those identified. The determination of free acid is, however, one of the most difficult procedures, because free acids do not pass through most conventional GC columns.

3.6 Miscellaneous Compounds

Table VI.X shows volatile miscellaneous compounds found in papayas. Some thiophene derivatives, which have a sulphurous odour, have been reported in mountain papaya from Chile (Idstein *et al.,* 1985). These sulphur-containing heterocyclic compounds had not been found in tropical fruits until recently (Idstein and Schreier, 1985). A unique feature of papaya is the presence of benzyl isothiocyanate (BITC), a well-known constituent of the Caricaceae (Tang *et al.,* 1972). BITC, which has a green, pungent, powerful cress-like odour, comprised 23% of the extract of papaya from Hawaii (Yamaguchi *et al.,* 1983). On the other hand, it was not found in the canned puree (Yamaguchi *et al.,* 1983). Tang (1971) reported that the BITC content in papaya depended on fruit maturity. It was highest in green, immature papaya, and had almost disappeared in ripened papaya. Green papayas are commonly used as salad in Southeast Asia. BITC is the major characteristic flavour component in this use. In addition, the high content of BITC in mature papaya seeds leads to the use of crushed papaya seeds as an ingredient of commercial papaya-seed salad dressing.

Tang (1971) also reported that immature papaya with a high BITC content exhibited stronger resistance to the fungal pathogen *Phytophthora parasitica* than did ripened papaya with a low BITC content. It is obvious that BITC plays an important rôle in the biochemical aspects of papaya.

4. PASSION FRUIT

Passion fruit, which originated in tropical America, has about 400 known species, of which some 30 bear edible fruit. All are probably indigenous to the American tropics, and many are known only to the native markets in South American countries, Mexico, and the West Indies. There are two major species, yellow (*Passiflora edulis flavicarpa*) and purple (*Passiflora edulis Sims*). Although in some locales the purple passion fruit is preferred for its flavour over the yellow passion fruit, the latter has been more productive in commercial plantings. Casimir *et al.* (1981) published a comprehensive review of passion fruit that includes volatile constituents reported up to 1981.

4.1 Flavour Characteristics

As in other fruits, there is no single compound that represents passion fruit aroma. Casimir and Whitfield (1978) reported an interesting result on the flavour impact values of passion fruit volatiles. According to their results,

TABLE VI.XI

Volatile hydrocarbons found in passion fruit

Compound	Cultivar	Source	Form	Odour quality	References
Pentane	Purple/yellow hybrid[1]	Taiwan	Fruit juice	Kerosene-like	d
Heptane	Purple	Australia	Fruit juice	Kerosene-like	a
	Purple/yellow hybrid	Taiwan	Fruit juice		d
Toluene	Purple	Australia	Fruit juice	Ethereal	a
o-Xylene	Hybrid	Taiwan	Fresh fruit/canned juice	Gassy, kerosene-like	f
m-Xylene	Hybrid	Taiwan	Fresh fruit/canned juice	Gassy, kerosene-like	f
p-Xylene	Hybrid	Hawaii	Fresh fruit/canned juice	Cold meat fat-like	f
1-Isopropyl-3-methylbenzene	Yellow	Hawaii	Fruit juice	Herbal, gassy, mint-like	c
1-Isopropenyl-4-methylbenzene	Yellow	Hawaii	Fruit juice	Citrus-, lemon-like	c
Naphthalene	Yellow	Hawaii	Fruit juice	Pungent, choking, dry tarry	c
1,1,6-Trimethyl-1,2-dihydro-naphthalene	Purple	Australia	Fruit juice		a
α-Pinene	Yellow	Hawaii	Fruit juice	Floral, pine-, cedar wood-like	c
Myrcene	Purple/yellow hybrid	Taiwan	Fruit juice	Fresh green grass-like	d
	Yellow	Hawaii	Fruit juice		c
	Yellow	Brazil	Fruit pulp		e
(E)-Ocimene	Purple	Australia	Fruit juice	Warm, herbaceous	a
	Yellow	Brazil	Fruit pulp		e
(Z)-Ocimene	Purple/yellow hybrid	Taiwan	Fruit juice	Warm, herbaceous	d
	Purple	Australia	Fruit juice		a
	Yellow	Brazil	Fruit pulp		e
Limonene	Purple/yellow hybrid	Taiwan	Fruit juice	Fresh, sweet, lemon-like	d
	Yellow	Hawaii	Fruit juice		c
	Yellow	Brazil	Fruit pulp		e
Car-3-ene	Yellow	Hawaii	Fruit juice	Sweet, diffusive, penetrating	c
α-Terpinene	Yellow	Brazil	Fruit pulp	Refreshing, lemon-, citrus-like	e
γ-Terpinene	Yellow	Brazil	Fruit pulp	Flat, dull, slightly citrus-like	e
Terpinolene	Yellow	Hawaii	Fruit juice		c
	Purple/yellow hybrid	Taiwan	Fruit juice	Floral, sweet, pine-like	d
	Yellow	Brazil	Fruit pulp		e
α-Humulene	Hybrid	Taiwan	Fresh fruit	Fresh, green, floral	f

[1] Hybrid F$_1$; P. edulis Sims. × P. edulis forma flavicarpa.
[a] Murray et al. (1972, 1973).
[c] Winter and Kloti (1972).
[d] Chen et al. (1982).
[e] Engel and Tressl (1983b).
[f] Yamaguchi et al. (1983).

6-(but-2'-enylidene)-1,5,5-trimethylcyclohex-1-ene and (Z)-hex-3-enyl butanoate contributed 30 and 11%, respectively, of the passion fruit flavour profile. Murray *et al.* (1972) pointed out that the biosynthesis of a number of the volatile flavouring constituents of passion fruit is associated with the production or degradation of carotenoid pigments. Those compounds are linalool, β-ionone, dihydro-β-ionone, and the lactone of 2-hydroxy-2,6,6-trimethylcyclohexylidene acetic acid. Engel and Tressl (1983b) found a pool of nonvolatile polar precursors of passion fruit volatiles, such as glycosides of monoterpene alcohols and hydroxylated linalool derivatives, which could be transformed into important aroma components by chemical or enzymic reactions.

4.2   Hydrocarbons

Table VI.XI shows the volatile hydrocarbons found in passion fruit. The low molecular weight aliphatic or aromatic hydrocarbons, which have a kerosene-like odour, may come from solvent contaminations. Their quantities are, however, not significant to the fruit flavours. α-Pinene, which has a floral, pine-, and cedar wood-like aroma, has been found only in the yellow passion fruit from Hawaii (Winter and Kloti, 1972). In addition to α-pinene, several other monoterpene hydrocarbons, such as (*E*)- and (*Z*)-ocimene, α- and γ- terpinene, limonene, myrcene, car-3-ene and terpinolene, have been reported in various passion fruits. These monoterpene hydrocarbons are important ingredients for artificial flavours and fragrances.

4.3   Esters

Table VI.XII shows the volatile esters found in passion fruits. Esters are the major group of volatiles found in passion fruits. Eight methyl esters and 29 ethyl esters have been reported in various passion fruits. Chen *et al.* (1982) identified 33 esters in the headspace of passion fruit juice from Taiwan. Ethyl butanoate, which has a fruity, banana-like aroma, was found as a major component in almost all varieties. Ethyl hexanoate, which possesses a floral, fruity, and wine-like aroma, is also one of the most abundant esters in passion fruit. Butyl acetate, butyl butanoate, hexyl butanoate, and hexyl hexanoate, which have somewhat mixed flavours of banana, pineapple, and pear, have also been found in all varieties of passion fruit. Hexyl butanoate was the major component of passion fruit from Taiwan, and it comprised 14% of the fruit extract (Yamaguchi *et al.,* 1983). The esters of $C_2 - C_6$ saturated and unsaturated acids give a floral, sweet, fruity flavour to passion fruits. Citronellyl acetate, which has a fresh rose-like aroma, is the only terpene ester so far found in passion fruit (Murray *et al.,* 1972). Eight γ-lactones and one δ-lactone have been reported, which give creamy and fruity flavours to passion fruits.

TABLE VI.XII

Volatile esters found in passion fruit

Compound	Cultivar	Source	Form	Odour quality	References
Methyl acetate	Purple	Australia	Fruit juice	Sweet, ethereal, diffusive	e
	Purple/yellow hybrid[1]	Taiwan	Fruit juice		d
Methyl butanoate	Purple/yellow hybrid	Taiwan	Fruit juice	Banana-, pineapple-like	d
Methyl hexanoate	Yellow	Hawaii	Fruit juice	Pineapple-, apricot-like	c
	Purple/yellow hybrid	Taiwan	Fruit juice		d
Methyl 3-hydroxyhexanoate	Purple	Australia	Fruit juice	Oily, fruity, wine-like	a
Methyl salicylate	Yellow	Hawaii	Fruit juice	Pungent, sweet, fruity	c
	Hybrid	Taiwan	Fresh fruit		f
Methyl 4-hydroxybenzoate					
Dimethyl carbonate	Yellow	Hawaii	Fruit juice		c
Methyl furoate	Yellow	Hawaii	Fruit juice	Berry-, wine-like	c
Ethyl acetate	Purple	Australia	Fruit juice	Ethereal, fruity, banana-, brandy-like	a
	Purple	New Guinea	Fruit juice		b
	Yellow	Hawaii	Fruit juice		c
	Purple/yellow hybrid	Taiwan	Fruit juice		d
Ethyl propanoate	Purple	Australia	Fruit juice	Fruity, ethereal, rum-like	a
	Purple/yellow hybrid	Taiwan	Fruit juice		d
Ethyl butanoate	Purple	Australia	Fruit juice	Ethereal, fruity, banana-like	a
	Purple	New Guinea	Fruit juice		b
	Yellow	Hawaii	Fruit juice		c
	Purple/yellow hybrid	Taiwan	Fruit juice		d
	Yellow	Brazil	Fruit pulp		e
	Hybrid	Taiwan	Fresh fruit		f
Ethyl (E)-but-2-enoate	Yellow	Hawaii	Fruit juice	Sour, caramel, fruity	c
	Purple/yellow hybrid	Taiwan	Fruit juice		d
Ethyl 3-hydroxybutanoate	Yellow	Hawaii	Fruit juice	Slightly nutty, coconut-like	c
	Hybrid	Taiwan	Fresh fruit		f
Ethyl acetoacetate	Yellow	Hawaii	Fruit juice	Sweet, fruity, fermented	c
Ethyl pentanoate	Purple	Australia	Fruit juice	Ethereal, fruity, apple-like	a
	Purple/yellow hybrid	Taiwan	Fruit juice		d

(continued)

TABLE VI.XII (*continued*)

Compound	Cultivar	Source	Form	Odour quality	References
Ethyl hexanoate	Purple	Australia	Fruit juice	Floral, fruity, wine-like	a
	Purple	New Guinea	Fruit juice		b
	Yellow	Hawaii	Fruit juice		c
	Purple/yellow hybrid	Taiwan	Fruit juice		d
	Yellow	Brazil	Fruit pulp		e
	Hybrid	Taiwan	Fresh fruit		f
Ethyl hex-2-enoate	Purple	Australia	Fruit juice		a
	Yellow	Hawaii	Fruit juice		c
	Purple/yellow hybrid	Taiwan	Fruit juice		d
Ethyl (Z)-hex-3-enoate	Purple	Australia	Fruit juice		a
	Yellow	Hawaii	Fruit juice		c
	Purple/yellow hybrid	Taiwan	Fruit juice		d
Ethyl 3-hydroxyhexanoate	Purple	Australia	Fruit juice		a
	Purple/yellow hybrid	Taiwan	Fruit juice		d
Ethyl heptanoate	Purple	Australia	Fruit juice	Fruity, green, berry-like	a
Ethyl (Z)-hept-4-enoate	Yellow	Hawaii	Fruit juice		c
Ethyl (E)-hept-4-enoate	Yellow	Hawaii	Fruit juice		c
Ethyl octanoate	Yellow	Hawaii	Fruit juice	Wine, apricot-, banana-like	c
	Purple/yellow hybrid	Taiwan	Fruit juice		d
Ethyl (Z)-oct-3-enoate	Purple	Australia	Fruit juice		a
	Purple/yellow hybrid	Taiwan	Fruit juice		d
Ethyl (E)-oct-3-enoate	Purple	Australia	Fruit juice		a
Ethyl (Z)-oct-4-enoate	Purple	Australia	Fruit juice		a
Ethyl (E)-oct-4-enoate	Yellow	Hawaii	Fruit juice		c
Ethyl (Z)-octa-4,7-dienoate	Purple	Australia	Fruit juice		a
	Purple/yellow hybrid	Taiwan	Fruit juice		d
Ethyl decanoate	Yellow	Hawaii	Fruit juice	Floral, oily, nut-like	c
Ethyl furoate	Yellow	Hawaii	Fruit juice	Warm, fruity, plum-like	c
Diethyl malonate	Yellow	Hawaii	Fruit juice	Slightly balsamic, apple-like	c
Diethyl carbonate	Yellow	Hawaii	Fruit juice		c
Ethyl cinnamate	Yellow	Hawaii	Fruit juice	Sweet, balsamic, honey-like	c
	Hybrid	Taiwan	Canned juice		f
Ethyl α-acetylcinnamate	Yellow	Hawaii	Fruit juice		c
2-Phenethyl acetate	Purple	Australia	Fruit juice	Fruity, rose-, honey-like	a

Compound	Variety	Origin	Product	Aroma	Ref.
2-Phenethyl butanoate	Yellow	Hawaii	Fruit juice	Dry, sweet, musty, rose-like	c
2-Phenethyl hexanoate	Yellow	Hawaii	Fruit juice	Fresh, rose-, pineapple-like	c
Propyl butanoate	Purple/yellow hybrid	Taiwan	Fruit juice	Sweet, fruity, pineapple-like	d
Propyl hexanoate	Purple	Taiwan	Fruit juice	Pineapple-, blackberry-like	d
Butyl acetate	Yellow	Hawaii	Fruit juice	Ethereal, fruity, pear-like	c
	Purple/yellow hybrid	Taiwan	Fruit juice		d
	Hybrid	Taiwan	Fresh fruit/canned juice		f
Butyl butanoate	Purple	Australia	Fruit juice	Fresh, sweet, fruity	a
	Purple	New Guinea	Fruit juice		b
	Yellow	Hawaii	Fruit juice		c
	Purple/yellow hybrid	Taiwan	Fruit juice		d
	Hybrid	Taiwan	Fresh fruit		f
Ethyl hexanoate	Purple	Australia	Fruit juice	Heavy, vinous-fruity	a
	Yellow	Taiwan	Fruit juice		d
2-Methylpropyl acetate	Purple/yellow hybrid	Taiwan	Fruit juice	Pineapple-, pear-like	d
1,2-Dimethylpropyl acetate	Purple/hybrid	Taiwan	Fruit juice	Pear-, banana-, apple-like	d
1-Methylbutyl butanoate	Purple	Australia	Fruit juice	Heavy sweet, banana-like	a
Pentyl hexanoate	Yellow	Taiwan	Fruit juice	Pungent, fruity, apple-like	d
3-Methylbutyl acetate	Yellow	Hawaii	Fruit juice	Fresh, pear-, banana-like	c
	Hybrid	Taiwan	Fresh fruit/canned juice		f
3-Methylbutyl hexanoate	Hybrid	Taiwan	Fresh fruit	Fresh, fruity, pineapple-like	f
Hexyl acetate	Yellow	Hawaii	Fruit juice	Sweet, berry-, pear-like	c
	Purple/yellow hybrid	Taiwan	Fruit juice		d
	Hybrid	Taiwan	Fresh fruit		f
Hexyl propanoate	Purple	Australia	Fruit juice	Sweet, earthly, herbaceous	a
Hexyl butanoate	Purple	Australia	Fruit juice	Powerful fruity	a
	Purple	New Guinea	Fruit juice		b
	Yellow	Hawaii	Fruit pulp		c
	Yellow	Brazil	Fruit juice		e
	Hybrid	Taiwan	Fresh fruit/canned juice		f
(Z)-Hex-3-enyl acetate	Purple	Australia	Fruit juice	Green, sharp fruity	a
	Yellow	Hawaii	Fruit juice		c
	Purple/yellow hybrid	Taiwan	Fruit juice		d
	Hybrid	Taiwan	Fresh fruit		f
(E)-Hex-3-enyl acetate	Purple	Australia	Fruit juice		a
(Z)-Hex-3-enyl butanoate	Purple	Australia	Fruit juice	Green, banana-, wine-like	a
	Purple	New Guinea	Fruit juice		b
	Purple/yellow hybrid	Taiwan	Fruit juice		d
	Hybrid	Taiwan	Fresh fruit		f

(continued)

258

TABLE VI.XII (continued)

Compound	Cultivar	Source	Form	Odour quality	References
(E)-Hex-3-enyl butanoate	Purple	Australia	Fruit juice	Fruity, wine-, cognac-like	a
(Z)-Hex-4-enyl butanoate	Purple	Australia	Fruit juice		a
(E)-Hex-4-enyl butanoate	Purple	Australia	Fruit juice		a
(Z)-Hexa-3,5-dienyl butanoate	Purple	Australia	Fruit juice		a
Hexyl hexanoate	Purple	Australia	Fruit juice	Fresh vegetable-like	a
	Purple	New Guinea	Fruit juice		b
	Yellow	Hawaii	Fruit juice		c
	Purple/yellow hybrid	Taiwan	Fruit juice		d
	Hybrid	Taiwan	Fresh fruit/canned juice		f
(Z)-Hex-3-enyl hexanoate	Purple	Australia	Fruit juice	Powerful fruity, green	a
	Purple	New Guinea	Fruit juice		b
	Yellow	Hawaii	Fruit juice		c
	Purple/yellow hybrid	Taiwan	Fruit juice		d
	Hybrid	Taiwan	Fresh fruit/canned juice		f
(E)-Hex-3-enyl hexanoate	Purple	Australia	Fruit juice		a
(Z)-Hex-4-enyl hexanoate	Purple	Australia	Fruit juice		a
(E)-Hex-4-enyl hexanoate	Purple	Australia	Fruit juice		a
1-Methylhexyl acetate	Purple	Australia	Fruit juice	Fruity, fatty, green	a
	Purple/yellow hybrid	Taiwan	Fruit juice		d
1-Methylhexyl butanoate	Purple	Australia	Fruit juice		a
	Purple	New Guinea	Fruit juice		b
	Purple/yellow hybrid	Taiwan	Fruit juice		d
	Purple/yellow hybrid	Taiwan	Fruit juice		d
1-Methylhexyl hexanoate	Purple	Australia	Fruit juice		a
	Purple	New Guinea	Fruit juice		b
	Purple/yellow hybrid	Taiwan	Fruit juice		d
Octyl acetate	Purple	Australia	Fruit juice	Fruity, fatty, apple-like	a
Octyl hexanoate	Yellow	Hawaii	Fruit juice	Fruity, mild herbaceous	c
Benzyl formate	Yellow	Hawaii	Fruit juice	Fruity, green, herbaceous	c
Benzyl acetate	Purple	Australia	Fruit juice	Floral, jasmin-, lily-like	a
	Purple	New Guinea	Fruit juice		b
	Yellow	Hawaii	Fruit juice		c
	Purple/yellow hybrid	Taiwan	Fruit juice		d
	Hybrid	Taiwan	Fresh fruit		f

Benzyl butanoate	Purple	Australia	Fruit juice	Sweet, floral, jasmin-like	a
	Yellow	Hawaii	Fruit juice		c
Benzyl methylpropanoate	Hybrid	Taiwan	Fresh fruit	Fruity, floral, berry-like	f
Benzyl hexanoate	Yellow	Hawaii	Fruit juice	Mild fruity, apricot-like	c
Citronellyl acetate	Purple	Australia	Fruit juice	Fresh rose-like	a
γ-Butyrolactone	Yellow	Hawaii	Fruit juice	Slightly caramel	c
2-Methyl-γ-butyrolactone	Yellow	Hawaii	Fruit juice		c
γ-Hexalactone	Purple	Australia	Fruit juice	Herbaceous, coumarin-like	a
	Yellow	Hawaii	Fruit juice		c
	Hybrid	Taiwan	Canned juice		f
γ-Heptalactone	Yellow	Hawaii	Fruit juice	Malty, caramel, nut-like	c
γ-Octalactone	Purple	Australia	Fruit juice	Coconut-, tonka bean-like	a
	Hybrid	Taiwan	Canned juice		f
δ-Octalactone	Yellow	Hawaii	Fruit juice		c
γ-Decalactone	Yellow	Hawaii	Fruit juice		c
	Hybrid	Taiwan	Fresh fruit/canned juice	Creamy, fruity, peach-like	f
γ-Undecalactone	Yellow	Hawaii	Fruit juice	Oily, fruity, peach-like	c
γ-Dodecalactone	Yellow	Hawaii	Fruit juice	Oily, pear, peach, plum-like	c
Lactone of 2,6,6-trimethyl-2-hydroxycyclohexylacetic acid	Purple	Australia	Fruit juice		a
	Yellow	Hawaii	Fruit juice		c

[1] See Table VI.XI.
[a] Murray et al. (1972, 1973).
[b] Parliment (1972).
[c] Winter and Kloti (1972).
[d] Chen et al. (1982).
[e] Engel and Tressl (1983b).
[f] Yamaguchi et al. (1983).

TABLE VI.XIII

Volatile aldehydes and ketones found in passion fruit

Compound	Cultivar	Source	Form	Odour quality	References
Acetaldehyde	Purple	Australia	Fruit juice	Pungent, ethereal, nauseating	a
Hexanal	Yellow	Taiwan	Fruit juice	Green grass-like	d
	Purple/yellow hybrid[1]	Taiwan	Fruit juice		d
	Yellow	Brazil	Fruit pulp		e
	Hybrid	Taiwan	Fresh fruit/canned juice		f
4-Hydroxy-3-methoxy benzaldehyde (vanillin)	Yellow	Hawaii	Fruit juice	Intensely sweet, vanilla-like	c
Furfural (2-furaldehyde)	Yellow	Hawaii	Fruit juice	Cold meat-, gravy-, coconut-like	c
	Yellow	Brazil	Fruit pulp		e
	Hybrid	Taiwan	Canned juice		f
5-Methylfurfural	Yellow	Hawaii	Fruit juice	Warm, sweet, caramel	c
Benzaldehyde	Yellow	Hawaii	Fruit juice	Green wood bark-, twigs-like	c
Acetone	Purple	Australia	Fruit juice	Light ethereal, nauseating	a
	Purple/yellow hybrid	Taiwan	Fruit juice		d
	Hybrid	Taiwan	Fresh fruit/canned juice		f
1-Phenylpropan-2-one	Yellow	Hawaii	Fruit juice	Green	c
4-(4-Hydroxyphenyl)-3-methyl-butan-2-one	Yellow	Hawaii	Fruit juice		c
Phenylbutadione	Yellow	Hawaii	Fruit juice	Faint balsamic, vanilla-like	c
Pentan-2-one	Purple	Australia	Fruit juice	Sweet, ethereal, banana-like	a
	Yellow	Hawaii	Fruit juice		c
	Purple/yellow hybrid	Taiwan	Fruit juice		d
Pentan-3-one	Yellow	Hawaii	Fruit juice	Ethereal, fruity	c
Penta-2,4-dione	Yellow	Hawaii	Fruit juice		c
Cyclopentanone	Yellow	Hawaii	Fruit juice		c
	Yellow	Taiwan	Fruit juice		d
	Hybrid	Taiwan	Fresh fruit		f
3,5,5-Trimethylcyclopent-1-enone	Hybrid	Taiwan	Canned juice		f
2,2,6-Trimethylcyclohexanone	Yellow	Hawaii	Fruit juice		c
2,2,6-Trimethyl-2-hydroxy-cyclohexanone	Hybrid	Taiwan	Canned juice		f
Heptan-2-one	Purple	Australia	Fruit juice	Fruity, cinnamon bark-like	a
	Yellow	Hawaii	Fruit juice		c

Compound	Type	Origin	Source	Aroma	Ref.
6-Methylhept-5-en-2-one	Purple	Australia	Fruit juice	Oily, green, banana-like	a
	Yellow	Hawaii	Fruit juice		c
Nonan-2-one	Purple	Australia	Fruit juice	Fruity, floral, herbaceous	a
	Yellow	Hawaii	Fruit juice		c
	Purple/yellow hybrid	Taiwan	Fruit juice		d
Nonan-3-one	Yellow	Hawaii	Fruit juice	Pungent, gassy, herbal	c
Nonan-4-one	Yellow	Hawaii	Fruit juice		c
Undecan-2-one	Purple	Australia	Fruit juice	Fruity, rose-, orange-like	a
1-Oxa-8-oxo-2,6,10,10-tetra-methylspiro[4,5]dec-6-ene	Yellow	Hawaii	Fruit juice		c
7,9,9-Trimethylbicyclo-[4.4.0]-deca-1,6-dien-3-one	Yellow	Hawaii	Fruit juice		c
5,5,9-Trimethylbicyclo-[4.4.0]-deca-1,9-dien-3-one	Yellow	Hawaii	Fruit juice		c
(E)-6,10-Dimethylundeca-5,9-dien-2-one	Yellow	Hawaii	Fruit juice		c
(Z)- and (E)-6,10-Dimethylun-deca-3,5,9-trien-2-ones	Yellow	Hawaii	Fruit juice		c
β-Ionone	Purple	Australia	Fruit juice	Warm, woody, violet-like	a
	Purple	New Guinea	Fruit juice		b
	Yellow	Hawaii	Fruit juice		c
	Purple/yellow hybrid	Taiwan	Fruit juice		d
	Hybrid	Taiwan	Fresh fruit/canned juice		f
Dihydro-β-ionone	Purple	Australia	Fruit juice	Slightly violet-like	a
	Yellow	Hawaii	Fruit juice		c
	Hybrid	Taiwan	Fresh fruit/canned juice		f
Acetylfuran	Yellow	Hawaii	Fruit juice	Green, resinous, woody	c
	Hybrid	Taiwan	Canned juice		f
2-Methyltetrahydrofuran-3-one	Hybrid	Taiwan	Canned juice		f
2,5-Dimethyl-2H-furan-3-one	Hybrid	Taiwan	Canned juice		f

[1] See Table VI.XI.
[a] Murray et al. (1972, 1973).
[b] Parliment (1972).
[c] Winter and Kloti (1972).
[d] Chen et al. (1982).
[e] Engel and Tressl (1983b).
[f] Yamaguchi et al. (1983).

4.4   Aldehydes and Ketones

Table VI.XIII shows the volatile aldehydes and ketones found in passion fruits. Only two aliphatic aldehydes have been reported. Hexanal, which has a green grass-like odour, was found in almost all cultivars. 4-Hydroxy-3-methoxybenzaldehyde (vanillin), with an intensely sweet and vanilla-like aroma, has only been reported in yellow passion fruit from Hawaii (Winter and Kloti, 1972). Furfural has been found in large quantity (21.09%) in canned juice from Taiwan (Yamaguchi *et al.,* 1983). On the other hand, it was not found in samples obtained from the headspace of fresh fruits (Chen *et al.,* 1982) or a solvent extract from fresh fruits (Yamaguchi *et al.,* 1983). Heat treatment during the canning process may produce furfural from sugars. Chan *et al.* (1973) reported that formation of sugar degradation products, such as acetylfuran and 2-methyltetrahydrofuran-3-one, caused an off-flavour in passion fruit products. It is interesting that a series of odd-numbered carbon alkan-2-ones ($C_{3, 5, 7, 9, 11}$) was found in passion fruits. C_6 ketones were not reported in any passion fruits, even though C_6 alcohols and esters were the most abundant volatiles in passion fruits. β-Ionone has been reported in many cultivars of passion fruit. It has a warm, woody, and violet-like aroma, and is used in licorice candy and in chewing gum (Arctander, 1969).

4.5   Alcohols and Acids

Table VI.XIV shows the volatile alcohols and acids found in passion fruits. The simplest alcohol, methanol, has been found only in purple passion fruit from Australia (Murray *et al.,* 1972). Methanol has been reported in only a few natural products. Ethanol was a major component of volatiles of passion fruit juice from Australia (Murray *et al.,* 1972). Chen *et al.* (1982) found, however, only trace amounts of ethanol in fruit juice from Taiwan, and they proposed that the failure of ethanol recovery was due to the poor retention ability of ethanol on Tenax-GC. Hexan-1-ol, which has a fatty, fruity, and wine-like flavour, has been reported in all types of passion fruits. It was one of the major constituents of the fruit juice extract of Taiwan passion fruit (Yamaguchi *et al.,* 1983). Hexanoic acid content (13%) was greatest in the same sample. Hexan-1-ol and hexanoic acid comprised over 20% of the extract of canned fruit juice from Taiwan (Yamaguchi *et al.,* 1983). The presence of hex-3-enols, which have an intensely green and foliage-like odour, may give the characteristic green aroma to passion fruit. Many terpene alcohols have been reported in passion fruits, but none of them has been found in significant amounts. A series of aliphatic acids has been reported, particularly in yellow passion fruit from Hawaii (Winter and Kloti, 1972). $C_2 - C_8$ acids may play some rôle in passion fruit flavour, but the odour of the acids with higher molecular weight than nonanoic acid becomes weaker and less important. These high molecular weight fatty acids may come from the wax layer of fruits.

TABLE VI.XIV

Volatile alcohols and acids found in passion fruit

Compound	Cultivar	Source	Form	Odour quality	References
Methanol	Purple	Australia	Fruit juice	Ethereal	a
Ethanol	Purple	Australia	Fruit juice	Sickly, sweet, ethereal	a
	Purple/yellow/hybrid	Taiwan	Fruit juice		d
Propan-1-ol	Yellow	Hawaii	Fruit juice	Alcoholic, nauseating	c
2-Methylpropan-1-ol	Yellow	Hawaii	Fruit juice	Pungent	c
	Purple/hybrid	Taiwan	Fruit juice		d
Butan-1-ol	Purple	Australia	Fruit juice	Sweet, sickly	a
	Yellow	Hawaii	Fruit juice		c
	Yellow	Brazil	Fruit pulp		e
	Hybrid	Taiwan	Fresh fruit/canned juice		f
Butane-2,3-diol	Yellow	Hawaii	Fruit juice		c
2-Methylbutan-1-ol	Yellow	Hawaii	Fruit juice	Sweet, valeric	c
3-Methylbutan-1-ol	Yellow	Hawaii	Fruit juice	Choking, cough provoking	c
2-Methylbut-3-en-2-ol	Yellow	Hawaii	Fruit juice	Oily, herbaceous, fungus-like	c
	Yellow	Brazil	Fruit pulp		e
3-Methylbut-2-en-1-ol	Yellow	Hawaii	Fruit juice	Green, oily, gassy	c
	Yellow	Brazil	Fruit pulp		e
	Hybrid	Taiwan	Canned juice		f
3-Methylbut-3-en-1-ol	Yellow	Hawaii	Fruit juice		c
Pentan-1-ol	Yellow	Hawaii	Fruit juice	Harsh, fusel oil-like	c
	Purple/yellow/hybrid	Taiwan	Fruit juice		d
		Taiwan	Fresh fruit/canned juice		f
Pentan-2-ol	Yellow	Hawaii	Fruit juice	Ethereal, wine-like	c
	Purple/yellow	Taiwan	Fruit juice		d
Cyclopentanol	Yellow	Hawaii	Fruit juice		c
Hexan-1-ol	Purple	Australia	Fruit juice	Fatty, fruity, wine-like	a
	Purple	New Guinea	Fruit juice		b
	Yellow	Hawaii	Fruit juice		c
	Purple/yellow/hybrid	Taiwan	Fruit juice		d
	Yellow	Brazil	Fruit pulp		e
	Hybrid	Taiwan	Fresh fruit/canned juice		f

(continued)

264

TABLE VI.XIV (*continued*)

Compound	Cultivar	Source	Form	Odour quality	References
(Z)-Hex-3-en-1-ol	Purple	Australia	Fruit juice	Intensely green, gassy	a
	Purple	New Guinea	Fruit juice		b
	Yellow	Hawaii	Fruit juice		c
	Yellow	Brazil	Fruit pulp		e
	Hybrid	Taiwan	Fresh fruit/canned juice		f
(E)-Hex-3-en-1-ol	Purple	Australia	Fruit juice	Intensively green, foliage-like	a
	Hybrid	Taiwan	Fresh fruit/canned juice		f
(E)-Hex-4-en-1-ol	Yellow	Hawaii	Fruit juice		c
Heptan-1-ol	Yellow	Hawaii	Fruit juice	Fresh, light green, nut-like	c
	Hybrid	Taiwan	Fresh fruit		f
Heptan-2-ol	Purple	New Guinea	Fruit juice	Fresh, herbaceous, lemon-like	b
	Purple/yellow/hybrid	Taiwan	Fruit juice		d
	Hybrid	Taiwan	Fresh fruit		f
Octan-1-ol	Purple	Australia	Fruit juice	Powerful, orange-, rose-like	a
	Yellow	Hawaii	Fruit juice		c
	Purple/yellow/hybrid	Taiwan	Fruit juice		d
	Hybrid	Taiwan	Fresh fruit		f
(Z)-Oct-3-en-1-ol	Purple	Australia	Fruit juice	Sweet, fermented	a
(E)-Oct-3-en-1-ol	Purple	Australia	Fruit juice	Sweet, fermented	a
Nonan-2-ol	Purple	Australia	Fruit juice	Powerful fruity green	a
	Purple	Taiwan	Fruit juice		d
(Z)-Non-2-en-1-ol	Yellow	Hawaii	Fruit juice	Diffusive, fatty, orris-like	c
Benzyl alcohol	Yellow	Hawaii	Fruit juice	Faint sweet	c
	Hybrid	Taiwan	Fresh fruit/canned juice		f
2-Phenylethanol	Yellow	Hawaii	Fruit juice	Mild, warm, rose-, honey-like	c
	Hybrid	Taiwan	Fresh fruit/canned juice		f
3-Phenylpropan-1-ol	Yellow	Hawaii	Fruit juice	Warm, mild balsamic, floral	c
Linalool	Purple	Australia	Fruit juice	Refreshing, floral, woody	a
	Yellow	Hawaii	Fruit juice		c
	Yellow	Brazil	Fruit pulp		e
	Hybrid	Taiwan	Fresh fruit		f
α-Terpineol	Purple	Australia	Fruit juice	Floral, lilac-like	a
	Yellow	Hawaii	Fruit juice		c
	Yellow	Brazil	Fruit pulp		e
	Hybrid	Taiwan	Fresh fruit		f

Compound	Colour	Origin	Sample	Odour description	Ref.
Terpinen-4-ol	Yellow	Hawaii	Fruit juice		c
4-Terpineol	Yellow	Hawaii	Fruit juice		c
	Yellow	Brazil	Fruit pulp		e
1,8-Cineol	Hybrid	Taiwan	Fresh fruit	Fresh, camphor-like	f
	Purple	Australia	Fruit juice		a
Citronellol	Yellow	Hawaii	Fruit juice	Fresh rose-like	c
Geraniol	Purple	Hawaii	Fruit juice	Mild sweet, floral, rose-like	a
	Purple	Australia	Fruit juice		a
	Yellow	Hawaii	Fruit juice		c
	Yellow	Brazil	Fruit pulp		e
	Hybrid	Taiwan	Fresh fruit		f
Hotrienol	Yellow	Brazil	Fruit pulp	Fresh, floral, lime-like	e
Myrcenol	Yellow	Brazil	Fruit pulp	Camphor-, lime-like	e
(Z)-Ocimenol	Yellow	Brazil	Fruit pulp	Camphor-, lime-like	e
(E)-Ocimenol	Yellow	Brazil	Fruit pulp	Fresh, sweet, rose-like	e
Nerol	Yellow	Brazil	Fruit pulp		e
Nerol oxide	Yellow	Brazil	Fruit juice		e
(Z)-Linalool oxide (furanoid)	Purple	Australia	Fruit juice	Sweet, woody, floral, earthy	a
	Yellow	Hawaii	Fruit juice		c
	Yellow	Brazil	Fruit pulp		e
	Hybrid	Taiwan	Canned juice		f
(E)-Linalool oxide (furanoid)	Purple	Australia	Fruit juice	Sweet, woody, floral, earthy	a
	Yellow	Hawaii	Fruit juice		c
	Yellow	Brazil	Fruit pulp		e
	Hybrid	Taiwan	Canned juice		f
(E)-Anhydrolinalool oxide A	Yellow	Brazil	Fruit pulp		e
(Z)-Anhydrolinalool oxide B	Yellow	Brazil	Fruit pulp		e
Furfuryl alcohol	Yellow	Hawaii	Fruit juice	Warm, oily, burning	c
Acetic acid	Purple	New Guinea	Fruit juice	Pungent, stingingly sour	b
	Hybrid	Taiwan	Canned juice		f
Methylpropanoic acid	Yellow	Hawaii	Fruit juice	Powerful diffusive, sour	c
	Hybrid	Taiwan	Canned juice		f
Butanoic acid	Purple	New Guinea	Fruit juice	Sour, rancid butter-like	b
	Hybrid	Taiwan	Canned juice		f
But-2-enoic acid	Yellow	Hawaii	Fruit juice		c
2-Methylbutanoic acid	Yellow	Hawaii	Fruit juice	Pungent, cheese-like	c
	Hybrid	Taiwan	Canned juice		f

(continued)

TABLE VI.XIV (continued)

Compound	Cultivar	Source	Form	Odour quality	References
Pentanoic acid	Yellow	Hawaii	Fruit juice	Animal perspiration-like	c
Hexanoic acid	Purple	New Guinea	Fruit juice	Rancid fat-, sweat-like	b
	Yellow	Hawaii	Fruit juice		c
A hexenoic acid	Purple	New Guinea	Fruit juice	Heavy, rancid fat-like	b
	Hybrid	Taiwan	Fresh fruit/canned juice		f
(Z)-Hex-3-enoic acid	Yellow	Hawaii	Fruit juice		c
Heptanoic acid	Yellow	Hawaii	Fruit juice		c
	Hybrid	Taiwan	Canned juice		f
(E)-Hept-3-enoic acid	Yellow	Hawaii	Fruit juice		c
Octanoic acid	Purple	New Guinea	Fruit juice	Burning, sour cheese-like	b
	Yellow	Hawaii	Fruit juice		c
An octenoic acid	Purple	New Guinea	Fruit juice		b
	Hybrid	Taiwan	Canned juice		f
(Z)-Oct-3-enoic acid	Yellow	Hawaii	Fruit juice		c
(E)-Oct-3-enoic acid	Yellow	Hawaii	Fruit juice		c
Nonanoic acid	Yellow	Hawaii	Fruit juice	Waxy, nut-, brandy-like	c
Decanoic acid	Yellow	Hawaii	Fruit juice	Fatty, waxy	c
Dodecanoic acid	Yellow	Hawaii	Fruit juice	Fatty, waxy, refreshing	c
Tetradecanoic acid	Yellow	Hawaii	Fruit juice	Faint waxy, oily	c
Pentadecanoic acid	Yellow	Hawaii	Fruit juice		c
Hexadecanoic acid	Yellow	Hawaii	Fruit juice	Virtually odourless	c
Benzoic acid	Yellow	Hawaii	Fruit juice	Urine-like	c
Phenylacetic acid	Yellow	Hawaii	Fruit juice	Sweet, animal honey-like	c
Cinnamic acid	Yellow	Hawaii	Fruit juice	Very faint honey-like	c
Furoic acid	Yellow	Hawaii	Fruit juice	Mild caramel	c

[1] See Table VI.XI.
[a] Murray et al. (1972, 1973).
[b] Parliment (1972).
[c] Winter and Kloti (1972).
[d] Chen et al. (1982).
[e] Engel and Tressl (1983b).
[f] Yamaguchi et al. (1983).

4.6 Miscellaneous Compounds

Table VI.XV shows the miscellaneous volatile compounds found in passion fruits. 4-Allyl-2-methoxyphenol (eugenol), which has a strong clove oil-like odour, has only been reported in the yellow passion fruit from Hawaii (Winter and Kloti, 1972). A series of other phenol derivatives was also found in the same sample, but these may be implicated in off-flavour of fruits.

Casimir *et al.* (1981) described the details of ionone-related flavour compounds, such as edulans I, II, III, IV and megastigmatrienes, in their review. Edulans I and II have strongly floral, rose-like aromas and make a specific contribution to the flavour of fresh passion fruit juice. Some isomers of megastigmatrienes have a rose-like or raspberry-like odour in dilute aqueous solution. These compounds are not listed in Table VI.XV.

5. GUAVA

Guava (*Psidium guajava* L.) is a native of Central America. It was distributed in Asia by the early 17th century and is now cultivated in nearly 60 countries. Guava is particularly important in Florida, Hawaii, Egypt, South Africa, Brazil, Colombia, and the West Indies. Guava production is still much lower than that of major tropical fruits, but it is increasing gradually. In addition to consumption as fresh fruit, guava has been used in many different ways: jellies, jams, cheese, ketchup, puree, juice powder, nectar, and juices. Mature guavas do not keep well, so they are processed soon after ripening.

5.1 Flavour Characteristics

Fresh guava fruit has a quince, banana-like aroma (Herrmann, 1983). Wilson and Shaw (1978) suggested that β-caryophyllene plays an important rôle in guava aroma. Later, MacLeod and de Troconis (1982b) also found β-caryophyllene in guava, and reported its odour quality as guava-like and slightly burnt. They also reported that 2-methylpropyl acetate, hexyl acetate, and benzaldehyde had guava-like odour among 40 volatiles identified in guava. Stevens *et al.* (1970) suggested that β-ionone, which has a low odour threshold and an intense violet aroma, contributed floral flavour to guava puree. They also isolated several aromatic compounds such as methyl benzoate, 2-phenethyl acetate, methyl cinnamate, and cinnamyl acetate, which probably contribute a pleasant fruity aroma to the fruits. Cinnamyl acetate reportedly has the most guava-like aroma. On the other hand, MacLeod and de Troconis (1982b) did not find any of these aromatic compounds in an essence of fresh guava fruit. They reported myrcene as a constituent with a significant guava character.

TABLE VI.XV

Miscellaneous volatile compounds found in passion fruit

Compound	Cultivar	Source	Form	Odour quality	References
Acetol (hydroxypropanone)	Hybrid	Taiwan	Canned juice	Sweet, caramel	f
Acetoin (3-hydroxybutanone)	Yellow	Hawaii	Fruit juice	Creamy, fatty, butter-like	c
2-Ethoxyethanol	Hybrid	Taiwan	Fresh fruit/canned juice	Faint earthy, mushroom-like	f
Diethyleneglycol monoethyl-ether	Yellow	Hawaii	Fruit juice	Faint, dry musty, ethereal	c
Phenol	Yellow	Hawaii	Fruit juice	Medicinal	c
2-Methylphenol	Yellow	Hawaii	Fruit juice		c
3-Methylphenol	Yellow	Hawaii	Fruit juice	Dry tarry, medicinal, leathery	c
4-Methylphenol	Yellow	Hawaii	Fruit juice	Tarry, smoky, medicinal	c
4-Ethylphenol	Yellow	Hawaii	Fruit juice	Woody, phenolic, medicinal	c
4-Allylphenol	Yellow	Hawaii	Fruit juice	Dry, tarry, medicinal	c
2,4-Dimethylphenol	Yellow	Hawaii	Fruit juice		c
2,4,5-Trimethylphenol	Yellow	Hawaii	Fruit juice		c
2-Methyoxyphenol	Yellow	Hawaii	Fruit juice	Sweet, powerful smoke-like	c
4-Allyl-2-methoxyphenol	Yellow	Hawaii	Fruit juice	Warm, spicy, clove oil-like	c
1-(3,4,5-Trimethoxyphenyl)-prop-2-ene	Yellow	Hawaii	Fruit juice		c
2,6,6-Trimethyl-2-vinyltetra-hydropyran	Yellow	Brazil	Fruit pulp		e
2,2-Dimethyl-5-(1-methylprop-enyl)tetrahydrofuran	Yellow	Brazil	Fruit pulp		e
1-Oxa-2,6,10,10-tetramethyl-spiro[4,5]dec-6-ene	Yellow	Hawaii	Fruit juice		c
Dimethylmaleic anhydride	Yellow	Hawaii	Fruit juice	Diffusive, gassy green	c
(Z)-Rose oxide	Yellow	Hawaii	Fruit juice		c
2-Methylquinoxaline	Yellow	Hawaii	Fruit juice		c
Ethyl 3-methylthiopropanoate	Yellow	Hawaii	Fruit juice		c
4-Methyl-5-vinylthiazole	Yellow	Hawaii	Fruit juice		c
3-Methylthiohexan-1-ol	Yellow	Hawaii	Fruit juice		c
2-Methyl-4-propyl-1,3-oxa-thiane	Yellow	Hawaii	Fruit juice		c
N,N-Dimethylformamide	Hybrid	Taiwan	Fresh fruit		f

[1] See Table VI.XI.

[e] Engel and Tressl (1983b).

TABLE VI.XVI

Volatile hydrocarbons found in guava

Compound	Source	Form	Odour quality	References
Pentane	South Africa	Puree	Kerosene-like	e
Hexane	South Africa	Puree	Kerosene-like	e
Methylcyclohexane	Brazil	Fruit pulp		a
3,5-Dimethylheptane	South Africa	Puree		e
Octane	Brazil	Fruit pulp		a
	Venezuela	Fruit pulp		b
Nonane	Brazil	Fruit pulp		a
Decane	Brazil	Fruit pulp		a
	Venezuela	Fruit pulp		b
Undecane	Brazil	Fruit pulp		a
Dodecane	Brazil	Fruit pulp		a
Tridecane	Brazil	Fruit pulp		a
Tetradecane	Brazil	Fruit pulp		a
Pentadecane	Brazil	Fruit pulp		a
Hexadecane	Brazil	Fruit pulp		a
Benzene	South Africa	Puree	Choking, diffusive, gassy	e
Toluene	Brazil	Fruit pulp	Ethereal	a
Ethylbenzene	Brazil	Fruit pulp	Sweet, gassy	a
p-Xylene	Brazil	Fruit pulp	Fatty, cold meat-like	a
Propylbenzene	Brazil	Fruit pulp		a
Vinylbenzene	Brazil	Fruit pulp	Extremely diffusive, sweet, gassy	a
1-Ethyl-2-methylbenzene	Brazil	Fruit pulp		a
Butylbenzene	Brazil	Fruit pulp		a
2,5-Dimethylstyrene	Brazil	Fruit pulp		a
p-Cymene	Brazil	Fruit pulp	Dry woody, resinous, pine-like	a
β-Myrcene	Brazil	Fruit pulp	Fresh green grass-like, warm, balsamic	a
	Venezuela	Fruit pulp		b
Limonene	Brazil	Fruit pulp	Fresh, sweet, lemon-like	a
	Venezuela	Fruit pulp		b
	Hawaii	Puree		c
	Taiwan	Puree		d
	South Africa	Puree		e
	Florida	Puree		f
(Z)-Ocimene	Brazil	Fruit pulp	Warm, herbaceous	a
α-Pinene	Taiwan	Peel/leaf/puree	Floral, resinous, pine-, cedar wood-like	d
β-Pinene	Florida	Puree	Dry woody, resinous, pine-like	f
Terpinolene	Brazil	Fruit pulp	Floral, sweet, pine-like	a
α-Copaene	Venezuela	Fruit pulp		b
β-Copaene	Florida	Puree		f
α-Cubebene	Taiwan	Puree		d

(continued)

TABLE VI.XVI (*continued*)

Compound	Source	Form	Odour quality	References
β-Caryophyllene	Venezuela	Fruit pulp	Floral, dry woody, clove leaf oil-like	b
	Hawaii	Puree		c
	Taiwan	Peel/puree		d
Farnesene	Florida	Puree	Mild sweet, warm	f
Aromadendrene	Taiwan	Peel/puree		d
α-Humulene	Venezuela	Fruit pulp	Fresh, green, floral	b
	Taiwan	Peel/puree		d
	Florida	Puree		f
β-Humulene	Florida	Puree	Fresh, green, floral, clove-like	f
β-Bisabolene	Venezuela	Fruit pulp	Warm, sweet, spicy, balsamic	b
	Taiwan	Peel/puree		d
	Florida	Puree		f
α-Selinene	Venezuela	Fruit pulp	Pepper-like	b
	Florida	Puree		f
β-Selinene	Florida	Puree	Floral, woody, almond-like	f
δ-Cadinene	Taiwan	Puree	Mild, dry woody	d
	Florida	Puree		f
γ-Muurolene	Taiwan	Puree		d
Curcumene	Florida	Puree		f

[a] Idstein and Schreier (1985).
[b] MacLeod and de Troconis (1982b).
[c] Stevens *et al.* (1970).
[d] Shiota (1978).
[e] Torline and Ballschmieter (1973).
[f] Wilson and Shaw (1978).

5.2 Hydrocarbons

Table VI.XVI shows the volatile hydrocarbons found in guava. Low molecular weight aliphatic and aromatic hydrocarbons have been reported, but the source of these compounds was not clear, and may be solvent contamination. Limonene, which has a fresh, sweet, and lemon-like aroma, was found in many varieties of guavas. Wilson and Shaw (1978) determined 11 terpene hydrocarbons in Florida guava puree extract. They found that β-caryophyllene was by far the largest single component in the hydrocarbon fraction obtained by thin layer chromatography. β-Caryophyllene and limonene comprised over 95% of the total hydrocarbons present in the same sample. However, MacLeod and de Troconis (1982b) found slightly more α-humulene than β-caryophyllene, but neither of them was a major component of the essence obtained from Venezuelan guava. Another sesquiterpene hydrocarbon, bisabolene, has also been reported in many cultivars of guava. This compound has received much

TABLE VI.XVII

Volatile esters found in guava

Compound	Source	Form	Odour quality	References
Methyl acetate	South Africa	Puree	Sweet, ethereal, diffusive	e
Methyl propanoate	South Africa	Puree	Ethereal, rum-like	e
Methyl butanoate	South Africa	Puree	Apple, banana-, pine-apple-like	e
Methyl 4-hydroxybutanoate	Brazil	Fruit pulp		a
Methyl hexanoate	Brazil	Fruit pulp	Ethereal, pineapple-, apricot-like	a
Methyl (Z)-hex-3-enoate	Brazil	Fruit pulp		a
Methyl octanoate	Brazil	Fruit pulp	Fruity, wine-, orange-like	a
Methyl 2-furoate	Brazil	Fruit pulp	Wine-, berry-like	a
Methyl benzoate	Brazil	Fruit pulp	Pungent, heavy sweet, floral	a
	Hawaii	Puree		c
Methyl (Z)-cinnamate	Brazil	Fruit pulp	Balsamic, strawberry-like	a
Methyl (E)-cinnamate	Brazil	Fruit pulp	Balsamic, strawberry-like	a
	Hawaii	Puree		c
Methyl hexadecanoate	Brazil	Fruit pulp		a
Methyl non-3-enoate	Taiwan	Puree		d
Methyl nicotinoate (methyl pyridine-3-carboxylate)	Brazil	Fruit pulp	Warm, sweet, herba-ceous	a
Ethyl acetate	Brazil	Fruit pulp	Ethereal, fruity, banana-, brandy-like	a
	Venezuela	Fruit pulp		b
	South Africa	Puree		e
Ethyl propanoate	Brazil	Fruit pulp	Fruity, ethereal, rum-like	a
Ethyl butanoate	Brazil	Fruit pulp	Ethereal, fruity, banana-like	a
	Venezuela	Fruit pulp		b
	Taiwan	Peel/leaf		d
	South Africa	Puree		e
Ethyl (E)-but-2-enoate	Brazil	Fruit pulp	Sour, caramel, fruity	a
Ethyl pentanoate	Brazil	Fruit pulp	Ethereal, fruity, apple-like	a
	Taiwan	Puree		d
Ethyl hexanoate	Brazil	Fruit pulp	Fruity, floral, wine-like	a
	Venezuela	Fruit pulp		b
Ethyl (E)-hex-2-enoate				a
Ethyl octanoate	Brazil	Fruit pulp	Fruity, wine-, apricot-, banana-like	a
	Venezuela	Fruit pulp		b
Ethyl decanoate	Venezuela	Fruit pulp	Floral, sweet, oily, nut-like	b
Ethyl dodecanoate	Venezuela	Fruit pulp	Slightly mango-like	b

(*continued*)

TABLE VI.XVII (*continued*)

Compound	Source	Form	Odour quality	References
Ethyl tetradecanoate	Venezuela	Fruit pulp	Oily, ethereal, violet-like	b
Ethyl hexadecanoate	Venezuela	Fruit pulp	Faint waxy, mild sweet	b
Ethyl 2-furoate	Brazil	Fruit pulp	Warm, fruity, plum-like	a
Ethyl benzoate	Brazil	Fruit pulp	Blackcurrant-like	a
Ethyl phenylacetate	Brazil	Fruit pulp	Powerful sweet, honey-like	a
Ethyl (*E*)-cinnamate	Brazil	Fruit pulp	Sweet, balsamic, fruity, honey-like	a
Propyl acetate	Brazil	Fruit pulp	Diffusive, ethereal, pear, raspberry-like	a
2-Methylpropyl acetate	Venezuela	Fruit pulp	Diffusive, ethereal, rum-like	b
3-Phenylpropyl acetate	Brazil	Fruit pulp	Sweet, floral, fruity, balsamic	a
	Taiwan	Puree		d
Butyl acetate	Venezuela	Fruit pulp	Ethereal, fruity, pear-like	b
3-Methylbutyl acetate	Venezuela	Fruit pulp	Fruity, pear, banana-like	b
3-Methylbutyl butanoate	Brazil	Fruit pulp	Sweet, apricot-, banana-like	a
A pentenyl benzoate	Taiwan	Puree		d
Hexyl acetate	Brazil	Fruit pulp	Sweet, fruity, berry-, pear-like	a
	Venezuela	Fruit pulp		b
	Taiwan	Peel/puree		d
(*E*)-Hex-2-enyl acetate	Brazil	Fruit pulp	Powerful fresh, banana-, apple-, pear-like	a
(*Z*)-Hex-3-enyl acetate	Brazil	Fruit pulp	Green, sharp fruity	a
	Venezuela	Fruit pulp		b
	Taiwan	Peel/leaf/puree		d
	South Africa	Puree		e
	Hawaii	Puree		c
(*Z*)-Hex-3-enyl methyl-propanoate	Brazil	Fruit pulp		a
Cyclohexyl acetate	Brazil	Fruit pulp	Sweet, fruity, candy-like	a
(*Z*)-Hex-3-enyl hexanoate	Brazil	Fruit pulp	Powerful fruity, green	a
(*Z*)-Hex-2-enyl hexanoate	Brazil	Fruit pulp		a
(*E*)-Hex-2-enyl hexanoate	Brazil	Fruit pulp	Powerful fruity, apple-, pear-, pineapple-like	a
(*E*)-Hex-3-enyl hexanoate	Brazil	Fruit pulp		a
Octyl acetate	Venezuela	Fruit pulp	Fruity, fatty, apple-like	b
Cinnamyl acetate	Brazil	Fruit pulp	Sweet, mild balsamic, floral	a
	Hawaii	Puree		c
	Taiwan	Peel/leaf/puree		d
Benzyl acetate	Taiwan	Peel/leaf/puree	Floral, jasmin-, lily-like	d
2-Phenethyl acetate	Venezuela	Fruit pulp	Sweet, rose-, honey-like	c

TABLE VI.XVII (*continued*)

Compound	Source	Form	Odour quality	References
γ-Hexalactone	Taiwan	Leaf	Herbaceous, coumarin-like	d
γ-Decalactone	Brazil	Fruit pulp	Creamy, fruity, peach-like	a

[a] Idstein and Schreier (1985).
[b] MacLeod and de Troconis (1982b).
[c] Stevens *et al.* (1970).
[d] Shiota (1978).
[e] Torline and Ballschmieter (1973).

attention as a fragrance ingredient, because it is a very attractive, colourless, and stable material, which forms an important part of the fragrance picture of certain perfumes (Arctander, 1969).

5.3 Esters

Table VI.XVII shows the volatile esters found in guava. The main esters reported in guavas are simple esters, that is 14 methyl esters and 15 ethyl esters. It is obvious that esters are the major group of volatiles. MacLeod and de Troconis (1982b) found that esters comprised over 55% of the components identified in the essence of guava from Venezuela. Among the esters identified, the major representatives were ethyl acetate (*ca.* 26%), which has an ethereal, fruity, banana-like, and brandy-like aroma; ethyl hexanoate (15.5%), which has a fruity, floral, almond-like, and wine-like odour; and ethyl butanoate, which has an ethereal, fruity, and banana-like flavour (8.7%). (*Z*)-Hex-3-enyl acetate, which possesses a green and sharp, fruity aroma, was one of the major constituents of guava from Brazil (Idstein and Schreier, 1985). A relatively uncommon compound, methyl nicotinate (methyl pyridine-3-carboxylate), which has a warm, sweet, and herbaceous aroma, was detected by the same authors. Many esters of unsaturated C_6 alcohols, which have a powerful fruity, apple-like, banana-like, and pineapple-like flavour, have been found in various guavas. These compounds may contribute strong fruity aromas to guavas.

5.4 Aldehydes and Ketones

Table VI.XVIII shows the volatile aldehydes and ketones found in guava. Idstein and Schreier (1985) reported that 50% of the total amounts of volatiles from Brazilian guava were aldehydes. They proposed that most of them were typical products of peroxidation of unsaturated fatty acids, and that some ketones, such as pent-1-en-3-one and oct-1-en-3-one, were lipid degradation products. Hexanal, which has a fatty green and green grass-like aroma, and

TABLE VI.XVIII

Volatile aldehydes and ketones found in guava

Compound	Source	Form	Odour quality	References
Acetaldehyde	Venezuela	Fruit pulp	Pungent, ethereal, nauseating	b
3-Phenylpropenal	Brazil	Fruit pulp	Warm, spicy, cassia-, cinnamon-like	a
Butanal	South Africa	Puree	Pungent, irritating	e
(E)-But-2-enal	Brazil	Fruit pulp		a
Pentanal	Brazil	Fruit pulp	Powerful diffusive, choking, penetrating	a
2-Methylpent-4-enal	Brazil	Fruit pulp		a
2-Methylpent-2-enal	South Africa	Puree	Powerful gassy, green	e
(Z)-Pent-2-enal	Brazil	Fruit pulp		a
(E)-Pent-2-enal	Brazil	Fruit pulp		a
Hexanal	Brazil	Fruit pulp	Fatty green, green grass-like	a
	Venezuela	Fruit pulp		b
	Hawai	Puree		c
(Z)-Hex-3-enal	Brazil	Fruit pulp	Deep green, strawberry leaf-like	a
(E)-Hex-3-enal	Brazil	Fruit pulp		a
(E)-Hex-2-enal	Brazil	Fruit pulp		a
	Taiwan	Peel/leaf/puree	Green, fruity, pungent vegetable-like	d
1-Methylcyclohex-3-en-1-carboxaldehyde	Brazil	Fruit pulp		a
(E, E)-Hexa-2,4-dienal	Brazil	Fruit pulp		a
Heptanal	Brazil	Fruit pulp	Powerful diffusive, oily, fatty	a
(E, Z)-Hepta-2,4-dienal	Brazil	Fruit pulp		a
(E, E)-Hepta-2,4-dienal	Brazil	Fruit pulp		a
(E)-Oct-2-enal	Brazil	Fruit pulp	Green leaf-, foliage-like	a
(E, E)-Octa-2,4-dienal	Brazil	Fruit pulp		a
(E, Z)-Nona-2,6-dienal	Brazil	Fruit pulp	Extremely diffusive, green vegetable-like	a
(E, E)-Nona-2,6-dienal	Brazil	Fruit pulp		a
Decanal	Brazil	Fruit pulp	Herbaceous, lily-, violet-like	a
(E)-Dec-2-enal	Brazil	Fruit pulp	Powerful, waxy, orange-like	a
(Z, E)-Deca-2,4-dienal	Brazil	Fruit pulp	Extremely powerful, orange-like	a
(E, E)-Deca-2,4-dienal	Brazil	Fruit pulp		a
Benzaldehyde	Venezuela	Fruit pulp	Almond-, wood bark-like	b
	Hawai	Puree		c
	Taiwan	Peel/leaf/puree		d
Furfural (2-furaldehyde)	Venezuela	Fruit pulp	Cold meat-, gravy-, coconut-like	b
	Taiwan	Puree		d
5-Methylfurfural	Venezuela	Fruit pulp	Green wood bark-, twigs-like	b

TABLE VI.XVIII (*continued*)

Compound	Source	Form	Odour quality	References
5,6-Dihydro-2*H*-pyran-2-carboxaldehyde	Brazil	Fruit pulp		a
Vanillin (4-hydroxy-3-methoxybenzaldehyde)	Brazil	Fruit pulp	Intensely sweet, vanilla-like	a
Citral	Hawaii	Puree	Intensely lemon-like	c
Neral	Taiwan	Puree	Intensely lemon-like	d
Geranial	Taiwan	Puree	Intensely lemon-like	d
Perillaldehyde	Taiwan	Puree	Fatty, spicy, ginger-grass-like	d
Acetone	Venezuela	Fruit pulp	Light, ethereal, nauseating	b
Methylbutanone	South Africa	Puree	Ethereal, camphor-like	e
Butanone	Venezuela	Fruit pulp	Ethereal	b
Butanedione	Venezuela	Fruit pulp	Pungent, oily, butter-like	b
Pentan-3-one	Brazil	Fruit pulp	Ethereal, fruity	a
3-Hydroxybutanone (acetoin)	Brazil	Fruit pulp	Intensely creamy, fatty, butter-like	a
Pent-1-en-3-one	Brazil	Fruit pulp		a
4-Methylheptan-3-one	Brazil	Fruit pulp		a
A methylheptenone	Taiwan	Peel/leaf/puree		d
6-Methylhept-5-en-2-one	Brazil	Fruit pulp		a
Octan-3-one	Brazil	Fruit pulp	Spicy, butter-, lavender oil-like	a
Oct-3-en-3-one	Brazil	Fruit pulp		a
4-Hydroxy-5-methyl-2*H*-furan-3-one	Brazil	Fruit pulp		a
2,5-Dimethyl-4-hydroxy-2*H*-furan-3-one	Brazil	Fruit pulp		a
3,4-Dihydro-8-hydroxy-3-methyl-2-benzo-1*H*-pyran-1-one	Brazil	Fruit pulp		a
Acetylfuran	Venezuela	Fruit pulp	Green resinous, woody	b
β-Ionone	Hawaii	Puree	Warm woody, violet-like	c
Carvone	Brazil	Fruit pulp	Herbaceous, bread-, dill seed-like	a

[a] Idstein and Schreier (1985).
[b] MacLeod and de Troconis (1982b).
[c] Stevens *et al.* (1970).
[d] Shiota (1978).
[e] Torline and Ballschmieter (1973).

(*E*)-hex-2-enal, which has a green, fruity, pungent, and vegetable-like odour, were the major components of guava from Brazil (Idstein and Schreier, 1985). The characteristic green fruity aroma of guava may be due to the presence of C_6-C_{10} dienals, because they have powerful diffusive, green vegetable-like, lily-like, and violet-like aromas. The presence of carvone, a major component

TABLE VI.XIX

Volatile alcohols and acids found in guava

Compound	Source	Form	Odour quality	References
Methanol	South Africa	Puree	Pungent, ethereal	e
Ethanol	South Africa	Puree	Sickly, ethereal	e
	Venezuela	Fruit pulp		b
2-Phenylethanol	Hawaii	Puree	Mild, warm, rose-, honey-like	c
	Brazil	Fruit pulp		a
3-Phenylpropanol	Brazil	Fruit pulp	Warm, lilac-, rose-, hyacinth-like	a
	Taiwan	Puree		d
2-Methylpropan-1-ol	Hawaii	Puree	Pungent, ethereal	c
2-Methylpropan-2-ol	Venezuela	Fruit pulp		b
Butan-1-ol	Brazil	Fruit pulp	Sweet, sickly	a
Butan-2-ol	Brazil	Fruit pulp	Oily vinous	a
3-Methylbutan-1-ol	Venezuela	Fruit pulp	Sweet, valeric	b
Pentan-2-ol	Brazil	Fruit pulp	Ethereal, wine-like	a
Pent-1-en-3-ol	Hawaii	Puree	Powerful gassy green	c
	Brazil	Fruit pulp		a
Pentan-1-ol	Hawaii	Puree	Harsh, fusel oil-like	c
	Brazil	Fruit pulp		a
(E)-Pent-2-enol	Brazil	Fruit pulp		a
(Z)-Pent-2-enol	Brazil	Fruit pulp		a
Hexan-1-ol	Hawaii	Puree	Chemical, fatty, fruity, wine-like	c
	Brazil	Fruit pulp		a
	Taiwan	Peel/leaf/puree		d
(E)-Hex-3-en-1-ol	Venezuela	Fruit pulp	Intensely green, foliage-like	b
	Hawaii	Puree		c
	Brazil	Fruit pulp		a
(Z)-Hex-3-en-1-ol	Brazil	Fruit pulp	Intensely green, gassy	a
	Taiwan	Peel/leaf/puree		d
(E)-Hex-2-en-1-ol	Brazil	Fruit pulp	Fruity green, caramel	a
Heptan-1-ol	Brazil	Fruit pulp	Fresh, light green, nut-like	a
Octan-1-ol	South Africa	Puree	Powerful, orange-, rose-like	e
	Venezuela	Fruit pulp		b
	Hawaii	Puree		c
	Brazil	Fruit pulp		a
Oct-1-en-3-ol	Brazil	Fruit pulp	Sweet, ethereal	a
(E)-Oct-2-en-1-ol	Brazil	Fruit pulp		a
Nonan-1-ol	Hawaii	Puree	Oily, petal-like	c
Decan-3-ol	Brazil	Fruit pulp		a
Undecan-5-ol	Brazil	Fruit pulp		a
Tetradecanol	Brazil	Fruit pulp	Very faint coconut oil-like	a
Hexadecanol	Brazil	Fruit pulp	Virtually odourless	a
Octadecanol	Brazil	Fruit pulp	Virtually odourless	a
Eugenol	Brazil	Fruit pulp	Warm spicy, clove leaf-like	a

TABLE VI.XIX (*continued*)

Compound	Source	Form	Odour quality	References
Cinnamyl alcohol	Brazil	Fruit pulp	Warm balsamic, sweet	a
α-Terpineol	Hawaii	Puree	Floral, lilac-like	c
	Taiwan	Peel/leaf/puree		d
1,8-Cineol	Taiwan	Peel/leaf/puree	Fresh, camphor-like	d
Terpinen-4-ol	Taiwan	Peel/leaf/puree		d
Linalool	Taiwan	Peel/leaf/puree	Refreshing, floral, woody	d
Borneol	Taiwan	Leaf/puree	Dry woody, slightly camphor-like	d
Cadineol	Taiwan	Puree		d
δ-Cadinol	Taiwan	Peel/leaf/puree		d
4-Oxodihydro-β-ionol	Brazil	Fruit pulp		a
Acetic acid	Brazil	Fruit pulp	Pungent, stingingly sour	a
Phenylpropanoic acid	Brazil	Fruit pulp	Sweet floral, honey-like	a
Butanoic acid	Brazil	Fruit pulp	Sour, rancid butter-like	a
Pentanoic acid	Taiwan	Puree	Animal perspiration-like	d
Hexanoic acid	Brazil	Fruit pulp	Heavy, rancid fat-like	a
(E)-Hex-2-enoic acid	Brazil	Fruit pulp		a
Heptanoic acid	Brazil	Fruit pulp	Sour, sweet, fatty	a
Octanoic acid	Brazil	Fruit pulp	Burning, sour cheese-like	a
Nonanoic acid	Brazil	Fruit pulp	Waxy, nut-, brandy-like	a
Decanoic acid	Brazil	Fruit pulp	Sour, rancid fat-like	a
Tetradecanoic acid	Brazil	Fruit pulp	Faint waxy, oily	a
Hexadecanoic acid	Brazil	Fruit pulp	Virtually odourless	a
Benzoic acid	Brazil	Fruit pulp	Sweet, honey-like	a
Cinnamic acid	Brazil	Fruit pulp	Very faint honey-like	a

[a] Idstein and Schreier (1985).
[b] MacLeod and de Troconis (1982b).
[c] Stevens *et al.* (1970).
[d] Shiota (1978).
[e] Torline and Ballschmeiter (1973).

of spearmint, is somewhat surprising, because no related compounds were reported in guava (Idstein and Schreier, 1985). Some furan derivatives, which are known as sugar degradation products, were found in the fruit puree of Brazilian guava (Idstein and Schreier, 1985). They may cause some off-flavour of fruit, but their quantities are not significant for the overall flavour quality of guavas.

5.5 Alcohols and Acids

Table VI.XIX shows the volatile alcohols and acids found in guava. Nearly 40 different alcohols have been reported in guava. Guava may be one of the most alcohol-rich tropical fruits. Ethanol was reported in the fruit pulp of Venezuelan guava as one of the major components (MacLeod and de Troconis,

278

Volatile miscellaneous compounds found in guava

Compound	Source	Form	Odour quality	References
1-Phenoxybutane	Brazil	Fruit pulp		a
1,2-Dimethoxybenzene	Brazil	Fruit pulp		a
(Z)-Theaspirane	Brazil	Fruit pulp		a
(E)-Theaspirane	Brazil	Fruit pulp		a
Dimethyl disulphide	Brazil	Fruit pulp	Sulphurous, pungent	a
Dimethyl trisulphide	Brazil	Fruit pulp	Pungent, sulphurous	a
Pentane-3-thiol	Brazil	Fruit pulp	Sulphurous	a
2-Methylthiophene	Brazil	Fruit pulp	Sulphurous, pungent	a
3-Methylthiophene	Brazil	Fruit pulp	Sulphurous, pungent	a
2-Ethylthiophene	Brazil	Fruit pulp	Sulphurous, pungent	a
Methylpyrazine	Brazil	Fruit pulp	Roasted, toasted, burning	a
Trimethylpyrazine	Brazil	Fruit pulp	Roasted, toasted, burning	a
Acetylpyrazine	Brazil	Fruit pulp	Roasted, toasted, nut-like	a
5-Ethoxythiazole	Brazil	Fruit pulp	Green vegetable-like	a
Benzothiazole	Brazil	Fruit pulp	Faint sulphurous	a
	Taiwan	Puree		d
2-Methylthiobenzothiazole	Brazil	Fruit pulp		a

[a] Idstein and Schreier (1985).
[d] Shiota (1978).

1982b). High ethanol content may be related to fermentation. Butan-1-ol, which is the major constituent of papaya essence, was found only in the fruit pulp of Brazilian guava, which contained hexan-1-ol and (Z)-hex-3-en-1-ol as major components (Idstein and Schreier, 1985). (Z)-Hex-3-en-1-ol, which has an intensely green and grassy odour, was the major constituent of fruit puree obtained from Taiwan guava (Shiota, 1978). 1,8-Cineol, which has a fresh and camphor-like odour, was present in the leaves and peel of Taiwan guava in large quantities, but was present in only trace amounts in fruits (Shiota, 1978). A series of aliphatic acids has been reported in guava from Brazil (Idstein and Schreier, 1985). The low molecular weight acids, $C_2 - C_8$, may contribute some characteristic flavours to guava, but acids higher than nonanoic have only a faint odour, or no odour.

5.6 Miscellaneous Compounds

Table VI.XX shows the volatile miscellaneous compounds found in guava. Idstein and Schreier (1985) reported many S- and N-containing heterocyclic compounds for the first time in guava. Thiophenes, which generally have a pungent, sulphurous odour, give a faint green, pleasant aroma on dilution. The sensory properties of the thiophenes have been summarized by Ohloff and Flament (1979). Pyrazines are well known as Maillard browning reaction products, and have been used widely in artificial cooked flavours. Alkylpyrazines generally have a roasted or toasted flavour, and the alkoxypyrazines often have an earthy, vegetable-like odour. Recently, pyrazines have also received much attention as fragrance ingredients. Thiazoles and pyrazines have somewhat similar sensory properties. The alkylthiazoles have green, nutty, roasted, vegetable, or meaty flavours (Pittet and Hruza, 1974; Ho and Jin, 1985). Recently, these heterocyclic compounds have been found to be present in various fruits at a very low level. They may play an important rôle in the fruit flavours because of their low odour thresholds.

REFERENCES

Arctander, S., 1969. Perfume and Flavor Chemicals. Published by the author, Montclair, NJ.

Ashraf, M.A., Khan, N., Ahmad, M. and Elahi, M., 1981. Studies on the pectinesterase activity and some chemical constituents of some Pakistani mango varieties during storage ripening. *J. Agric. Food Chem.*, 29: 526 – 528.

Casimir, D.J. and Whitfield, F.B., 1978. Flavour impact values: A new concept for assigning numerical values for the potency of individual flavour components and their contribution to the overall flavour profile. *Ber. Int. Fruchtsaft.-Union, Wiss.-Techn. Komm.*, 15: 325 – 347.

Casimir, D.J., Kefford, J.F. and Whitfield, F.B., 1981. Technology and flavor chemistry of passion fruit juices and concentrates. *Adv. Food Res.*, 27: 243 – 295.

Chan, H.T., Flath, R.A., Forrey, R.R., Cavaletto, C.G., Nakayama, T.O.M. and Brekke, J.E., 1973. Development of off-odors and off-flavors in papaya puree. *J. Agric. Food Chem.*, 21: 566 – 570.

Chen, C.-C., Kuo, M.-C., Hwang, L.S., Wu, J.S.-B. and Wu, C.-M., 1982. Headspace components of passion fruit juice. *J. Agric. Food Chem.*, 30: 1211 – 1215.

Engel, K.-H. and Tressl, R., 1983a. Studies on the volatile components of two mango varieties. *J. Agric. Food Chem.*, 31: 796 – 801.

Engel, K.-H. and Tressl, R., 1983b. Formation of aroma components from nonvolatile precursors in passion fruit. *J. Agric. Food Chem.*, 31: 998 – 1002.

Flath, R.A. and Forrey, R.R., 1977. Volatile components of papaya (*Carica papaya* L., Solo variety). *J. Agric. Food Chem.*, 25: 103 – 109.

Gholap, A.S. and Bandyopadhyay, C., 1976. Fatty acid composition as a quality index of ripe mango (*Mangifera indica* L.) pulp. *Indian Food Packer*, 30: 63 – 64.

Gholap, A.S. and Bandyopadhyay, C., 1980. Fatty acid biogenesis in ripening mango (*Mangifera indica* L. var. Alphonso). *J. Agric. Food Chem.*, 28: 839 – 841.

Herrmann, K., 1983. Exotische Lebensmittel. Springer, Berlin, Heidelberg, New York.

Ho, C.T. and Jin, Q.Z., 1985. Aroma properties of some alkylthiazoles. *Perf. Flavorist*, 9(6): 15 – 18.

Hunter, G.L.K., Bucek, W.A. and Radford, T., 1974. Volatile components of canned alphonso mango. *J. Food Sci.*, 39: 900 – 903.

Idstein, H. and Schreier, P., 1985. Volatile constituents from Guava (*Psidium guajava*, L.) fruit. *J. Agric. Food Chem.*, 33: 138 – 143.

Idstein, H., Keller, T. and Schreier, P., 1985. Volatile constituents of Mountain Papaya (*Carica candamarcensis*, syn. *C. pubescens* Lenne et Koch) fruit. *J. Agric. Food Chem.*, 33: 663 – 666.

MacLeod, A.J. and de Troconis, N.G., 1982a. Volatile flavour components of mango fruit. *Phytochemistry*, 21: 2523 – 2526.

MacLeod, A.J. and de Troconis, N.G., 1982b. Volatile flavour components of guava. *Phytochemistry*, 6: 1339 – 1342.

280

MacLeod, A.J. and Pieris, N.M., 1983. Volatile components of papaya (*Carica papaya* L.) with particular reference to glucosinolate products. *J. Agric. Food Chem.*, 31: 1005 – 1008.

MacLeod, A.J. and Snyder, C.H., 1985. Volatile components of two cultivars of mango from Florida. *J. Agric. Food Chem.*, 33: 380 – 384.

Morales, A.L. and Duque, C., 1987. Aroma constituents of the fruit of the mountain papaya (*Carica pubescens*) from Colombia. *J. Agric. Food Chem.*, 35: 538 – 540.

Murray, K.E., Shipton, J. and Whitfield, F.B., 1972. The chemistry of food flavor. I. Volatile constituents of passion fruit, *Passiflora edulis. Aust. J. Chem.*, 25: 1921 – 1933.

Murray, K.E., Shipton, J. and Whitfield, F.B., 1973. The flavour of purple passion fruit. *Food Technol. Aust.*, 25: 446 – 448.

Ohloff, G. and Flament, I., 1979. The rôle of heteroatomic substances in the aroma compounds of foodstuffs. *Forstschr. Chem. Org. Naturst.*, 36: 231 – 283.

Parliment, T., 1972. Volatile constituents of passion fruit. *J. Agric. Food Chem.*, 20: 1043 – 1045.

Pittet, A.O. and Hruza, D.E., 1974. Comparative study of flavor properties of thiazole derivatives. *J. Agric. Food Chem.*, 22: 264 – 269.

Shiota, H., 1978. Review and study on guava fruits flavor. *Koryo*, 121: 23 – 30.

Stevens, K.L., Brekke, J.E. and Stern, D.J., 1970. Volatile constituents in guava. *J. Agric. Food Chem.*, 18: 598 – 599.

Tang, C.S., 1971. Benzyl isothiocyanate of papaya fruit. *Phytochemistry*, 10: 117 – 121.

Tang, C.S., Bhothipaksa, K. and Frank, H.A., 1972. Bacterial degradation of benzyl isothiocyanate. *Appl. Microbiol.*, 6: 1145 – 1148.

Torline, P. and Ballschmieter, H.M.B., 1973. Volatile constituents from guava, I. A comparison of extraction methods. *Lebensm.-Wiss. Technol.*, 6: 32 – 33.

Wilson, G.W. III and Shaw, P.E., 1978. Terpene hydrocarbons from *Psidium guajava. Phytochemistry*, 17: 1435 – 1436.

Winter, M. and Kloti, R., 1972. Uber das Aroma der gelben Passionsfrucht. *Helv. Chim. Acta*, 55: 1916 – 1921.

Wolfe, H.S., 1962. The mango in Florida – 1887 to 1962. *Proc. Fla. State Hortic. Soc.*, 75: 387 – 391.

Yamaguchi, K., Nishimura, O., Toda, H., Mihara, S. and Shibamoto, T., 1983. Chemical studies on tropical fruits. In: G. Charalambous and G. Inglett (Editors), Instrumental Analysis of Foods, Recent Progress, Vol. 2. Academic Press, New York, pp. 93 – 117.

Chapter VII

THE FLAVOUR OF EXOTIC FRUIT

HARRY YOUNG and VIVIENNE J. PATERSON
Division of Horticulture and Processing, Department of Scientific and Industrial Research, Auckland (New Zealand)

1. INTRODUCTION

In recent years there has been an expansion of horticultural crops in international trade. In addition to the traditional fruit such as apples and citrus, there is a growing consumer interest in exotic fruit. We should define what we mean by exotic. Fruit that are considered to be exotic by some people may be common fruit in another culture. For the purpose of this chapter we will use the term 'exotic' to include unusual fruit that are unknown or have only recently begun to be traded on major world markets.

Good flavour is a combination of taste, mainly attributed to the non-volatile components, e.g. the sugars and non-volatile acids, texture, and the aroma, which is due to the volatile compounds present. This latter gives the fruit its particular flavour character, and is particularly important in determining consumer appeal of exotic fruit. It is these volatile flavour compounds that this review will concentrate on, as there are very few reports on the other flavour aspects.

Investigations into the nature of the volatile aroma compounds of fruit are motivated by many reasons. Most often there is a need to understand flavour changes and off-flavours developing during processing. There is also a need to learn the flavour changes that occur during maturing, ripening and storage of fruit to determine the best harvest maturity and post-harvest handling methods. Selection of consumer-preferred cultivars is also very important, especially when developing a new fruit industry.

In recent years there have been many publications on the flavour of exotic fruit, but no recent comprehensive review. Dupaigne (1978) reviewed published work on the composition and aroma of little-known tropical fruit including durian, jackfruit, and feijoa, but concluded that there had been little published work on the aroma of these fruit. Herrmann (1981) also reviewed a number of exotic fruit, but most of the references were for studies on composition of non-volatile components rather than flavour.

Volatile flavour compounds from a wide range of fruit and other foods have been listed by van Straten *et al.* (1983). We have not included tables of compounds for those fruit that were cited by van Straten *et al.* unless new results have been reported.

2. KIWIFRUIT

Kiwifruit, *Actinidia deliciosa* (A. Chevalier) C. F. Liang et A. R. Ferguson var *deliciosa* (syn. *A. chinensis* Planchon var. *hispida* C. F. Liang) is a deciduous vine native to China. It was first developed as a commercial crop in New Zealand during the 1960s, but is now grown in quantity in several countries, including New Zealand, the U.S.A., France, Italy, and Chile. Initially it was promoted as a new fruit on the luxury markets of Europe, but it is now widely available.

This fruit is harvested when it is still unripe and very hard. In this state it can be stored at 0°C for several months. When ripened at room temperature the fruit softens and develops its unique flavour. Kiwifruit is usually used as fresh fruit, although there is growing interest in processed products, e.g. frozen pulp, wine and beverages. Almost all reports on the flavour of this fruit have been on fresh fruit.

The flavour components of kiwifruit (Table VII.I) have been investigated by Shiota (1982), Young *et al.* (1983), Young and Paterson (1985), and more recently by Takeoka *et al.* (1986). Pfannhauser (1987) has investigated a processed product.

Shiota (1982) used simultaneous distillation and extraction, using pentane/dichloromethane (7 : 3 w/w) as the extraction solvent, to isolate the volatile flavour components. In this extract, ethyl butanoate was found to be the major component. Relatively high levels of methyl and ethyl benzoate were present. Apart from H_2S (see below) this is the only report of the occurrence of sulphur compounds (methyl and ethyl methylthioacetate, and ethyl 3-methylthiopropanoate) and long-chain hydrocarbons (C-20 to C-23).

Young *et al.* (1983) used low-temperature vacuum distillation to isolate the flavour components of kiwifruit. They considered it important to avoid elevated temperature, since this fruit exhibits considerable flavour degradation on heating. In their work the aqueous concentrate was analysed by collecting the organic components on Chromosorb 105 traps or by extraction of the aqueous concentrate with Freon. Similar results were obtained with either method. The major component found by these authors was (*E*)-hex-2-enal.

Takeoka *et al.* (1986) investigated the volatile components of kiwifruit flavour using vacuum distillation at temperatures below 30°C followed by extraction into Freon 11. Twenty-two additional minor components were identified. (*E*)-Hex-2-enal was again the main component. The identification of (*Z*)-pent-2-en-1-ol is of interest, since this compound is only rarely found in fruit volatiles.

The delicate flavour of kiwifruit is affected by the maturity at harvest, the post-harvest handling (e.g. storage time) and, more especially, the degree of ripeness of the fruit at the time of sampling (Young and Paterson, 1985). In the under-ripe fruit the volatiles were dominated by aldehydes, giving the fruit an intense green note. With over-ripe fruit, esters dominated the aroma profile (increasing from 330 to 7870 µg/kg). Fruit harvested immature developed less

TABLE VII.I

Volatile flavour components of kiwifruit[a]

Esters

Ethyl acetate	S, Y1, Y2, T, P
Methyl acetate	T
Ethyl benzoate	S, Y2, T, P
Methyl benzoate	S, Y1, Y2, T, P
Ethyl butanoate	S, Y1, Y2, T, P
Methyl butanoate	S, Y1, Y2, T, P
Propyl butanoate	Y2
2-Methylpropyl butanoate	S
Ethyl furoate	S, T
Methyl furoate	S, T, P
Ethyl hexanoate	S, Y2, T
Methyl hexanoate	S, Y2
Ethyl 3-hydroxybutanoate	S
Ethyl 2-methylbutanoate	T
Ethyl 3-methylbutanoate	P
Ethyl 2-methylpropanoate	S, Y2, T
Methyl 2-methylpropanoate	Y2, T
Ethyl methylthioacetate	S
Methyl methylthioacetate	S
Ethyl 3-methylthiopropanoate	S
Ethyl octanoate	S
Methyl octanoate	S
Ethyl pentanoate	S, Y2
Methyl pentanoate	Y2, T, P
Dibutyl phthalate	S
Ethyl propanoate	S, Y2, T
Methyl propanoate	Y2, T
Methyl salicylate	T

Aldehydes and ketones

Benzaldehyde	T, P
2,5-Dimethyl-4-methoxy-2H-furan-3-one	T
2-Furaldehyde	S, P
(E, E)-Hepta-2,4-dienal	T
(E, Z)-Hepta-2,4-dienal (i)	T
(E)-Hept-2-enal	T
Hexanal	Y1, Y2, T, P
Hexa-2,4-dienal	T
(Z)-Hex-2-enal	Y1, Y2, T, P
(E)-Hex-2-enal	S, Y1, Y2, T, P
(E)-Hex-3-enal	T, P
(Z)-Hex-3-enal	T, P
5-Methyl-2-furaldehyde	S, P
Pentanal	T, P
(E)-Pent-2-enal	Y1, Y2, T
Pent-1-en-3-one	Y1, Y2, T, P
Phenylacetaldehyde	S

(*continued*)

284

TABLE VII.I (*continued*)

Alcohols	
Ethanol	Y1, Y2
2-Ethylhexan-1-ol	T
Heptan-1-ol	T, P
Hexadecan-1-ol	S
Hexan-1-ol	S, Y1, Y2, T, P
(*E*)-Hex-3-en-1-ol	Y1, Y2, T, P
(*Z*)-Hex-3-en-1-ol	S, Y1, Y2, T, P
(*E*)-Hex-2-en-1-ol	S, Y1, Y2, T
Hex-2-en-1-ol	P
(*Z*)-Hex-2-en-1-ol (i)	T
2-Methylpropan-1-ol	S
3-Methylbut-2-en-1-ol	P
Pentan-1-ol	Y1, Y2, T, P
Pent-1-en-3-ol	Y1, Y2, T
(*Z*)-Pent-2-en-1-ol	Y1, T
2-Phenylethanol	S
Hydrocarbons	
Biphenyl	S
Docosane	S
Eicosane	S
Ethylbenzene	T
Heneicosane	S
Naphthalene	T
Nonadecane	S
Styrene	T
Toluene	S, T
Trieicosane	S
m-Xylene	T
o-Xylene	T
p-Xylene	T
Terpenes	
Car-3-ene	T
Carvone	S
p-Cymene	S, T, P
β-Damascenone	S
(*E*)-Hotrienol	S
Limonene	S, Y2, P
Linalool	S
Linalool oxide	S
α-Pinene	S, T, P
β-Pinene	T, P
Terpinen-4-ol	S
α-Terpineol	S
Verbenone	S

[a] S = Shiota (1982); Y1 = Young *et al.* (1983); Y2 = Young and Paterson (1985); T = Takeoka *et al.* (1986); P = Pfannhauser (1987).
(i) = Tentative identification.

volatiles. There is a marked reduction of volatile levels with storage. Fruit harvested at optimal maturity (soluble solids content 8.0%) were best at maintaining their flavour during storage.

The differences found in the volatiles reported by the three groups are most probably due to the methods of isolation, the source of the fruit, and the ripeness of the fruit. Although Shiota (1982) did not give any indication of the ripeness of the fruit used, the very high level of ethyl butanoate found was probably the result of using over-ripe fruit. The presence of the long-chain hydrocarbons could be due to the high-temperature distillation method employed to isolate the volatiles.

By monitoring at an odour port during GC of kiwifruit volatiles, Young *et al.* (1983) deduced that the important contributors to kiwifruit flavour were ethyl butanoate, hexanal and (*E*)-hex-2-enal. A strong sulphury note that is released when kiwifruit tissue is disrupted has been shown to be H_2S (Young and Paterson, unpublished). Takeoka *et al.* (1986) did not comment on the flavour significance of the compounds identified.

The component considered by Shiota (1982) to be an important contributor to the unique flavour of this fruit was β-damascenone. This compound was not detected in other reports on kiwifruit flavour. Rather surprisingly, he did not consider ethyl butanoate, with its low sensory threshold (Flath *et al.*, 1967), to be an important contributor, even though it is the major component in the volatiles collected (25%).

2-Furaldehyde and 5-methyl-2-furaldehyde, identified in kiwifruit puree (Pfannhauser, 1987), were also found in the fresh fruit volatiles isolated at elevated temperatures (Shiota, 1982). These could be heat induced artifacts (Sieso and Crouzet, 1977).

Two compounds, undeca-1-(*E, Z*)-3,5-triene and undeca-1-(*E, Z, Z*)-3,5,8-tetraene, which have extremely low odour thresholds, have been detected in kiwifruit homogenised in methanol in the ratio 1 : 33. However, the significance of these two compounds to kiwifruit flavour has not been proven (Berger *et al.*, 1985).

With the increasing international production of kiwifruit there is a growing interest in the manufacture of processed products from kiwifruit. However, there is an off-flavour problem associated with processed kiwifruit products (Young *et al.*, 1983; Pfannhauser, 1987). The problem of flavour degradation in the processed products needs to be solved if these products are to gain high consumer acceptance.

3. FEIJOA

Feijoa, *Feijoa sellowiana* Berg, is a member of the Myrtaceae family, which has at least 32 edible species (Maggs, 1984). It is a small evergreen shrub native to South America, but is becoming a commercially important crop in many countries (Klein and Thorp, 1987). The fruit, which is egg-shaped and has a

TABLE VII.II

Volatile flavour components of feijoa fruit[a]

Esters	
Ethyl acetate	A, B
3-Methylbutyl acetate	B
Hexenyl acetate	A
(Z)-Hex-3-enyl acetate	B
Benzyl benzoate	B
Ethyl benzoate	A, B, C
Hexenyl benzoate (i)	A
(Z)-Hex-3-enyl benzoate	B
Hexyl benzoate	B
Methyl benzoate	A, B, C
Ethyl butanoate	A, B, C
Hept-2-yl butanoate	C
(Z)-Hex-3-enyl butanoate	B, C
Methyl butanoate	B
3-Methylbutyl butanoate	B
Ethyl cinnamate	A, B
Methyl cinnamate	B
Ethyl hexadecanoate	B
Ethyl hexanoate	B, C
(Z)-Hex-3-enyl hexanoate	B, C
Ethyl 3-hydroxybutanoate	B
Ethyl lactate	B
Ethyl 4-methoxybenzoate	A, B
Methyl 4-methoxybenzoate	A, B
3-Methylbutyl methylpropanoate	B
Ethyl octanoate	B
Ethyl phenylpropanoate	B
Diethyl phthalate	B
Ethyl propanoate	B
Hexenyl propanoate (i)	A
Ethyl salicylate	B
Methyl salicylate	B
Aldehydes and ketones	
2,5-Dimethyl-2H-furan-3-one	B
2-Furaldehyde	B
Heptan-2-one	A, C
Hexenal	A
(E)-Hex-2-enal	A, B
3-Hydroxybutanone	B
Nonan-2-one	A, B, C
Octan-3-one	A, B
Phenylacetaldehyde	B
p-Tolualdehyde	B
Undecan-2-one	A, B
Vanillin (4-hydroxy-3-methoxybenzaldehyde)	B

TABLE VII.II (*continued*)

Alcohols	
Benzyl alcohol	B
Heptan-2-ol	B
Hexan-1-ol	B
(Z)-Hex-3-en-1-ol	B
3-Methylbutan-1-ol	B
Octan-3-ol	B
2-Phenylethanol	B
Terpenes and sesquiterpenes	
β-Copaene	B
β-Elemene	B
α-Farnesene	B
(Z)-Farnesol	B
Geraniol	B
Linalool	B
Methyl geranate	B
Myrcene	C
Nerol	B
(E)-Ocimene	B
(E)-β-Ocimene	C
Terpinen-4-ol	B
α-Terpineol	B
p-Vinylguaiacol	B

[a] A = Hardy and Michael (1970); B = Shiota *et al.* (1980); C = Shaw *et al.* (1983).
(i) = Tentative identification.

green skin even when ripe, produces an intense distinctive aroma when ripe. Table VII.II lists the volatile flavour components found in feijoa.

Hardy and Michael (1970) identified 16 compounds in the vacuum distillate from homogenised whole feijoa fruit. The major components in the flavour volatiles from this fruit were the methyl and ethyl esters of benzoic acid. Other aromatic esters identified included hexenyl benzoate, ethyl cinnamate, methyl 4-methoxybenzoate and ethyl 4-methoxybenzoate. Octan-3-one, at a level of 5%, was the third largest peak in the gas chromatogram.

Shiota *et al.* (1980) extended the list of compounds found in the volatiles of feijoa. Using simultaneous distillation/extraction they obtained a concentrate in which aromatic esters made up over 30% of the volatiles, methyl and ethyl benzoate were 11 and 17%, respectively. The extract was also rich in terpenoids and sesquiterpenoids. Quantitatively, the most important terpenoid identified was linalool, found at a level similar to that of methyl and ethyl benzoate.

The aroma volatiles released by intact feijoa fruit during ripening were studied by Shaw *et al.* (1983). As the fruit ripened the relative amounts of benzoate esters increased. The report did not indicate whether this increase was accompanied by decreases in the absolute levels of the other components.

Feijoa is unusual in that ethyl and methyl benzoate dominate the aroma profile. By monitoring an odour port during GC of feijoa volatiles, Hardy and Michael (1970) linked the characteristic feijoa aroma with these aromatic esters. Shiota *et al.* (1980) suggested that feijoa has a guava-like flavour as a consequence of the sesquiterpene content (Wilson and Shaw, 1978; MacLeod and de Troconis, 1982a). Pattabhiraman *et al.* (1968), however, considered carbonyl compounds were the main contributors to the flavour of guava, and indeed octan-3-one (5 – 8%) and several other carbonyl compounds were found in feijoa volatiles. Shaw *et al.* (1983) considered methyl benzoate, ethyl benzoate, and ethyl butanoate, among the 11 compounds identified, to be important contributors to feijoa flavour. Starodubtseva and Kharebava (1986) concurred with the opinion that ethyl butanoate is an important component of feijoa flavour.

Surprisingly, linalool, found at very high levels by Shiota *et al.* (1980), was not detected by Hardy and Michael (1970) or by Shaw *et al.* (1983). This highly odoriferous compound would be expected to make a significant contribution to the overall flavour.

The flavour of feijoa fruit was improved by treatment with acetaldehyde or with 100% nitrogen (Pesis *et al.,* 1987). These treatments increased production of volatiles, and treated fruit were preferred by taste panelists.

4. ROSEAPPLE

Roseapple (*Syzygium jambos* L. syn *Eugenia jambos* L.) is another member of the Myrtaceae family. Also known as malabar plum, it is native to South East Asia and is widely cultivated. Apart from the distinctive rose water flavour, the fruit are almost tasteless (Lee *et al.,* 1975a). They are eaten fresh or may be preserved (Chin and Yong, 1980).

Lee *et al.* (1975b) extracted the volatiles from stored (0°C) roseapple pulp by vacuum distillation using a rotary evaporator, followed by dichloromethane extraction of the aqueous distillate. Aliphatic or terpene alcohols accounted for 12 of the 17 compounds identified. Although no quantitative data were given, the alcohols comprised most of the total peak area in the chromatogram, with (Z)-hex-3-en-1-ol being the major peak. Other compounds were (Z)- and (E)-linalool oxide, hexanal, 3-hydroxybutanone and cinnamaldehyde. Although the lack of esters in the volatiles would be consistent with the bland flavour of roseapple, many of the less intense GC peaks were not identified. Also, the pulp was stored for several months before sampling, during which time it is possible that degradation of esters may have occurred.

The rosy aroma of the fruit was attributed to linalool, 3-phenylpropan-1-ol and the two linalool oxides.

5. CHERIMOYA

The cherimoya is a member of the Annonaceae, which is a large family with over 2000 species, many of which have edible fruit. The best known are species of *Annona,* which includes custard apple*, soursop*, sweetsop and ilama, and species of *Asimina* (pawpaw), *Rollinia* and *Cananga* (Leboeff *et al.,* 1982).

Cherimoya *(Annona cherimolia)* probably originates in Central America (Gardiazabal and Rosenberg, 1986) and is grown commercially in a number of subtropical regions (George, 1984). Considered one of the finest of the *Annona* fruit, it has a sweet, delicate and delicious flavour (Tankard, 1987). It can be eaten fresh or can be processed into drinks, ice cream or sherbets.

In an investigation of the cherimoya flavour volatiles, Idstein *et al.* (1984) isolated the volatiles from cherimoya pulp using high vacuum distillation followed by liquid/liquid extraction. This study demonstrated the usefulness of on-line FT-IR (Fourier transform infrared) as a complement to GC-MS in the identification of flavour volatiles.

The 203 compounds identified are listed in Table VII.III, and include 23 hydrocarbons, 58 esters, 54 alcohols and 47 carbonyl compounds. Hexan-1-ol and 3-methylbutan-1-ol, present at > 1000 µg/kg, and butan-1-ol and linalool, at 500 – 1000 µg/kg, were quantitatively the most important components. Butyl, 3-methylbutyl, and hexyl esters of butanoic acid were found at levels between 100 and 500 µg/kg. Other esters identified included a series of 3-methylbutanoates and 3-methylbutyl esters. The authors suggested that the numerous 3-methylbutyl esters and 3-methylbutanoates present indicate a forced leucine metabolism in this fruit.

In addition to linalool, a number of terpene hydrocarbons and oxygenated terpenes was also identified. The presence of (*Z*)- and (*E*)-pent-2-en-1-ol, both uncommon plant volatiles, and their esters was noted.

Further sensory studies will be necessary to identify the significant contributors to cherimoya flavour.

6. CUSTARD APPLE

The name 'custard apple' is often associated with several species of *Annona,* however in commerce it refers to *Annona atemoya.* This species is considered to be a hybrid of *A. cherimolia* and *A. squamosa* (sweetsop) (Tankard, 1987). Originating from South America, it is grown in many tropical regions including parts of Australia. It has an aromatic, juicy white flesh with a characteristic flavour (Wyllie *et al.,* 1987). The fruit may be eaten fresh, or in ice cream, sherbet, fruit drinks and pies.

The flavour volatiles of *A. atemoya* cv. African Pride have been reported by Wyllie *et al.* (1987). Custard apple pulp was blended with water and extracted with a Likens – Nickerson apparatus using pentane, followed by concentration at 50°C. All of the 37 compounds that were identified (Table VII.IV) are terpenoid in nature. This makes an interesting contrast to two other related fruit, cherimoya and soursop, that have been investigated and contain

* Included in this review.

TABLE VII.III

Volatile flavour components of cherimoya fruit[a]

Esters
  Benzyl acetate
  Butyl acetate
  Ethyl acetate
  3-Methylbut-2-enyl acetate
  3-Methylbutyl acetate
  Methyl benzoate
  3-Methylbutyl benzoate
  Benzyl butanoate
  Butyl butanoate
  But-2-enyl butanoate
  Ethyl butanoate
  (E)-Hex-2-enyl butanoate
  (Z)-Hex-3-enyl butanoate
  Hexyl butanoate
  Methyl butanoate
  3-Methylbutyl butanoate
  2-Methylpropyl butanoate
  (E)-Pent-2-enyl butanoate
  (Z)-Pent-2-enyl butanoate
  Pent-3-enyl butanoate
  Pentyl butanoate
  Butyl but-2-enoate
  Hexyl but-2-enoate
  3-Methylbutyl but-2-enoate
  Diethyl carbonate
  Ethyl decanoate
  Methyl decanoate
  Hexyl hexanoate
  Methyl hexanoate
  3-Methylbutyl hexanoate
  (E)-Pent-2-enyl hexanoate
  Ethyl 2-hydroxy-4-methylpentanoate
  3-Methylbutyl 2-hydroxypropanoate
  Propyl 2-hydroxypropanoate
  Hexyl 2-methylbutanoate
  Methyl 2-methylbutanoate
  Benzyl 3-methylbutanoate
  Butyl 3-methylbutanoate
  (E)-Hex-3-enyl 3-methylbutanoate
  (Z)-Hex-3-enyl 3-methylbutanoate
  3-Methylbut-3-enyl 3-methylbutanoate
  3-Methylbutyl 3-methylbutanoate
  Hexyl 3-methylbutanoate
  (Z)-Pent-2-enyl 3-methylbutanoate
  Pentyl 3-methylbutanoate
  Benzyl 2-methylpropanoate
  Hexyl 2-methylpropanoate
  Methyl octadecanoate
  Methyl octanoate

  Ethyl 4-oxopentanoate
  Methyl pentadecanoate
  Hexyl pentanoate
  Butyl propanoate
  3-Methylbutyl propanoate

Aldehydes and ketones
  Benzaldehyde
  Benzophenone
  Butyrophenone
  Cyclohexanone
  2,5-Dimethyl-4-methoxy-2H-furan-3-one
  Dodecanal
  Eicosan-2-one
  2-Ethyloctanal
  2-Furaldehyde
  Heneicosan-3-one
  Heptadecan-2-one
  Heptadecan-3-one
  (E, E)-Hepta-2,4-dienal
  (Z, E)-Hepta-2,4-dienal
  (E)-Hex-2-enal
  3-Hydroxybutanone
  3-Methylacetophenone
  4-Methylacetophenone
  2-Methylbenzaldehyde
  3-Methylbenzaldehyde
  4-Methylbenzaldehyde
  3-Methylbutylpenta-2,4-dione
  6-Methylhept-5-en-2-one
  2-Methyltetrahydrofuran-3-one
  Nonadecan-3-one
  Nonan-2-one
  Octadecan-2-one
  Pentadecanal
  Pentadecan-2-one
  Pentadecan-3-one
  Pentan-2-one
  Pent-3-en-2-one
  Propiophenone
  Tetradecanal
  Tetradecan-2-one
  Tridecan-2-one
  Tridecan-3-one
  3,3,5-Trimethylcyclohexanone

Alcohols
  Benzyl alcohol
  Butan-1-ol
  Butan-2-ol

TABLE VII.III (*continued*)

Cyclopentanol
Decan-1-ol
Dodecan-1-ol
2-Ethylhexan-1-ol
Heptan-1-ol
Hexadecan-1-ol
Hexan-1-ol
Hex-1-en-3-ol
(*E*)-Hex-2-en-1-ol
(*E*)-Hex-3-en-1-ol
(*Z*)-Hex-3-en-1-ol
2-Methylbutan-2-ol
3-Methylbutan-1-ol
2-Methylbut-3-en-2-ol
4-(Methylethyl)benzyl alcohol
2-Methylpentan-2-ol
2-Methylpropan-1-ol
Nonadecan-1-ol
Nonan-1-ol
Nonan-3-ol
Octadecan-1-ol
Octan-1-ol
(*Z*)-Oct-2-en-1-ol
Pentan-1-ol
Pentan-2-ol
Pentan-3-ol
Pent-1-en-3-ol
(*E*)-Pent-2-en-1-ol
(*Z*)-Pent-2-en-1-ol
(*E*)-Pent-3-en-1-ol
Pent-4-en-2-ol
2-Phenylethanol
Tetradecan-1-ol
Tridecan-1-ol

Terpenes
Bornyl acetate
α-Cadinol
Camphene
Campholenealdehyde
Camphor
Carvone
Caryolan-1-ol
Chrysanthenone
1,8-Cineol
p-Cymene
Geraniol
Geranylacetone
Limonene
Linalool
(*E*)-Linalool oxide, furanoid
(*Z*)-Linalool oxide, furanoid

(*E*)-Linalool oxide, pyranoid
(*Z*)-Linalool oxide, pyranoid
Linalyl acetate
Linalyl propanoate
Mentha-2,8-dienol
p-Menth-1-ene
Menthol
Menthone
β-Myrcene
Myrtenal
Myrtenol
Neo-alloocimene
Nerol
(*E*)-Ocimene
(*Z*)-Ocimene
Pinocamphone
α-Terpinene
γ-Terpinene
Terpinen-4-ol
α-Terpineol
Terpinolene
Terpinyl acetate
α-Thujene
Thymol
Verbenone

Hydrocarbons
1,3-Diethylbenzene
2,5-Dimethylstyrene
Ethylbenzene
Propylbenzene
Styrene
Toluene
m-Xylene
o-Xylene
p-Xylene

Miscellaneous
Acetic acid
1-Methoxy-4-(prop-1-enyl)benzene
Benzonitrile
Benzothiazole
m-Cresol
1,4-Dimethoxybenzene
2,2-Dimethyl-5-(1-methylpropenyl)tetrahydrofuran
Dodecanoic acid
2-Ethylthiophene
Heptanoic acid
γ-Hexalactone
Hexanoic acid
3-Methylbutanoic acid
5-(Methylethyl)-2-methyl-2-vinyltetrahydrofuran

(continued)

TABLE VII.III (*continued*)

Methylpyrazine	Octanoic acid
2-Methylthiophene	Phenol
3-Methylthiophene	1-Phenoxybutene
γ-Nonalactone	(*E*)-Theaspirane
Nonanoic acid	2,6,6-Trimethyl-2-vinyltetrahydropyran

[a] Idstein *et al.* (1984).

TABLE VII.IV

Volatile flavour components of custard apple fruit[a]

Terpenes	Sesquiterpenes
Borneol	Aromadendrene
Bornyl acetate	Bicyclogermacrene
Camphene	δ-Cadinene
1,8-Cineol	α-Cadinol
p-Cymene	T-Cardinol
Humulene	Caryophyllene
Isopinocarveol	α-Copaene
Limonene	α-Cubebene
Linalool	β-Cubebene
Myrcene	β-Elemene
β-(*E*)-Ocimene	Germacrene D
β-(*Z*)-Ocimene	Globulol
α-Phellandrene	T-Muurrolol
β-Phellandrene	Spathulenol
α-Pinene	Viridifluorene
β-Pinene	Viridifluol
α-Terpinene	
Terpinen-4-ol	Hydrocarbons
α-Terpineol	α-*p*-Dimethylstyrene
Terpinolene	

[a] Wyllie *et al.* (1987).

predominately esters and other non-terpenoid compounds (MacLeod and Pieris, 1981b; Paull *et al.*, 1983; Idstein *et al.*, 1984).

Over 80% of the extract was comprised of β-pinene (26%), α-pinene (21%), germacrene D (11%), bicyclogermacrene (15%), limonene (7%) and bornyl acetate (3%). Also present were oxygenated compounds in minor amounts, particularly the sesquiterpene alcohols globulol, spathulenol and cardinols. No sensory data were given on the contribution any of these compounds may make to the flavour of custard apples.

Wyllie *et al.* (1987) found no marked difference in the relative levels of the individual components or in the total amount of volatiles produced by mature fruit ranging from under-ripe to over-ripe.

Soursop (*Annona muricata*) is from tropical America and the West Indies, and has the largest fruit of the *Annona* species, weighing up to 4 kg. It is slightly more acid than other species (MacLeod and Pieris, 1981b), although the name 'soursop' is said to refer to the sour odour of the skin (Johns and Stevenson, 1979).

The fruit is pleasantly aromatic and juicy, but can be fibrous. The rich creamy juice is popular as a drink on its own or mixed with milk to make 'champola' (Tankard, 1987).

MacLeod and Pieris (1981b) studied the volatile components extracted from fruit pulp using a modified Likens – Nickerson apparatus and 2-methylbutane. The extract, concentrated by low-temperature high-vacuum distillation, yielded an essence with characteristic soursop aroma.

Esters comprised nearly 80% of the sample, of which methyl hexanoate (31%) and methyl hex-2-enoate (27%) were the major components. A series of methyl and ethyl esters of C-4, C-6 and C-8 straight-chain carboxylic acids and the six corresponding 2-enoates were found. Other compounds found at levels greater than 1% were β-farnesene (6%), dichloromethane (6%), hexan-1-ol (3%), methyl nicotinate (methyl pyridine-3-carboxylate, 2%), along with minor amounts of sesquiterpenes and some branched-chain hydrocarbons. A sulphur/nitrogen heterocyclic compound, which is unusual for flavour volatiles of fruit, was tentatively identified. No individual component was found to be characteristic of soursop aroma. The unique aroma probably results from the blend of the 15 esters and β-farnesene.

The postharvest physiological changes in soursop fruit during ripening were studied by Paull (1982). Fruit became more bland and developed a slight off-flavour as they approached the over-ripe state. This correlated with a decline in the titratable acidity and total phenols. Ester production in soursop began to rise 3 days after harvest, peaking 2 – 3 days later, followed by a dramatic fall and with a gradual increase in the levels of a number of unidentified peaks in the gas chromatogram (Paull *et al.*, 1983). They suggested that some of these peaks may be responsible for the off-odour detected in over-ripe fruit, and speculated that one of the peaks may correspond to an unidentified peak reported as having a rancid odour by MacLeod and Pieris (1981b).

A list of the volatiles identified in soursop is given in Table VII.V. Paull *et al.* (1983) found that methyl butanoate was the most abundant ester, whereas MacLeod and Pieris (1981b) found methyl hexanoate to be about five times more abundant. Bearing in mind the different fruit and the different techniques used to study the volatiles, the composition of the volatiles and hence the flavour may be critically dependent on the degree of ripeness, as is the case with kiwifruit (Young and Paterson, 1985).

Propyl 4-hydroxybutanoate has been shown to enhance the flavour of natural soursop nectar. Sensory evaluation clearly recognised the stronger and more 'aromatic' flavour of samples to which the compound had been added (Natara-

TABLE VII.V

Volatile flavour components of soursop fruit[a]

Esters	Alcohols
Ethyl butanoate	Hexan-1-ol
Methyl butanoate	(Z)-Hex-3-en-1-ol
Ethyl but-2-enoate	2-Methylhexan-1-ol
Methyl but-2-enoate	
Methyl furoate	Terpenes
Ethyl hexanoate	(E)-β-Farnesene
Methyl hexanoate	
Ethyl hex-2-enoate	Hydrocarbons
Methyl hex-2-enoate	Styrene
Methyl hex-3-enoate	Toluene
Methyl nicotinate (methyl pyridine-3-carboxylate)	
Ethyl octanoate	Miscellaneous
Methyl octanoate	Chloroform
Ethyl oct-2-enoate	Dichloromethane
Methyl oct-2-enoate	1,1-Diethoxyethane

[a] MacLeod and Pieris (1981b).

jan and Karunanithy, 1973). Although propyl 4-hydroxybutanoate has a strong soursop-like aroma, it has not been identified in soursop volatiles.

8. CARAMBOLA

Carambola, *Averrhoa carambola* L. (Oxalidaceae), is originally from S.E. Asia, but it is also grown in warmer regions of China, India, South America and Australia (Johns and Stevenson, 1979). A highly ornamental fruit, it is about 10 cm in length and star-shaped in cross section. Because of its unusual shape, it is also known as starfruit or five corners. The fruit is eaten fresh or used as a juice, and has a sweet refreshing taste, described variously as apple- or apricot-like or grape-like (Wilson *et al.,* 1985). The fruit contain up to 0.7% oxalic acid, depending on cultivar (Wagner *et al.,* 1975). Removing the 'wing' edges, which contain most of the oxalic acid, was found to improve the flavour (Tankard, 1987).

The colour of different cultivars ranges from white to deep yellow. The yellow-coloured fruit are more attractive, but the white varieties are reported to be sweeter. Wagner *et al.* (1975) found wide variability in the flavour qualities and composition between cultivars grown in Florida.

Shiota (1984) used the Likens–Nickerson technique to extract the volatile components from carambola. Ethyl anthranilate (ethyl *o*-aminobenzoate) was the major component. There was a high level of ethyl *N*-methylanthranilate, a compound which had not previously been identified from a botanical source.

Among the methyl and ethyl esters found were benzoates, diethyl succinate, methyl salicylate, ethyl hexa-2, 4-dienoate (ethyl sorbate), ethyl cinnamate and methyl *N*-methylanthranilate. Terpenes were detected in minor amounts.

An essence of carambola was prepared by dichloromethane extraction of the aqueous filtrate of pureed fruit by Wilson *et al.* (1985). Of the 40 components identified, methyl anthranilate was the most abundant compound (21%). The compounds identified in the two investigations are listed in Table VII.VI.

Several components were considered by Wilson *et al.* (1985) to be important contributors to the characteristic carambola aroma notes. The strong grape-like note in the fresh fruit and the solvent extract was consistent with the presence of methyl anthranilate (methyl *o*-aminobenzoate). Other esters, especially ethyl acetate and ethyl butanoate, were important, imparting the warm fruity character to carambola fruit. The green grassy note was attributed to 6-methyl-hept-5-en-2-one and the corresponding alcohol, (*Z*)-, and (*E*)-hex-3-en-1-ol and hexanal. They found that the green grassy aroma was less apparent in the extract than in the fruit. Phenylethanol (3.4%) and the acetate (4.6%), with their strong floral rose-like aroma (Arctander, 1969), were also considered to be important.

There are major differences between the results of the two investigations. Wilson *et al.* (1985) did not detect the presence of ethyl anthranilate in their extract, whereas it was the major component in the extract prepared by Shiota (1984). Methyl anthranilate was a minor component (*ca.* 1%) in the study by Shiota (1984), but was the major component (21%) in the extract of Wilson *et al.* (1985). The differences observed may be due to the cultivars involved, the condition of the fruit, or the methods used in preparing the extracts. For example, it is possible that the ethyl anthranilate is an artifact of exposure of the fruit to elevated temperature.

9. CASHEW APPLE

The cashew, *Anacardium occidentale* (Anacardiaceae) is a native of tropical America. The tree is better known for the popular cashew nut, of which major producing countries are India, Brazil, Mozambique, Tanzania and Kenya (Sturtz, 1984). The cashew apple is the peduncle to which the cashew nut is attached. It is consumed as fresh fruit, or used in pickles, jellies, jams, chutneys and wine (Johns and Stevenson, 1979), and in the preparation of an alcoholic beverage called 'Fenny' in Goa (Ramteke *et al.*, 1984). Until fully ripe the cashew apple is very sour and astringent (Maciel *et al.*, 1986), but ripe fruit are edible and develop a pleasant characteristic flavour on cooking (MacLeod and de Troconis, 1982c). A cashew apple brandy is non-astringent, unlike the wine, and still retains the 'exotic' flavour of cashew apple (Rao, 1985).

MacLeod and de Troconis (1982c) used the Likens – Nickerson technique to isolate the flavour volatiles from fresh Venezuelan fruit. The yield of volatiles at 3.6 µg/kg fresh weight is consistent with the weak flavour of this fruit. The major compounds identified in the extract were terpene hydrocarbons (38%), especially car-3-ene, and aldehydes (26%). Esters accounted for only 0.7% of the volatiles.

TABLE VII.VI

Volatile flavour components of carambola fruit[a]

Esters	
Cinnamyl acetate	W
Ethyl acetate	W
(E)-Hex-2-enyl acetate	S
3-Methylbutyl acetate	S
Phenethyl acetate	W
Ethyl anthranilate (ethyl o-aminobenzoate)	S
Methyl anthranilate	S, W
Ethyl benzoate	S, W
Methyl benzoate	S, W
Ethyl butanoate	S, W
Ethyl (E)-but-2-enoate	S
Ethyl cinnamate	S
Ethyl decanoate	S
Diethyl glutarate	W
Ethyl hexanoate	S, W
Methyl hexanoate	S
Ethyl hexa-2,4-dienoate	S, W
Ethyl N-methylanthranilate	S
Methyl N-methylanthranilate	S
Ethyl nicotinate (ethyl pyridine-3-carboxylate)	W
Ethyl nonanoate	S
Ethyl octanoate	S
Methyl salicylate	S, W
Diethyl succinate	S, W
Aldehydes and ketones	
Acetaldehyde	W
Acetophenone	S, W
Benzaldehyde	W
Cinnamaldehyde	W
Heptanal	S
Hexanal	W
2-Methylacetophenone	W
3-Methylbutan-2-one	W
6-Methylhept-5-en-2-one	W
Nonanal	S
Octanal	S
Alcohols	
Benzyl alcohol	W
Decan-1-ol	S
Hexan-1-ol	W
(E)-Hex-2-en-1-ol	S
(Z)-Hex-3-en-1-ol	S, W
(E)-Hex-3-en-1-ol	W
3-Methylbutan-1-ol	W
6-Methylhept-5-en-2-ol	W
2-Methylpropan-1-ol	W
Nonan-1-ol	S

TABLE VII.VI (*continued*)

Octan-1-ol	S, W
Pentan-1-ol	W
Pent-1-en-3-ol	W
Phenylethanol	W
Terpenes	
Borneol	W
Carvone	W
1,8-Cineol	W
3,4-Dihydro-β-ionone	S
β-Ionone	S, W
Limonene	W
(*E*)-Ocimene	S
α-Pinene	W
β-Pinene	W
Terpinen-4-ol	W
Hydrocarbons	
Eicosane	S
Heptadecane	S
Nonadecane	S
Pentacosane	S
Pentadecane	S
Theaspirane	S
Tricosane	S
Tridecane	S
Miscellaneous	
Benzothiazole	W
1,2-Dimethoxybenzene	W
3,5,5-Trimethylcyclohex-2-enone	S
Quinoline	W

[a] S = Shiota (1984); W = Wilson *et al.* (1985).

Car-3-ene was the major component (24%). This terpene is not a common fruit volatile, but was also found in high levels in mango, another Anacardiaceae member (MacLeod and de Troconis, 1982b; MacLeod and Snyder, 1985). In contrast to the results of MacLeod and de Troconis (1982c), Maciel *et al.* (1986), using headspace concentration, found esters were the major components in the volatiles of both processed and unprocessed juice. The main ones were ethyl esters of acetic, 3-methylbutanoic and hexanoic acids. Aldehydes and carene were only minor components in the juice. Interestingly, relatively large amounts of methylpropanoic acid and 3-methylbutanoic acid were present. Dimethyl sulphide, dimethyl disulphide and dimethyl trisulphide were detected in both processed juice, which had been treated with SO_2, and unprocessed juice. A list of the volatile components identified in cashew apple by MacLeod and de Troconis (1982c), and Maciel *et al.* (1986) is given in Table VII.VII.

298

TABLE VII.VII

Volatile flavour components of cashew apples[a]

Esters	
Ethyl acetate	A, B
3-Methylbutyl acetate	B
Pentyl acetate	B
Ethyl benzoate	B
Methyl benzoate	B
Ethyl butanoate	B
Methyl butanoate	B
3-Methylbutyl butanoate	B
Ethyl but-2-enoate	B
Ethyl heptanoate	B
Ethyl hexanoate	B
Methyl hexanoate	B
Ethyl hex-2-enoate	B
Ethyl 2-mercaptopropanoate (i)	B
Methyl 2-methylenebutanoate	B
Methyl 2-methylbutanoate	B
Ethyl 3-methylbutanoate	B
3-Methylbut-3-enyl 3-methylbutanoate	B
Propyl 3-methylbutanoate	B
Ethyl 3-methylbut-2-enoate	B
Ethyl 4-methyloctanoate	B
Butyl 2-methylpentanoate	B
Ethyl 3-methylpentanoate	B
Ethyl octanoate	A, B
Ethyl pentanoate	B
1-Methylethyl pentanoate	B
2-Methylpropyl pentanoate	B
Methyl pent-2-enoate	B
Aldehydes and ketones	
Acetophenone	A, B
Benzaldehyde	A, B
2-Furaldehyde	A, B
Hexanal	A
(E)-Hex-2-enal	A
3-Hydroxybutanone	B
3-Methylbutanal	A
Nonanal	A
(E)-Non-2-enal	A
Pentan-2-one	B
Phenylacetaldehyde	A
Alcohols	
Benzyl alcohol	B
3-Ethylpentan-3-ol	B
Hex-3-en-1-ol	B
(Z)-Hex-3-en-1-ol	A
3-Methylbutan-1-ol	B
6-Methylheptan-1-ol	B

TABLE VII.VII (*continued*)

2-Methylpentan-1-ol	A
4-Methylpentan-1-ol	B
Octan-1-ol	B
2-Phenylethanol	B
Terpenes	
Carene	B
Car-3-ene	A
Caryophyllene	A, B
Limonene	A, B
α-Phellandrene	A
β-Phellandrene	B
α-Selinene	A
α-Terpinene	A
γ-Terpinene	B
Hydrocarbons	
Ethylcyclohexane	A
Methylcyclohexane	A
Octane	A
Toluene	A
m and/or *p*-Xylene	A
o-Xylene	A
Miscellaneous	
Dimethyl disulphide	B
Dimethyl trisulphide	B
γ-Hexalactone	B
3-Methylbutanoic acid	B
Methylpropanoic acid	B
Pyridine	B

[a] A = MacLeod and de Troconis (1982c); B = Maciel *et al.* (1986).
(i) = Tentative identification.

When cashew apple juice is concentrated, most of the aroma is lost. The recovery of the aroma compounds during juice concentration was investigated by Ramteke *et al.* (1984). They prepared an aroma concentrate with a typical strong cashew apple character which could be used to improve the flavour of concentrated juice.

On the basis of the levels found and/or odour qualities, MacLeod and de Troconis (1982c) suggested that important flavour compounds include hexanal, car-3-ene, limonene, (*E*)-hex-2-enal and benzaldehyde. However, they were not able to make positive deductions as to which compounds are important contributors to the characteristic cashew apple flavour. With cashew apple juice, Maciel *et al.* (1986) considered that various esters contribute sweet, fruity flavour characters, while the two volatile acids contribute the pungent/sour character. Assessments obtained at an odour port during GC showed that individually none of these compounds duplicates the flavour of the juice.

There was a distinguishable flavour difference between processed and un-processed juices, but this was mainly attributed to the unpleasant odour of sulphur dioxide which was added to the processed juice as a preservative. Lower levels of SO_2 and methylpropanoic acid and 3-methylbutanoic acid could lead to improvement of the flavour of the juice (Maciel *et al.,* 1986).

It is difficult to determine if the very different volatile profiles obtained by MacLeod and de Troconis (1982c) and Maciel *et al.* (1986) are the result of the juicing process, since the later report gave little details on the process used. The differences may also be explained by the cultivars involved as again no details are available, or by the method of isolation. High temperature is known to liberate or alter terpene levels in grape and passion fruit (Engel and Tressl., 1983).

10. MOUNTAIN PAPAYA

Mountain papaya (*Carica pubescens* Lenée and Koch) from the Caricaceae family is also known as mountain pawpaw or papayuela. Originating in Colombia and Ecuador, mountain papaya is a popular fruit in Chile where it is grown commercially for canning (Idstein *et al.,* 1985). The fruit are also eaten fresh. It has a strong aromatic flavour which is stable to heating and storage (MacKenzie and Strachan, 1980).

There have been two reports on the volatile components of mountain papaya which differed in the methods used to extract the volatiles and the source of the fruit (Idstein *et al.,* 1985; Morales and Duque, 1987). A combined list of the volatiles identified in the two studies is given in Table VII.VIII.

From Chilean fruit, Idstein *et al.* (1985) used vacuum steam distillation followed by solvent extraction of the distillate to isolate the volatiles, which were separated into three fractions by column chromatography on silica gel. The 197 components identified included 103 esters, 28 hydrocarbons, 30 carbonyl compounds, 15 alcohols and 8 lactones. Esters were quantitatively the major components, especially methyl and ethyl butanoate and butyl acetate (which were at levels > 800 ppm). Apart from the esters, the other main components (50 – 300 ppm) were butan-1-ol, linalool, benzyl isothiocyanate, 3-hydroxybutanone and 3-acetoxybutanone.

Morales and Duque (1987) identified 45 components in the volatiles isolated from Colombian fruit using a modified Likens – Nickerson apparatus. Recovery of volatiles was high at 110 mg/kg. Esters comprised 63% of the total volatiles, and were dominated by ethyl and butyl esters of C-2 to C-8 acids, including several 3-hydroxy esters and 2-enoates. Alcohols accounted for 30% of the total volatiles, and were mainly even-carbon-numbered compounds. The major components were ethyl butanoate (33%) and butan-1-ol (26%).

Differences were found in the aroma volatiles of fruit grown in Chile (Idstein *et al.,* 1985) and Colombia (Morales and Duque, 1987). For example, methyl (*Z*)-hex-3-enoate, 3-methylbutyl acetate, methyl 3-hydroxyhexanoate, and ethyl

TABLE VII.VIII

Volatile flavour components of mountain papaya fruit[a]

Esters	
Benzyl acetate	I
Butyl acetate	I, M
Ethyl acetate	I, M
Hexyl acetate	I, M
2-Methylbutyl acetate	I
3-Methylbutyl acetate	I
Octyl acetate	I, M
Phenethyl acetate	I
Propyl acetate	I
Ethyl 4-acetoxybutanoate	I
Butyl benzoate	I, M
Ethyl benzoate	I, M
Methyl benzoate	I
Benzyl butanoate	I
Butyl butanoate	I, M
Ethyl butanoate	I, M
Hexyl butanoate	I
Methyl butanoate	I, M
2-Methylbutyl butanoate	I
3-Methylbutyl butanoate	I
Octyl butanoate	I
Phenethyl butanoate	I
Propyl butanoate	M
Butyl but-2-enoate	I
Ethyl but-2-enoate	I
Hexyl but-2-enoate	I
Methyl but-2-enoate	I
Butyl (E)-but-2-enoate	M
Ethyl (E)-but-2-enoate	M
Butyl decanoate	I, M
Ethyl decanoate	I
Hexyl decanoate	I
Methyl decanoate	I
2-Methylbutyl decanoate	I
Butyl dodecanoate	I
Ethyl dodecanoate	I
Methyl dodecanoate	I
2-Methylbutyl dodecanoate	I
Butyl 2-furanoate	I
Ethyl 2-furanoate	I
Methyl 2-furanoate	I
Ethyl heptanoate	M
Methyl heptanoate	I, M
Ethyl hexadecanoate	I
Methyl hexadecanoate	I
Butyl hexanoate	I, M
Ethyl hexanoate	I, M
Hexyl hexanoate	I
Methyl hexanoate	I, M

(continued)

TABLE VII.VIII (*continued*)

Propyl hexanoate	I
Ethyl (*E*)-hex-2-enoate	I
Methyl (*E*)-hex-2-enoate	I
Butyl (*Z*)-hex-3-enoate	I
Ethyl (*Z*)-hex-3-enoate	I
Hexyl (*Z*)-hex-3-enoate	I
Methyl (*Z*)-hex-3-enoate	I
Butyl 2-hydroxybenzoate	I, M
Ethyl 2-hydroxybenzoate	I, M
Methyl 2-hydroxybenzoate	I
Butyl 3-hydroxybutanoate	I, M
Ethyl 3-hydroxybutanoate	I, M
Hexyl 3-hydroxybutanoate	I
Methyl 3-hydroxybutanoate	I
Ethyl 4-hydroxybutanoate	I
Butyl 3-hydroxyhexanoate	I
Ethyl 3-hydroxyhexanoate	I
Hexyl 3-hydroxyhexanoate	I
Methyl 3-hydroxyhexanoate	I
Butyl 3-hydroxyoctanoate	I
Ethyl 3-hydroxyoctanoate	I
Methyl 3-hydroxyoctanoate	I
Methyl 4-methoxybenzoate	I
Ethyl 2-methylbenzoate	I
Methyl 2-methylbenzoate	I
Butyl 2-methylbutanoate	I
Ethyl 2-methylbutanoate	I, M
Methyl 2-methylbutanoate	I
Ethyl 3-methylbutanoate	I
Butyl methylpropanoate	I
Ethyl methylpropanoate	I
Butyl nicotinate (butyl pyridine-3-carboxylate)	I
Ethyl nicotinate	I
Methyl nicotinate	I
Methyl nonanoate	I
Butyl octanoate	I, M
Ethyl octanoate	I, M
Hexyl octanoate	I, M
Methyl octanoate	I, M
2-Methylbutyl octanoate	I
Butyl oct-2-enoate	M
Ethyl oct-2-enoate	M
Methyl (*E*)-oct-2-enoate	I
Methyl (*E*)-oct-3-enoate	I
Ethyl pentanoate	I
Methyl pentanoate	I
Methyl pentadecanoate	I
Butyl 2-phenylacetate	I
Ethyl 2-phenylacetate	I
Methyl 2-phenylacetate	I
Butyl propanoate	I, M
Ethyl propanoate	I, M

TABLE VII.VIII (*continued*)

Butyl tetradecanoate	I
Ethyl tetradecanoate	I
Methyl tetradecanoate	I
Ethyl thiopropanoate	I
Methyl tridecanoate	I
Methyl undecanoate	I
Aldehydes	
Acetaldehyde	M
Benzaldehyde	I
But-2-enal	I, M
Decanal	I
Dodecanal	I
2-Furaldehyde	I
4-Methylbenzaldehyde	I
2-Methyl-(E)-but-2-enal	I
(E)-Pent-2-enal	I
Phenylacetaldehyde	I
Tetradecanal	I
Ketones	
Acetone	M
Acetophenone	I
3-Acetoxybutanone	I
Benzophenone	I
Butane-2,3-dione	I
But-3-en-2-one	M
Cyclohexanone	I
Decan-2-one	I
Heptan-2-one	I, M
Hexadecan-2-one	I
Hexan-2-one	I
2-Hydroxyacetophenone	I
3-Hydroxybutanone	I
3-Methylacetophenone	I
6-Methylhept-5-en-2-one	I
Nonan-2-one	I
Octan-2-one	I
Oct-1-en-3-one	I, M
Tetradecan-2-one	I
Tridecan-2-one	I
Alcohols	
Benzyl alcohol	I, M
Butan-1-ol	I, M
Decan-1-ol	I
Ethanol	M
Hexan-1-ol	I, M
3-Methylbutan-1-ol	I
Octan-1-ol	I, M
Oct-1-en-3-ol	I
Pentan-1-ol	I

(*continued*)

TABLE VII.VIII (*continued*)

Pentan-2-ol	M
2-Phenylethanol	I
Terpenes	
Car-3-ene	I
1,8-Cineol	I, M
β-Damascone	I
β-Farnesene	I
Farnesol	I
Fenchone	I
β-Ionone	I
Linalool	I
(*E*)-Linalool oxide	M
(*E*)-Linalool oxide, furanoid	I
(*Z*)-Linalool oxide, furanoid	I
Linalyl acetate	I
Neryl acetate	I
Terpinen-4-ol	I
α-Terpineol	I
Hydrocarbons	
Alkanes C12 – C22	I
Alkenes C13, 15, 17, 19, 21, 23	I
Ethylbenzene	I
2-Methylnaphthalene	I
Naphthalene	I
Propenylbenzene	I
1,2,4,5-Tetramethylbenzene	I
1,2,3-Trimethylbenzene	I
o-, *m*-, *p*-Xylenes	I
Lactones	
γ-Heptalactone	I
δ-Heptalactone	I
γ-Hexalactone	I
γ-Nonalactone	I
δ-Nonalactone	I
γ-Octalactone	I
δ-Octalactone	I
γ-Undecalactone	I
Miscellaneous	
Acetic acid	I
Benzyl isothiocyanate	I
Butanoic acid	I
Decanoic acid	I
2,5-Dimethylthiophene	I
Hexanoic acid	I
2-Methylthiophene	I
3-Methylthiophene	I
Octanoic acid	I
Phenylacetonitrile	I
2-Phenethyl isothiocyanate	I

[a] I = Idstein *et al.* (1985); M = Morales and Duque (1987).

nicotinate (ethyl pyridine-3-carboxylate), major components in the Chilean fruit, were not detected in the Colombian fruit. Benzyl isothiocyanate, commonly found in Caricaceae, was detected at a level greater than 50 ppb in the Chilean mountain papaya (Idstein *et al.*, 1985), but was not reported in the Colombian fruit (Morale and Duque, 1987). Although the differences may be due to the different geographic regions in which the plants were grown (Morale and Duque, 1987), it should be noted that different methods of isolation were used.

The extract obtained by Idstein *et al.* (1985) had typical sensory properties of the fresh fruit, but when fractionated by chromatography on silica gel the individual fractions did not retain the fresh aroma. Fraction I (mainly hydrocarbons) was practically odourless, fraction II (esters, carbonyls and thiophene derivatives) had a strong fruity aroma, and fraction III (esters, lactones, alcohols, acids and oxygenated terpenes), with its floral-fruity notes, was considered to be closest to the original fruit aroma. Ethyl butanoate (32 mg/kg), ethyl acetate (29 mg/kg), butan-1-ol (20 mg/kg) and butyl acetate (8 mg/kg) were the major components found in the essence from Colombian fruit. A mixture of these chemicals in the proportions found in the essence had an aroma closely resembling that of the fresh fruit (Morales and Duque, 1987).

11. BABACO

Babaco, *Carica pentagona* Heilborn (Caricaceae), from Ecuador is a hybrid between *C. pubescens* Lenné and Koch (mountain pawpaw) and *C. stipulata* Badillo (Bollard, 1982). They are grown extensively in Ecuador and have been grown commercially in New Zealand where they were introduced in 1973. The fruit are larger than those of mountain pawpaw and are very juicy. They are either eaten fresh or canned.

The flavour is described as pleasant and fruity, but quite distinct from that of mountain pawpaw (Shaw *et al.*, 1985). In addition to their subtle and refreshing flavour, they have a comparatively low sugar content, *ca.* 6% SSC (Skinner, 1983).

Shaw *et al.* (1985) isolated the flavour volatiles from babaco using low-temperature high-vacuum distillation followed by ether extraction. The volatile components identified are given in Table VII.IX. This fruit is unique in that the volatiles are largely composed of aliphatic alcohols (75%), ranging from C-2 to C-10, with butan-1-ol (46%), hexan-1-ol (12%) and octan-1-ol (8%) the most dominant. Esters constitute 18% of the total volatiles, of which ethyl butanoate (8%) and ethyl hexanoate (5%) are the most abundant.

Babaco is similar to *C. pubescens* (Idstein *et al.*, 1985; Morales and Duque, 1987) in that the predominant volatiles are even-carbon-numbered alcohols and their esters derived from even-carbon-numbered acids. In addition, most of the compounds found in babaco are also found in mountain papaya. The main difference between the hybrid and its parent (*C. pubescens*) is in the relative amounts of the components, with esters predominating in the mountain papaya

306

and alcohols in the babaco. This is reflected in the much stronger aromatic flavour found in mountain papaya (MacKenzie and Strachan, 1980).

Although assessment at an odour port of the GC indicated that ethyl butanoate and ethyl hexanoate have babaco-like flavour (Shaw *et al.*, 1985),

TABLE VII.IX

Volatile flavour components of babaco fruit[a]

Esters	
Ethyl acetate	*
Methyl acetate	
Ethyl benzoate	*
Butyl butanoate	*
Ethyl butanoate	*
Heptyl butanoate	
Hexyl butanoate	*
Ethyl (*E*)-but-2-enoate	*
Butyl hexanoate	*
Ethyl hexanoate	*
Hexyl hexanoate	*
Octyl hexanoate	
Aldehydes and ketones	
Acetaldehyde	*
(*E*)-Hex-2-enal	
Octan-2-one	*
Tridecan-2-one	*
Alcohols	
Butan-1-ol	*
Decan-1-ol	*
Ethanol	*
Hexan-1-ol	*
(*E*)-Hex-2-en-1-ol	
(*Z*)-Hex-3-en-1-ol	
3-Methylbutan-1-ol	*
2-Methylpropan-1-ol	
Nonan-1-ol	
Octan-1-ol	*
Pentan-1-ol	*
Propan-1-ol	
Propan-2-ol	
Terpenes	
1,8-Cineol	*
Geraniol	
Linalool	*
α-Terpineol	*

[a] Shaw *et al.* (1985).
* Also identified in mountain papaya (Idstein *et al.*, 1985; Morales and Duque, 1987).

these esters are major components of many other fruit. The report did not comment on the significance of the alcohols to the overall flavour.

12. JACKFRUIT

A native of India, the jackfruit or jakfruit, *Artocarpus heterophyllus* Lam. syn. *A. integrifolia,* is an important fruit throughout tropical Asia, Africa and America (Johns and Stevenson, 1979). It is the largest cultivated fruit, up to 90 cm long (Sedgley, 1984) and up to 46 kg in weight (Tankard, 1987), and is in the same family (Moraceae) as the breadfruit. The ripe jackfruit are sweet and juicy, and have an aroma varying from pineapple-like to pungent, depending on the cultivar. Ripe fruit are usually eaten fresh or preserved, and immature fruit may be used as a vegetable similar to other starch plants. The seeds may also be eaten, either boiled or roasted.

The volatile components of jackfruit were isolated by Swords *et al.* (1978) using vacuum distillation followed by dichloromethane extraction of the distillate. No quantitative data were given for the 16 esters and 4 alcohols that were identified. However, the major peaks in the gas chromatogram were butyl butanoate, butyl 3-methylbutanoate, 3-methylbutyl 3-methylbutanoate, ethyl 3-methylbutanoate, butyl acetate and butan-1-ol. The majority of the volatiles were esters derived from C-4 and C-5, branched and straight chain acids and alcohols. No terpenoid compounds were detected. The mild fruity flavour of jackfruit was attributed to esters.

The flavour of processed jackfruit was found to be enhanced by the addition of ethyl 4-hydroxybutanoate and butyl 4-hydroxybutanoate at a rate of 120 and 100 ppm, respectively (Natarajan and Karunanithy, 1974). Neither of these compounds has been found in jackfruit.

13. DURIAN

Durian, *Durio zibethinus* Murr. (Bombacaceae), is one of the most unusual of tropical fruits. It releases an extremely unpleasant odour, yet the fruit are highly prized throughout South East Asia for their sweet and unique flavour. Most Europeans find the durian odour nauseating and repulsive. The durian has two distinct odour notes, one strong and onion-like, and the other delicate and fruity, while a fetid note develops if fruit are stored in a box (Baldry *et al.* 1972).

The fruit are very nutritious and sweet, developing a total soluble solids content of 36% with storage (Martin, 1980). Ripe durian are eaten fresh, or prepared in various ways, while unripe fruit may be cooked as a vegetable (Chin and Yong, 1980).

The first identification of durian volatiles was by Baldry *et al.* (1972). Using fruit from Singapore, they prepared a distillate which was analysed by various chemical tests and thin layer chromatography to measure thiols and thioethers, and by GC-MS to identify the other volatiles. A total of 25 compounds were identified, consisting of 7 sulphur compounds, 12 aliphatic esters, 2 aldehydes and 4 alcohols. Ethyl 2-methylbutanoate, ethanol and propan-1-ol were the major components identified by GC-MS. In an extract prepared in a similar way from Kuala Lumpur fruit, the major sulphur compound was propane-1-thiol and not thioethers as found in the Singapore fruit.

Moser *et al.* (1980) used headspace sampling to investigate the volatile sulphur components of durian from Thailand and steam distillation to identify the fruity aroma components. Ethyl 2-methylbutanoate, 1,1-diethoxyethane and ethyl acetate were identified in the distillate. In contrast to the distillate from durian fruit in which thioethers and propane-1-thiol were the major sulphur components (Baldry *et al.,* 1972), diethyl disulphide and diethyl trisulphide were the main sulphur compounds identified in the headspace. It is interesting to note that Baldry *et al.* (1972) found seven sulphur compounds, and while Moser *et al.* (1980) found eight, only two compounds were common to both (Table VII.X).

The levels of the main sulphur components along with H_2S were found to increase with maturity, with no H_2S detected in immature fruit. The major sulphur component of immature fruit was ethyl hydrodisulphide, the level of which decreased with maturity and was not detected in fully mature fruit (Moser *et al.,* 1980).

The onion-like odour of durian was found to be due mainly to thioethers in fruit from Singapore, and to thiols in fruit from Kuala Lumpur (Baldry *et al.,* 1972). Hydrogen sulphide and diethyl disulphide were thought to be responsible for the fetid odour. The fruity aroma note of durian was attributed to ethyl 2-methylbutanoate, which was the most abundant ester. This is in accord with

TABLE VII.X

Volatile sulphur compounds in durian fruit[a]

Hydrogen sulphide	B, M
Diethyl disulphide	B, M
Ethyl propyl disulphide	M
Diethyl trisulphide	M
Ethyl methyl trisulphide	M
Ethyl propyl trisulphide	M
Diethyl tetrasulphide	M
Diethyl thioether	B
Dimethyl thioether	B
Ethanethiol	B
Methanethiol	B
Propane-1-thiol	B
Ethyl hydrodisulphide (i)	M

[a] B = Baldry *et al.* (1972); M = Moser *et al.* (1980).
(i) In immature fruit only.

its low odour threshold (Flath *et al.*, 1967). An aqueous solution of 20 ppm of ethyl 2-methylbutanoate and 2.5 ppm of propane-1-thiol had an aroma similar, but not identical, to durian.

Moser *et al.* (1980) considered that hydrogen sulphide, hydrodisulphides, dialkyl polysulphides, ethyl esters and 1,1-diethoxyethane were the principal contributors to the aroma of durian.

14. MANGOSTEEN

The mangosteen (*Garcinia mangostana*) has been described as the 'Queen of tropical fruits', because it is considered one of the most delicious of fruit. This member of the Guttiferae family is native to the tropical jungles of Malaysia and Sumatra, and has long been cultivated in South East Asia (Tankard, 1987). The fruit is normally eaten fresh because while the flavour is stable under cool storage, the subtle aroma is lost during processing (Martin, 1980).

The volatile flavour components were investigated by MacLeod and Pieris (1982), who used the Likens – Nickerson technique to isolate the volatiles. The relatively low yield of volatiles (3 μg/kg) compared with other tropical fruit was not surprising because of the 'delicate and fugitive' flavour of mangosteen. Of the 52 components detected, 28 (91%) of the extract) were identified and are listed in Table VII.XI. Quantitatively, the major components were (Z)-hex-3-en-1-ol (27%), octane (15%), hexyl acetate (8%) and α-copaene (7%).

TABLE VII.XI

Volatile flavour components of mangosteen fruit[a]

Esters	Hydrocarbons
Hexyl acetate	Ethylcyclohexane
(*Z*)-Hex-3-enyl acetate	Heptane
	Octane
Aldehydes and ketones	Toluene
Acetone	*o*-, *m*-, *p*-Xylene
Benzaldehyde	
Hexanal	Terpenes
(*E*)-Hex-2-enal	α-Copaene
2-Furaldehyde	α-Terpineol
Furfuryl methyl ketone	Guaiene
5-Methyl-2-furaldehyde	Valencene
Nonanal	δ-Cadinene
Phenylacetaldehyde	γ-Cadinene
Alcohols	Miscellaneous
Hexan-1-ol	Dichloromethane
(*Z*)-Hex-3-en-1-ol	Pyridine

[a] MacLeod and Pieris (1982).

310

All of the compounds identified are common fruit volatiles and there appeared to be no unique character impact compound. Based on odour assessment, the main contributors to the mangosteen flavour were considered to be hexyl acetate, (Z)-hex-3-enyl acetate, and (Z)-hex-3-en-1-ol (as a consequence of the high level present). A mixture of these three compounds in the correct proportions had an aroma reminiscent of mangosteen, but harsher and more astringent (MacLeod and Pieris, 1982).

15. LOQUAT

The loquat (*Eriobotrya japonica* Lindl.), also known as the Japanese plum, is a member of the Roseaceae family, and is thought to have originated in China. It is an important fruit in Japan and Israel, but is grown in many parts of the world. Although they are normally eaten fresh, some are canned (Shaw, 1980). The flavour is described as mild, sub-acid, with some cultivars having an apple-like flavour.

The volatile components of loquat were investigated by Shaw and Wilson (1982), who used vacuum distillation on a rotary evaporator followed by dichloromethane extraction of the distillate. The volatile compounds identified are listed in Table VII.XII. Among the 17 components identified, the major ones were 2-phenylethanol, 3-hydroxybutanone, hexan-1-ol, hex-2-en-1-ol and diisobutyl phthalate.

Shaw and Wilson (1982) described the concentrated extract as having a strong aroma with a greenish, floral character and a harsh buttery note, whereas fresh loquat had a milder fruity aroma with an additional floral rose-like note. Using descriptors previously associated with the compounds found (Arctander, 1969),

TABLE VII.XII

Volatile flavour components of loquat fruit[a]

Esters	4-Hydroxy-4-methylpentan-2-one*
Ethyl acetate	Hexan-1-ol
Diisobutyl phthalate*	Hex-2-en-1-ol
Methyl cinnamate	Hex-3-en-1-ol
	3-Methylbutan-1-ol
Aldehydes and ketones	2-Phenylethanol
Acetone	Phenylpropanol
Benzaldehyde	
3-Hydroxybutanone	Terpenes
Phenylacetaldehyde	β-Ionone
Alcohols	Hydrocarbons
Benzyl alcohol	Cyclohexane

[a] Shaw and Wilson (1982).
* Probable artifact.

they attributed the floral note to 2-phenylethanol, the harsh buttery note to 3-hydroxybutanone, and the greenish note to hex-3-en-1-ol, hex-2-en-1-ol and phenylacetaldehyde.

Diisobutyl phthalate (a major component) and 4-hydroxy-4-methylpentan-2-one were thought to be artifacts.

16. BELI FRUIT

Beli or bael fruit, *Aegle marmelos* (Rutaceae), is a popular fruit in India and other parts of tropical Asia (Everett, 1981). The globular fruit has a hard rind surrounding the very sweet edible pulp. It is either eaten fresh or processed into a cream.

Concentrates of the volatile flavour compounds were prepared from fresh pulp using Soxhlet extraction and by the Likens – Nickerson technique (MacLeod and Pieris, 1981c). (Z)-Linalool oxide was the major component, comprising over 33% of the volatiles. Other high level compounds were 2-phenylethanol, 3-methylbutyl acetate, butyl 2-methylprop-2-enoate, and 3-methylbut-2-en-1-ol. A range of monoterpene and sesquiterpene hydrocarbons was identified.

MacLeod and Pieris (1981c) also prepared an extract from beli cream, a canned processed product. The cream produced about ten times more volatiles, but the relative amount of (Z)-linalool oxide was much lower (18%), while the proportion of the terpene and sesquiterpene hydrocarbons was markedly greater, 47 vs. 2% in the fresh fruit. There was also less 3-methylbut-2-en-1-ol and no 3-methylbutyl acetate or butyl 2-methylprop-2-enoate detected. The process used in the manufacture of the cream was not reported, but if heating was involved, the increase in terpene levels could be heat induced (cf. cashew apple).

Tokitomo *et al.* (1982) prepared aroma concentrates using several different methods, including with and without elevated temperatures. No quantitative data were reported, but from the gas chromatograms the major components were linalool oxide (furanoid), linalool, β-pinene and ethanol. Other compounds identified include 3,7-dimethylocta-1, 5, 7-trien-3-ol, β-ionone, and benzyl acetate. A list of the compounds identified in the two investigations is given in Table VII.XIII.

The volatiles isolated from ripe and unripe fruit showed quantitative and qualitative differences (Tokitomo *et al.,* 1982). A component (24%) tentatively identified as a steric isomer of 3,7-dimethylocta-1, 5, 7-trien-3-ol was almost not detectable in the unripe fruit.

Linalool oxide and linalool were considered in both reports to be important to the flavour of beli fruit. According to MacLeod and Pieris (1981c), α-phellandrene was the one compound that was consistently associated with the typical beli aroma when evaluated at an odour port during GC. However this compound was not detected by Tokitomo *et al.* (1982). Compounds shown in the proportions listed in Table VII.XIV were considered by Tokitomo *et al.*

312

TABLE VII.XIII

Volatile flavour components of beli fruit[a]

Esters	
Benzyl acetate	T
2-Cyclohexylethyl acetate (i)	M
Ethyl acetate	M
(Z)-Hex-3-enyl acetate	M
3-Methylbutyl acetate	M
Ethyl benzoate	M
Vinyl butanoate (i)	M
Methyl cinnamate	T
(Z)-Hex-3-enyl hexanoate	M
Butyl 2-methylprop-2-enoate	M
Ethyl phenylacetate	M
Methyl phenylacetate	M
Aldehydes and ketones	
Acetone	M
Benzaldehyde	M
Hexan-2-one	M
3-Methylbutanal	M
Nonanal	M
Pentane-2,4-dione (i)	M
Alcohols	
Benzyl alcohol	M, T
3,7-Dimethylocta-1,5,7-trien-3-ol	T
1,1-Dimethylprop-2-en-1-ol	M
Ethanol	T
3-Methylbutan-1-ol	M
3-Methylbut-2-en-1-ol	M
2-Phenylethanol	M, T
Terpenes	
Aromadendrene	T
γ-Cadinene	M
Camphene	M
Car-3-ene	T
Caryophyllene	M, T
α-Copaene	M
β-Copaene	T
p-Cymene	M, T
β-Elemene	M
α-Farnesene	T
β-Farnesene	M
α-Humulene	M
β-Ionone	T
β-Isothujene	M
Limonene	M, T
Linalool	M, T
(E)-Linalool oxide	M
(Z)-Linalool oxide	M

TABLE VII.XIII (*continued*)

Linalool oxides (furanoid)	T
Linalool oxides (pyranoid)	T
Menth-2-ene	M
Myrcene	M
α-Phellandrene	M
β-Phellandrene	M
α-Pinene	T
β-Pinene	T
α-Terpinene	M
α-Terpineol	M, T
Terpinen-4-ol	M
β-Thujene	M
α-Ylangene (i)	T
Hydrocarbons	
Benzene	M
Octadecane	M
Toluene	M
m- or *p*-Xylene	M
Miscellaneous	
Acetic acid	T
Carbon tetrachloride	M
Chloroform	M
Dichloromethane	M
Dimethylformamide (i)	M
γ-Pentalactone	M
Phenylacetonitrile	M
Pyridine	M

[a] M = MacLeod and Pieris (1981c); T = Tokitomo *et al.* (1982).
(i) = Tentative identification.

TABLE VII.XIV

Composition of synthetic beli fruit flavour[a]

Compound	Concentration (%)
Limonene	31.8
Car-3-ene	5.7
p-Cymene	2.0
Linalool oxide, I, II	19.2
Linalool	20.0
Benzyl acetate	5.2
Linalool oxide, III, IV	10.8
Benzyl alcohol	1.5
2-Phenylethanol	1.5
β-Ionone	2.5

[a] Tokitomo *et al.* (1982).

(1982) to have the fundamental flavour quality of beli fruit, while 3,7-di-methylocta-1,5,7-trien-3-ol, with its sweet floral note, and its isomers, enhance the attractiveness of the aroma. This trienol was not detected by MacLeod and Pieris (1981c). The lack of fresh notes in the one processed product studied could be due to the absence, or presence at very low levels only, of 3-methyl-butyl acetate, butyl 2-methylprop-2-enoate and 3-methylbut-2-enol (MacLeod and Pieris, 1981c).

17. WOOD APPLE

Wood apple (*Feronia limonia*) is a member of the Rutaceae family and is native to India and Sri Lanka. Despite the unattractive appearance of the brown fruit, wood apple have a delicious flavour. An underlying fatty note has been reported in the flavour (MacLeod and Pieris 1981a). The fruit are usually eaten fresh, but in Sri Lanka canned wood apple cream is economically important.

The volatile flavour components of fresh wood apple fruit and the canned cream have been investigated by MacLeod and Pieris (1981a). Two methods of extraction, a Soxhlet apparatus and a Likens – Nickerson method, were both found to be efficient, but the latter method gave a more concentrated extract. Trichlorofluoromethane was found to be the best extraction solvent in both methods when several solvents were compared.

In the fresh fruit, 29 components, comprising 96% of the total GC peak areas, were tentatively or positively identified, while in the cream there were 43 components (99% of GC peak area) tentatively or positively identified. Qualitatively, the fresh and canned fruit extracts were very similar, with esters being the main components (40 – 46%); however, quantitatively, the former gave five times more essence (21 mg/kg compared with 4.4 mg/kg). This was consistent with the weaker aroma intensity observed in the cream. In each extract ethyl butanoate was the major component (25%), and butanoic acid was the second most abundant compound, with 12% in the fruit and 21% in the cream. Acetone was found at a high level in both the fruit (15%) and the cream (12%).

A series of methyl and ethyl esters of butanoic acid, but-2-enoic acid and 3-hydroxybutanoic acid was identified, together with butan-1-ol and butanoic acid, and these compounds comprised 57 – 63% of the total volatiles. A similar series was identified for C-6 parent compounds, comprising 9 – 11% of the total volatiles, and there was some indication of a C-2 and a C-8 series. The presence of these even-carbon series led MacLeod and Pieris (1981a) to suggest that acetate precursors were involved in the biosynthesis of the flavour volatiles.

Some compounds were identified in the cream that were not found in the fresh fruit, the main ones being 2-furaldehyde (14%) and hept-2-en-4-one (1%). The additional compounds found in the cream were thought to be artifacts formed by thermal degradation, and contributed to the slight off-flavour in the cream. Although the cream had a much lower level of flavour volatiles, it still had the wood apple character.

Assessment of the odour quality of the GC peaks showed that the components with wood apple-like aroma were butanoic acid, methyl butanoate and ethyl 3-hydroxyhexanoate in both the fresh fruit and the wood apple cream, and also hexan-1-ol in the cream. However, these compounds do not provide the complete wood apple aroma. Butanoic acid contributed to the underlying fatty note, especially in the cream.

18. SAPODILLA

Sapodilla, *Manilkara achras* L.; syn *Achras sapota* L., (Sapotaceae) is a native of tropical America. It is known by several other names, including chiku and sapota, and it is commercially cultivated in India, Philippines, Sri Lanka, Mexico, Venezuela, Guatemala and other Central American countries. Many cultivars are recognised, and generally the fruit weigh 75 – 200 g, but some cultivars have fruit weighing up to 1 kg. It is best known for its chicle gum used in chewing gum manufacture.

The sweet tasting fruit have a delicate, characteristic aroma, but can sometimes be slightly astringent. Lakshminarayana and Rivera (1979) assessed a number of cultivars for composition and sensory quality, although no flavour chemistry analysis was reported. Several cultivars had excellent organoleptic qualities. The distribution, botany, cultivation, cultivars and post-harvest handling problems of sapodilla have been reviewed by Lakshminarayana (1980).

MacLeod and de Troconis (1982d) have identified the flavour volatiles which were extracted using a modified Likens – Nickerson apparatus with 2-methylbutane as solvent. The total volatiles recovered was very low, 5 μg/kg of fresh fruit. Sixty-five components were found, of which 47 were positively, and 14 partially, identified (Table VII.XV).

Nearly half the essence was comprised of benzyl-related compounds, with benzaldehyde (22%) and methyl benzoate (14%) being the main components. Benzyl alcohol, benzyl methyl ketone (1-phenylpropan-2-one) and ethyl and C-4 to C-6 alkyl benzoates were quantitatively also important. In addition to car-3-ene (1.5%), several other terpenes were found in trace amounts. Other high level components were methylcyclohexane and a branched C-8 hydrocarbon (co-eluting, 12%), toluene (6%), (*E*)-hex-2-enal (4%), octane (4%), 3-phenylpropan-1-ol (4%), and hexan-1-ol (3%). Hydrocarbons, alcohols, carbonyl compounds and two sulphur compounds accounted for most of the remaining components. No aliphatic esters were detected.

The components considered by odour assessment to have significant sapodilla character were methyl benzoate and methyl salicylate (1.2%). Other qualitatively important peaks were considered to be ethyl benzoate, 1-phenylbutan-1-one, benzaldehyde, hexan-1-ol and (*E*)-hex-2-enal. A mixture of these compounds in 2-methylbutane (inert solvent) had an aroma not entirely characteristic of fresh sapodilla fruit, but sufficiently similar for them to be considered as major contributors to the flavour.

316

TABLE VII.XV

Volatile flavour components of sapodilla fruit[a]

Esters	Methyleugenol
Hexyl acetate	3-Phenylpropan-1-ol
Butyl benzoate	
Ethyl benzoate	Hydrocarbons
Hexyl benzoate	Benzene
Methyl benzoate	Chloroform
Pentyl benzoate	Cyclohexane
Benzyl 3-methylbutanoate	Ethylcyclohexane
Methyl 4-methoxybenzoate	Heptadecane
Methyl salicylate	Isopropylbenzene
	Methylcyclohexane
Aldehydes and ketones	Nonadecane
Acetylfuran	Octadecane
Benzaldehyde	Octane
Butanedione	Styrene
Hexanal	Toluene
(E)-Hex-2-enal	m- and/or p-Xylene
2-Furaldehyde	o-Xylene
3-Methylbutanal	
1-Phenylbutan-1-one	Terpenes
1-Phenylpropan-2-one	Car-3-ene
	Caryophyllene
Alcohols	(Z)-Linalool oxide
Benzyl alcohol	(E)-Linalool oxide
Butan-1-ol	
Ethanol	Miscellaneous
Hexan-1-ol	Methoxybenzene
2-Methylbutan-2-ol	Dimethyl disulphide
3-Methylbutan-1-ol	Dimethyl trisulphide

[a] MacLeod and de Troconis (1982d).

19. TAMARIND

Tamarind, *Tamarindus indica* L. (Leguminaceae), is thought to be a native of Africa, but it is grown throughout the tropics, including India, which is the major producer (Bueso, 1980). All parts of the tree are used in a variety of ways including medicinal and industrial. The fruit are flattish, bean-like pods with a pulp that is high in sugar and has a pleasant acid taste making it a popular ingredient for meat seasonings and Worcestershire sauce. The pulp dehydrates naturally to a sticky paste which is highly valued as a delicacy. Other uses include beverages, syrups, frozen confectionery and canned products (Chapman, 1984).

Lee *et al.* (1975b) isolated the volatiles from tamarind pulp using vacuum steam distillation in a rotary evaporator at 40°C. The distillate was trapped with liquid nitrogen and extracted with dichloromethane to yield a concentrate which maintained the tamarind aroma.

The major component was 2-acetylfuran, to which was attributed a balsamic-cinnamic note. 2-Acetylfuran, together with 2-furaldehyde and 5-methyl-2-furaldehyde (which were present in minor amounts and which have sweet, caramel-like flavours), were considered to contribute significantly to the overall aroma.

Several compounds normally associated with roasted or fried food were identified in the extract. These were five simple pyrazines and two thiazoles, which were described as green, nutty, roasted or vegetable-like. Two linalool oxides and linalool were considered to contribute a slight rosy aroma. The citrus note was attributed to limonene, terpinen-4-ol, neral, α-terpineol, geranial and geraniol. Piperitone was noted for its fresh, minty, camphoraceous odour, and methyl salicylate, safrole, ionones, cinnamaldehyde and ethyl cinnamate for warm spice-like notes.

Some C-6 alcohols were quantitatively important. Other interesting components were 1-phenylpropan-1-one, 1-phenylpropan-2-one, 4-phenylpyrid-2-one, phenylacetaldehyde and 2-phenylethanol.

The sixty volatile components identified were believed to account adequately for the citrus, warm spice-like flavours and roasted notes which are characteristic of tamarind (Lee *et al.*, 1975b).

TABLE VII.XVI

Volatile flavour components of dalieb fruit[a]

Esters	Alcohols
Butyl acetate	Butan-1-ol
Ethyl acetate	Butan-2-ol
Phenethyl acetate	Ethanol
Butyl benzoate	Methanol
Ethyl benzoate	3-Methylbutan-1-ol
Butyl butanoate	6-Methylhept-5-en-2-ol
Ethyl butanoate	3-Methylpentan-1-ol
3-Methylbutyl butanoate	Pentan-1-ol
Ethyl but-2-enoate	Propan-1-ol
Ethyl hex-2-enoate	
Ethyl 3-methylpentanoate	Acids
Ethyl octanoate	Acetic acid
Ethyl pentanoate	Butanoic acid
	Hexanoic acid
Aldehydes and ketones	Methylpropanoic acid
6,10-Dimethylundeca-5,9-dien-2-one (geranylacetone)	Pentanoic acid
6-Methylhept-5-en-2-one	Propanoic acid
	Miscellaneous
	2-Propylfuran

[a] Harper *et al.* (1986).

20. DALIEB

Dalieb (*Borassus aethiopum* L.) is a palm which grows in the Sudan. It produces a large fruit, weighing $1 - 2$ kg, of which the fibrous but juicy orange pulp is edible. It has a sweet and pleasant aroma, and has been considered for processing into jam, squash and for flavouring (Harper *et al.*, 1986).

Both the steam distillate and direct headspace trapping of the volatiles from dalieb fruit have been studied by Harper *et al.* (1986). In each case the sample was introduced into the GC by the trap and purge method using Tenax. The volatile components identified in dalieb are given in Table VII.XVI.

The steam distillate contained mainly alcohols and esters, while direct sampling of the headspace gave additional compounds, including a series of C-2 to C-6 aliphatic acids and 2-propylfuran. The major peaks in the gas chromatogram of the steam distillate were ethanol, butan-1-ol and ethyl acetate. Other compounds identified included ethyl but-2-enoate and hex-2-enoate, 6-methylhept-5-en-2-one and 6,10-dimethylundeca-5,9-dien-2-one (geranylacetone). Quantitative estimation of the alcohols and volatile fatty acids showed that acetic acid and ethanol were present at high levels, approaching 0.1% of the fresh pulp.

21. PRICKLY PEAR

The cactus family (Cactaceae) includes many species with edible fruit, one of which is prickly pear (*Opuntia ficus indica* Mill.). This fruit, also known as tuna, Barbary fig or Indian fig, originates in the Americas, but grows extensively in many hot, semi-arid regions of the world. It is grown commercially in the Mediterranean, North America, Mexico, Peru and Central America (Johns and Stevenson, 1979). The fruit are generally eaten fresh, but some are dried or processed into juices and preserves.

The fruit can be up to 8 cm long, and have a very sweet flesh and aroma (Johns and Stevenson, 1979). Flath and Takahashi (1978) described the flavour as mildly sweet with little acid character and melon-like.

Vacuum steam distillation followed by ether extraction of the distillate gave an extract which showed about $140 - 150$ peaks in the gas chromatogram. Of these, sixty-one, representing 95% of the total peak areas, have been identified (Flath and Takahashi, 1978).

Alcohols were the main components (84%), in particular ethanol, which was responsible for 76% of the peak areas. Other alcohols included primary and secondary, saturated and unsaturated alcohols up to C-9, and cyclopentanol. Relatively low levels of esters were found (less than 3%). Three terpene hydrocarbons, myrcene, limonene and β-terpinene, and the terpene alcohol, linalool, were found at low levels.

Although ethanol quantitatively dominates the volatiles, it probably only makes a minor contribution to the flavour (Takeoka *et al.*, 1986). The presence of C-9 compounds, nonan-1-ol, nonenols, nonadienols, and non-2-enal, which are considered to have melon-like or cucumber-like notes (Forss *et al.*, 1962; Kemp *et al.*, 1974), is consistent with the light melon-like flavour of prickly pear.

22. PEPINO

The pepino (*Solanum muricatum* L.) is a perennial member of the Solanaceae and native to the Andes region of South America, where it is grown at altitudes up to 2700 m. It was recently developed as a commercial crop in New Zealand from a selection of plants collected from South America. The oval-shaped fruit has a very juicy flesh that is surrounded by a thin cream-coloured skin with irregular purple stripes. The flavour of the fruit from these original plants ranged from watery and bland to ones with a hot peppery taste (S. Dawes, personal communication, 1986). The pepino cultivars used in commercial production in New Zealand today have a flavour which is often described as melon-like or cucumber-like, with better cultivars having intense fruity aroma when ripe (Young and Paterson, 1986). At present the pepino is only used as the fresh fruit.

The word pepino, which is Spanish for cucumber, has been associated with several other fruit and vegetables, e.g. Pepino morado (*Solanum muricatum* Ait.) (Gomez, 1986). and *Cyclanthera pedata* (Benk, 1981).

The volatile flavour components of pepino have been studied by Shiota *et al.* (1988) and Young and Paterson (1986), and are listed in Table VII.XVII. Shiota *et al.* (1988) employed the Likens – Nickerson technique, which gave a concentrate from which 37 compounds were identified. The two most abundant were the isomeric 3-methylbut-2-enyl acetate and 3-methylbut-3-enyl acetate. Several isomeric 3-methylbutenyl 3-methylbutenoate esters were found. Of significance is the identification of straight-chain C-9 compounds, especially (Z)-non-6-enal and (Z)-non-6-en-1-ol. Unsaturated C-9 compounds are commonly associated with melon and cucumber flavour notes (Forss *et al.*, 1962; Kemp *et al.*, 1972a,b; Buttery *et al.*, 1982).

Large differences in the relative amounts of the volatile components were observed between cultivars and even between samples of the same cultivar.

Young and Paterson (1986) compared extracts prepared by the Likens – Nickerson technique and by vacuum distillation followed by ether extraction of the aqueous distillate. The low-temperature extract did not have the 'cooked' note, detected in the extract obtained by the elevated temperature method. The sensory quality of the low-temperature extract strongly resembled that of the fresh fruit. The main difference between the low- and high-temperature extracts was in the relative levels of the 3-methylbutenyl acetates. The former had a greater proportion of 3-methylbut-2-enyl acetate (31 vs. 15%). In addition, the low-temperature extract had higher levels of non-6-enal, nonanal and 3-methylbutenols.

320

TABLE VII.XVII

Volatile flavour components of pepino[a]

Esters	
Ethyl acetate	S, Y
Butyl acetate	S, Y
Hept-3-enyl acetate	Y
Hexyl acetate	S, Y
3-Methylbut-2-enyl acetate	S, Y
3-Methylbut-3-enyl acetate	S, Y
2-Methylbutyl acetate	S, Y
3-Methylbutyl acetate	S, Y
2-Methylpropyl acetate	S, Y
(Z)-Non-6-enyl acetate	S, Y
Nonyl acetate	S, Y
Pentyl acetate	S, Y
Propyl acetate	S, Y
Ethyl 3-methylbutanoate	Y
Butyl 3-methylbut-2-enoate	Y
Methyl 3-methylbut-2-enoate	S, Y
3-Methylbut-2-enyl 3-methylbutanoate	Y
3-Methylbut-3-enyl 3-methylbutanoate	S, Y
3-Methylbut-2-enyl 3-methylbut-2-enoate	S, Y
3-Methylbut-3-enyl 3-methylbut-2-enoate	S, Y
3-Methylbut-2-enyl 3-methylbut-3-enoate	Y
Aldehydes and ketones	
Butanone	Y
3-Methylbut-2-enal (i)	Y
3-Methylbut-3-enal (i)	Y
Hexanal	S, Y
(E)-Hex-2-enal	S, Y
Nonanal	S, Y
Nona-2,6-dienal	Y
(E)-Non-2-enal	S
(Z)-Non-6-enal	S, Y
Nonan-2-one	S
Pentan-3-one	Y
Penta-2,3-dione	S
Alcohols	
Butan-1-ol	S, Y
Ethanol	Y
2-Ethylhexan-1-ol	S
Hexan-1-ol	S, Y
3-Methylbut-2-en-1-ol	S, Y
3-Methylbut-3-en-1-ol	S, Y
3-Methylbut-3-en-2-ol	S, Y
Nonan-1-ol	S, Y
(Z)-Non-6-en-1-ol	S, Y

TABLE VII.XVII (*continued*)

Terpenes	
Terpinen-4-ol	S
α-Terpinene	S
(*E*)-Ocimene	S
Acids	
Dodecanoic acid	S
Hexadecanoic acid	S
Tetradecanoic acid	S

[a] S = Shiota *et al.* (1988); Y = Young and Paterson (1986).
(i) Tentative identification.

Flavour assessment at an odour port during GC indicated that 3-methylbut-2-enyl acetate was the most pepino-like component (Young and Paterson, 1986). The isomeric 3-methylbut-3-enyl acetate made only a small contribution to the estery note of pepino. The cucumber/melon note was easily attributed to the presence of (*Z*)-non-6-enal and to a lesser extent (*Z*)-non-6-en-1-ol.

There have been no investigations published on the effect of maturity and ripeness on the flavour of pepino.

Redgwell and Turner (1986) showed that the amino acids aspartic acid and glutamic acid increased several fold as the fruit matured and ripened. If branched-chain flavour compounds are assumed to be formed from amino acids (Nursten, 1970), fruit maturity at harvest could have a marked effect on the level of the important flavour compound 3-methylbut-2-enyl acetate, and hence the apparent intensity of the cucumber-note resulting from the C-9 compounds.

23. BACURI, CUPUACU, MURUCI AND TAPERABÁ

The Amazon region has a great number of edible fruit available in the local markets that have commercial significance. The volatile components of four popular dessert fruit from this region were investigated by Alves and Jennings (1979). Besides being eaten fresh, these four fruit are used in ice cream and juice manufacture. In each case the volatile flavour components were extracted from canned fruit pulp, using a Likens–Nickerson technique.

Bacuri (*Platonia insignis*) is a fruit the size of an orange with a bittersweet pulp which has a pleasant aroma and taste. Several components in the chromatogram of the volatile components were not identified, including the major peak. Linalool, which was a major peak, and heptan-2-one and (*Z*)-hex-3-enyl acetate, which are powerful odorants, were thought to be responsible for the flavour. Other components identified were heptane, (*Z*)- and (*E*)-linalool oxide, pentan-2-one, nonan-2-one, hept-2-ene, 2-furaldehyde, γ-terpinene and methyl dodecanoate.

Cupuacu (*Theobroma grandiflora*) has a pulp with an acid taste and a strong aroma. While many of the abundant components were not identified, ethyl butanoate and hexanoate were major components of the volatiles. Other compounds identified include ethyl and butyl 2-methylbutanoate, 2-methylbutyl 2-methylbutanoate, 2-furaldehyde and hexan-1-ol. The esters were thought to be important contributors to cupuacu aroma.

Muruci (*Brysonima crassifolia*) is a yellow fruit smaller than a cherry with a strong aroma. Ethyl hexanoate and butanone were two major peaks in the chromatogram, while a third unidentified peak was thought to be an artifact. Esters, especially ethyl esters, and ketones accounted for most of the volatiles identified, in addition to hexanal, 1,1-diethoxyethane, 1-methoxy-1-ethoxy-ethane, butan-2-ol and γ-terpinene. No individual compounds represented the aroma of muruci.

Tapereba (*Spondias lutea*) is a small fruit with a bittersweet juice and pleasant taste and aroma. Unfortunately, of the many peaks in the chromatogram, only methyl and ethyl benzoate, ethyl octanoate, ethyl cinnamate, hept-2-ene and ocimene were identified.

Alves and Jennings (1979) concluded that the aroma of each of these fruit was due to a combination of compounds rather than particular character impact compounds.

24. CONCLUSION

Most of the research on the flavours of the exotic fruit has been concerned with the fresh fruit. Unfortunately, much of the work has been done on fruit which have been bought at the local markets with little regard to the pre- and post-harvest history of the samples. There are several examples of quite large flavour differences found between fruit which are apparently of the same cultivar. Several studies, however, have revealed major flavour changes during fruit development, ripening and storage.

A wide variety of methods has been used to extract flavour volatiles. Each method favours a different range of components, and some may cause heat degradation or other reactions to the flavour compounds. Thus it is important to match the method with the objectives of the investigation.

2-Furaldehyde (furfural) and related compounds have been found in several processed products or in fresh fruit where elevated temperature was used in the isolation of the volatiles, suggesting that these compounds may be heat-induced artifacts. Kiwifruit is one example where low- and high-temperature techniques were used to isolate volatiles, but 2-furaldehyde was found only in the latter extract.

The formation of terpenes in some fruit (cashew apple and beli fruit) could be heat induced. It would be interesting to compare the volatiles of custard apple isolated by low-temperature techniques with the results reported in this review in which all compounds identified were terpenoid in nature.

Many of the fruit reviewed do not have easily identifiable flavour impact compounds, rather the flavour is a delicate balance of several components present. The flavour of these fruit could well be sensitive to agronomic and post-harvest practices.

Although the chemical identification of the flavour volatiles has advanced rapidly with the advent of modern sensitive instrumentation, such as the capillary GC-MS, very little is known about the importance of these compounds to the flavour of the fruit in which they are found. It will be an interesting challenge for the sensory specialist to define the significant flavour components of these fruit.

REFERENCES

Alves, S. and Jennings, W.G., 1979. Volatile composition of certain Amazonian fruits. *Food Chem.,* 4: 149–159.
Arctander, S. (Editor), 1969. Perfume and Flavor Chemicals. S. Arctander, Montclair, NJ.
Baldry, J., Dougan, J. and Howard, G.E., 1972. Volatile flavouring constituents of durian. *Phytochemistry,* 11: 2081–2084.
Benk, E., 1981. Weiterer beitrag zur kenntnis ausländischer obstfrüchte. *Riechst. Aromen Kosmet.,* 31(7): 234–235.
Berger, R.G., Drawert, F., Kollmannsberger, H. and Nitz, S., 1985. Natural occurrence of undecaenes in some fruits and vegetables. *J. Food Sci.,* 50: 1655–1656, 1667.
Bollard, E.G., 1981. Prospects for Horticulture: A Research Viewpoint. Department of Scientific and Industrial Research Discussion Paper No. 6, Publication No. ISSN 0110–5221, Wellington, New Zealand, p. 116.
Bueso, C.E., 1980. Soursop, tamarind and chironja. In: S. Nagy and P.E. Shaw (Editors). Tropical and Subtropical Fruits. AVI Publishing Co., Inc., Westport, CT, pp. 375–406.
Buttery, R.G., Seifert, R.M., Ling, L., Soderstrom, E.L., Ogawa, J.M. and Turnbaugh, J.G., 1982. Additional aroma components of honeydew melon. *J. Agric. Food Chem.,* 30: 1208–1211.
Chapman, K.R., 1984. Tamarind. In: P.E. Page (Editor), Tropical Tree Fruits for Australia. Queensland Dept. of Primary Industries, Brisbane, pp. 83–86.
Chin, H.F. and Yong, H.S., 1980. Malaysian Fruits in Colour. Tropical Press SDN, BHD, Kuala Lumpur, Malaysia.
Dupaigne, P., 1978. Mise au point sur la composition de l'arome des fruits tropicaux peu connus. *Fruits,* 33(6): 413–423.
Engel, K. and Tressl, R., 1983. Formation of aroma compounds from non volatile precursors in passion fruit. *J. Agric. Food Chem.,* 31: 998–1002.
Everett, T.H., 1981. New York Botanical Garden Illustrated Encyclopedia of Horticulture, Vol 1. Garland Publishing Inc., New York, pp. 62–63.
Flath, R.A. and Takahashi, J.M., 1978. Volatile constituents of prickly pear (*Opuntia ficus indica* Mill., de Castilla variety). *J. Agric. Food Chem.,* 26(4): 835–837.
Flath, R.A., Black, D.R., Guadagni, D.G., McFadden, W.H. and Schulz, T.H., 1967. Identification and organoleptic evaluation of compounds in Delicious apple essence. *J. Agric. Food Chem.,* 15: 29–35.
Forss, D.A., Dunstone, E.A., Ramshaw, E.H. and Stark, W., 1962. The flavor of cucumbers. *J. Food Sci.,* 27: 90–93.
Gardiazabal, F.I. and Rosenberg, G.M., 1986. Cultivation of Cherimoya. General University of Valpariso, Chile, pp. 1–2.
George, A.P., 1984. Annonaceae. In: P.E. Page (Editor), Tropical Tree Fruits for Australia. Queensland Department of Primary Industries, Brisbane, pp. 35–41.

324

Gomez, E.S., 1986. Frutas en Colombia. Ediciones Cultural, Bogota, p. 125.

Hardy, P.J. and Michael, B.J., 1970. Volatile components of feijoa fruits. *Phytochemistry,* 9: 1355 – 1357.

Harper, D.B., Nour, A.A.M. and Thompson, R.H., 1986. Volatile flavour components of dalieb (*Borassus aethiopum* L.). *J. Sci. Food Agric.,* 37: 685 – 688.

Herrmann, K., 1981. Übersicht über die chemische Zusammensetzung und die Inhaltsstoffe einer Reihe wichtiger exotischer Obstarten. *Z. Lebensm.-Unters.-Forsch.,* 173: 47 – 60.

Idstein, H., Herres, W. and Schreier, P., 1984. High-resolution gas chromatography-mass spectrometry and Fourier transform infrared analysis of cherimoya (*Annona cherimolia,* Mill.) volatiles. *J. Agric. Food Chem.,* 32: 383 – 389.

Idstein, H., Keller, T. and Schreier, P., 1985. Volatile constituents of mountain papaya (*Carica candamarcensis,* syn. *C. pubescens* Lenne et Koch) fruit. *J. Agric. Food Chem.,* 33: 663 – 666.

Johns, L. and Stevenson, V., 1979. The Complete Book of Fruit. Angus and Robertson, London.

Kemp, T.R., Knavel, D.E. and Stoltz, L.P., 1972a. Cis-6-nonenal: A flavour component of muskmelon fruit. *Phytochemistry,* 11: 3321 – 3322.

Kemp, T.R., Stoltz, L.P. and Knavel, D.E., 1972b. Volatile components of muskmelon fruit. *J. Agric. Food Chem.,* 20: 196 – 198.

Kemp, T.R., Knavel, D.E., Stoltz, L.P. and Lundin, R.E., 1974. 3,6-Nonadien-1-ol from *Citrullus vulgaris* and *Cucumis melo. Phytochemistry,* 13: 1167 – 1170.

Klein, J.D. and Thorp, T.G., 1987. Feijoa: post-harvesting handling and storage of fruit. *N.Z. J. Exp. Agric.,* 15: 217 – 221.

Lakshminarayana, S., 1980. Sapodilla and prickly pear, In: S. Nagy and P. Shaw (Editors), Tropical and Subtropical Fruits. AVI Publ. Inc., Westport, CT, pp. 415 – 441.

Lakshminarayana, S., and Rivera, M.A.M., 1979. Proximate characteristics and composition of sapodilla fruits grown in Mexico. *Proc. Fla. State Hortic. Soc.,* 92: 303 – 305.

Leboeuf, M., Cave, A., Bhaumik, P.K., Mukherjee, B. and Mukherjee, R., 1982. The phytochemistry of the Annonaceae. *Phytochemistry,* 21: 2783 – 2813.

Lee, P.L., Swords, G. and Hunter, G.L.K., 1975a. Volatile components of *Eugenia jambos* L., rose apple. *J. Food Sci.,* 40: 421 – 422.

Lee, P.L., Swords, G. and Hunter, G.L.K., 1975b. Volatile constituents of tamarind (*Tamarindus indica* L.). *J. Agric. Food Chem.,* 23: 1195 – 1199.

Maciel, M.I., Hansen, T.J., Aldinger, S.B. and Labows, J.N., 1986. Flavor chemistry of cashew apple juice. *J. Agric. Food Chem.,* 34: 923 – 927.

MacLeod, A.J. and de Troconis, N.G., 1982a. Volatile flavour components of guava. *Phytochemistry,* 21: 1339 – 1342.

MacLeod, A.J. and de Troconis, N.G., 1982b. Volatile flavour components of mango fruit. *Phytochemistry,* 21: 2523 – 2526.

MacLeod, A.J. and de Troconis, N.G., 1982c. Volatile flavour components of cashew 'apple' (*Anacardium occidentale*). *Phytochemistry,* 21: 2527 – 2530.

MacLeod, A.J. and de Troconis, N.G., 1982d. Volatile flavour components of sapodilla fruit (*Achras sapota* L.). *J. Agric. Food Chem.,* 30: 515 – 517.

MacLeod, A.J. and Pieris, N.M., 1981a. Volatile flavour components of wood apple (*Feronia limonia*) and a processed product. *J. Agric. Food Chem.,* 29: 49 – 53.

MacLeod, A.J. and Pieris, N.M., 1981b. Volatile flavor components of soursop (*Annona muricata*). *J. Agric. Food Chem.,* 29: 488 – 490.

MacLeod, A.J. and Pieris, N.M., 1981c. Volatile flavour components of beli fruit (*Aegle marmelos)* and a processed product. *J. Agric. Food Chem.,* 29: 1262 – 1264.

MacLeod, A.J. and Pieris, N.M., 1982. Volatile flavour components of mangosteen, *Garcinia mangostana. Phytochemistry,* 21: 117 – 119.

MacLeod, A.J. and Snyder, C.H., 1985. Volatile components of two cultivars of mango from Florida. *J. Agric. Food Chem.,* 33: 380 – 384.

Maggs, D.H., 1984. Myrtaceae. In: P.E. Page (Editor), Tropical Tree Fruits for Australia. Queensland Department of Primary Industries, Brisbane, pp. 20 – 24.

Martin, F.W., 1980. Durian and mangosteen, In: S. Nagy and P. Shaw (Editors), Tropical and Subtropical Fruits. AVI Publ. Inc., Westport, CT, pp. 407 – 414.

MacKenzie, R.A. and Strachan, G., 1980. Possibilities for processing mountain papayas in New Zealand. *Food Technol. N.Z.,* 15(8): 13 – 19.

Morales, A.L. and Duque, C., 1987. Aroma constituents of the fruit of the mountain papaya (*Carica pubescens)* from Colombia. *J. Agric. Food Chem.,* 35: 538 – 540.

Moser, R., Duvel, D. and Greve, R., 1980. Volatile constituents and fatty acid composition of lipids in *Durio zibethinus. Phytochemistry,* 19: 79 – 81.

Natarajan, P.N. and Karunanithy, R., 1973. Synthetic flavor enhancers – flavor enhancer of *Annona muricata. Can. Inst. Food Sci. Technol. J.,* 6: 248 – 249.

Natarajan, P.N. and Karunanithy, R., 1974. Synthetic flavour enhancers for *Artocarpus integrifolia* L. *Flavour Ind.,* 5(11/12): 282 – 283.

Nursten, H.E., 1970. Volatile compounds: The aroma of fruits. In: A.C. Hulme (Editor), The Biochemistry of Fruits and Their Products. Academic Press, London, pp. 239 – 268.

Pattabhiraman, T.R., Rao, P. and Sastry, L.V.L., 1968. Preliminary studies on the preparation of odor concentrates and identification of odorous ingredients in mango and guava. *Perfum. Essent. Oil Rec.,* 58: 733 – 736.

Paull, R.E., 1982. Postharvest variation in composition of soursop (*Annona muricata* L.) fruit in relation to respiration and ethylene production. *J. Am. Soc. Hortic. Sci.,* 107: 582 – 585.

Paull, R.E., Deputy, J. and Chen, N.J., 1983. Changes in organic acids, sugars, and headspace volatiles during fruit ripening of soursop (*Annona muricata* L.). *J. Am. Soc. Hortic. Sci.,* 108: 931 – 934.

Pesis, E., Avisar, I. and Zauberman, G., 1987. The effect of acetaldehyde vapors, anaerobic conditions, or ethylene on postharvest quality of feijoa fruit. *HortScience,* 22(5): 159.

Pfannhauser, W., 1987. Identification of kiwi aroma compounds by GC/MS and sniffing analysis. Proceedings of European Food Chem. IV: Rapid Analysis in Food Processing and Food Control. Euro. Food Chem. IV, Norwegian Food Research Institute, 1: 86 – 90.

Ramteke, R.S., Eipeson, W.E., Singh, N.S., Chikkaramu and Patwardhan, M.V., 1984. Studies on the preparation of aroma concentrate from cashew apple (*Anacardium occidentale). J. Food Sci. Technol. India,* 21: 248 – 249.

Rao, M.S.S., 1985. Scope for development of alcoholic beverage from cashew apple. *Acta Hortic.,* 108: 160 – 164.

Redgwell, R.J. and Turner, N.A., 1986. Pepino (*Solanum muricatum):* Chemical composition of ripe fruit. *J. Sci. Food Agric.,* 37: 1217 – 1222.

Sedgley, M., 1984. Moraceae. In: P.E. Page (Editor), Tropical Tree Fruits for Australia. Queensland Dept. of Primary Industries, Brisbane, pp. 100 – 107.

Shaw, G.J., Ellingham, P.J. and Birch, E.J., 1983. Volatile constituents of feijoa – headspace analysis of intact fruit. *J. Sci. Food Agric.,* 34: 743 – 747.

Shaw, G.J., Allen, J.M. and Visser, F.R., 1985. Volatile flavor components of babaco fruit (*Carica pentagona* Heilborn). *J. Agric. Food Chem.,* 33: 795 – 797.

Shaw, P.E., 1980. Loquat. In: S. Nagy and P. Shaw (Editors), Tropical and Subtropical Fruits. AVI Publ. Inc., Westport, CT, pp. 479 – 491.

Shaw, P.E. and Wilson, C.W., 1982. Volatile constituents of loquat (*Eriobotrya japonica* Lindl.) fruit. *J. Food Sci.,* 47: 1743 – 1744.

Shiota, H., 1982. Kiwifruit. *Koryo,* 137: 59 – 64.

Shiota, H., 1984. Starfruit. *Koryo,* 143: 37 – 43.

Shiota, H., Minami, T. and Tsuneya, T., 1980. The aroma constituents of strawberry guava (*Psidium cattelianum* Sabine), yellow guava (*Psidium cattelianum* Sabine var. lucidum Hort.) and ananas guava (*Feijoa sellowiana* Berg). *Koryo,* 128: 35 – 39.

Shiota, H., Young, H. and Paterson, V.J., 1988. Volatile flavour components of pepino fruit. *J. Sci. Food Agric.,* 43: 343 – 354.

Sieso V. and Crouzet J., 1977. Tomato volatile components: Effect of processing. *Food Chem.,* 2: 241 – 252.

Skinner, I., 1983. Babaco – The golden fruit with the hollow centre. *Aust. Hortic.,* 81(12): 34 – 47.

Starodubtseva, V.P. and Kharebava, L.G., 1986. Volatile complex of feijoa: 4. Content and importance of some essential oil components. *Subtrop. Kul't.,* 5: 118 – 123.

Straten, S. van, Maarse, H., de Beauveser, J.C. and Visscher, C.A., 1983. Volatile Compounds in Food: Qualitative Data, 5th edition. Division for Nutrition and Food Research T.N.O., Zeist, The Netherlands.

Sturtz, J.D., 1984. Anacardiaceae, In: P.E. Page (Editor), Tropical Tree Fruits for Australia. Queensland Department of Primary Industries, Brisbane, pp. 20 – 24.

Swords, G., Bobbio, P.A. and Hunter, G.L.K., 1978. Volatile constituents of jackfruit (*Arthocarpus heterophyllus*). *J. Food Sci.,* 43: 639 – 640.

Takeoka, G.R., Güntert, M., Flath, R.A., Wurz, R.E. and Jennings, W., 1986. Volatile constituents of kiwi fruit (*Actinidia chinensis* Planch.). *J. Agric. Food Chem.,* 34: 576 – 578.

Tankard, G.J., 1987. Exotic Tree Fruit for the Australian Home Garden. Thomas Nelson, Australia.

Tokitomo, Y., Shimono, Y., Kobayashi, A. and Yamanishi, T., 1982. Aroma components of baelfruit (*Aegle marmelos* Correa). *Agric. Biol. Chem.,* 46: 1873 – 1877.

Wagner, C.J., Bryan, W.L., Berry, R.E. and Knight, R.J., 1975. Carambola selection for commercial production. *Proc. Fla. State Hortic. Soc.,* 88: 466 – 469.

Wilson, C.W. and Shaw, P.E., 1978. Terpene hydrocarbons from *Psidium guajava*. *Phytochemistry,* 17: 1435 – 1436.

Wilson, C.W., Shaw, P.E., Knight, R.J., Nagy, S. and Klim, M., 1985. Volatile constituents of carambola (*Averrhoa carambola* L.). *J. Agric. Food Chem.,* 33: 199 – 201.

Wyllie, S.G., Cook, D., Brophy, J.J. and Richter, K.M., 1987. Volatile flavor components of *Annona atemoya* (custard apple). *J. Agric. Food Chem.,* 35: 768 – 770.

Young, H. and Paterson, V.J., 1985. The effects of harvest maturity, ripeness and storage on kiwifruit aroma. *J. Sci. Food Agric.,* 36: 352 – 358.

Young, H. and Paterson, V.J., 1986. The volatile flavour compounds of pepino. Abstracts of 10th Conf. Australia and New Zealand Society for Mass Spec., Dunedin, N.Z., p. 60.

Young, H., Paterson, V.J. and Burns, D.J.W., 1983. Volatile aroma constituents of kiwifruit. *J. Sci. Food Agric.,* 34: 81 – 85.

Chapter VIII

TOMATO, CUCUMBER AND GHERKIN

P.W. GOODENOUGH

AFRC Institute of Food Research, University of Reading, Shinfield, Reading RG2 9AT (Great Britain)

1. TOMATO

1.1 Importance and Origins

The tomato is a fruit almost universally treated as a vegetable, and a perennial plant that is almost universally cultivated as an annual. The tomato plant is the focus of a large agricultural industry. But the plant has a different historical pattern of development in farming communities than most of our traditional food plants. Unknown in the Old World of Europe, Asia and Africa, the tomato plant was first brought into cultivation in central and southern America. In particular, two of the species of the genus *Lycopersicon* (tomato) have fruit which change colour, while all other species have small green or purplish fruit. The two species are *L. esculentum* and *L. pimpinellifolium*. *L. pimpinellifolium,* or currant tomato, has small red fruit, but *L. esculentum* is the only tomato fruit ever to be found that is large and coloured. It was obvious from the first contact that Europeans had with the natives of the Andes and Mexico that the *esculentum* type had already been extensively hybridised and cultivated. As we shall see later, the tomato plant evolved in the western coastal plain of northern South America, but there are no archeological remains of cultivated tomatoes in the Andean region, nor is there any name for the plant in the native tongue. The comparison of hereditary enzyme variants reveals greater similarity between the older European cultivars and the primitive cultivar *L. esculentum* var. *cerasiforme,* or cherry tomatoes, of the Mexican area. In the Nahua tongue of Mexico, the plant is known as tomotl, and this is the origin of the modern name.

The first record of tomato plants in the Old World is credited to descriptions by Pier Andrea Mattiolus in 1558 (quoted by Sturtevant, 1889). The fruit was known as pomi d'oro or mala aurea (golden apple) and pomo amoris (love apple). Obviously the colour of some cultivars was golden rather than red. The plants are related to potatoes and other *Solanaceae,* the flower structure and fruit are very similar to all *Solanaceae*. Conclusions were erroneously drawn that the fruit were poisonous like deadly nightshade, and acceptance was very slowly gained. In 1581 a herbalist wrote 'These apples were eaten by Italians like melons, but the strong stinking smell gives one sufficient notice how unhealthful and evil they are to eat'. The link with deadly nightshade was to make superstitions persist until the 20th century, and there were only occasional written references until 1820. Linnaeus (1753) regarded the plant as part of the genus *Solanum,* but Miller (1754) separated an *esculentum* type cultivar into a genus *Lycopersicon.* The name *Lycopersicon* was from the Greek for 'wolf peach', as Anguillara writing in 1561 mistakenly thought Galen had so-named the tomato in the second century AD. Hill (1765) again re-classified the tomato plant as a species separate from *Solanum,* but with the generic name *Lycopersicum* (Latin derivation of *Lycopersicon),* and this classification was retained until 1914 when Druce pointed out that *Lycopersicon* Mill. was the original type-genus. This was adopted and has been used subsequently. The final subdivision of Muller into two genera was in 1940, and is shown in Table VIII.I.

As soon as the fruit was accepted and popularised it became heavily consumed in European countries and N. and S. America. An indication of the tomato fruit importance is seen in Table VIII.II, which shows that tomato fruit ranked sixteenth in nutrient concentration but first in contribution to the diet of the American population in 1980.

1.2 Botanical Details and Native Habitat

Even after the tomato came to be accepted as a food, debate persisted over its status as a fruit or vegetable. In 1893 the question came before the U.S. Supreme Court and the tomato was adjudged botanically to be a fruit but in the common language of the people was a vegetable. The fruit is a fleshy berry, or, in botanical terms, a swollen ovary. The body of the fruit, developed from the ovary wall which surrounds and encloses the seeds, is known as the pericarp, and consists of outer, radial and inner walls. The locular cavities appear as gaps

TABLE VIII.I

Classification of genus *Lycopersicon* (Mill.) into two subgenera

Subgenera	Eulycopersicon Glabrous, red or reddish-yellow fruit	Eriopersicon Hairy, whitish-green fruit with purplish stripes
Species	*L. pimpinellifolium*-currant tomato *L. esculentum* cultivated types *L. esculentum* var. *cerasiforme* cherry tomato, 1.5 – 2.5 cm dia.	*L. cheesmanii* var. *humifusum* *L. cheesmanii* from Galapago Island only *L. hirsutum* var. *glabratum* *L. chilense* originally *L. peruvianum* var. *dentatum* *L. chmielewski* and *L. parviflorum* are recognised by some authorities

TABLE VIII.II

Consumption vs nutritional content of some fruits and vegetables, including tomatoes, in the U.S.

Nutrient concentration		Contribution of nutrients to diet	
Crop	Rank	Crop	Rank
Broccoli	1	Tomatoes	1
Spinach	2	Oranges	2
Brussels sprouts	3	Potatoes	3
Lima beans	4	Lettuce	4
Peas	5	Sweet corn	5
Asparagus	6	Bananas	6
Artichokes	7	Carrots	7
Cauliflower	8	Cabbage	8
Sweet potatoes	9	Onions	9
Carrots	10	Sweet potatoes	10
Sweet corn	12	Peas	15
Potatoes	14	Spinach	18
Cabbage	15	Broccoli	21
Tomatoes	16	Lima beans	23
Bananas	18	Asparagus	25
Lettuce	26	Cauliflower	30
Onions	31	Brussels sprouts	34
Oranges	33	Artichokes	36

in the pericarp, and contain the seeds embedded in a jelly-like parenchymatous tissue originating from the placenta. The number of locules in normal fruit varies from two upwards, and is more or less characteristic for each cultivar. Prior to pollination, as well as during a relatively short period after anthesis, the growth of the fruit is mainly by cell division (Spurr, 1959), after which cell enlargement is responsible for growth of the fruit. Vacuoles appear in the cells, and differentiation in composition becomes apparent. The fruit is now ready to ripen, which, although having no scientific meaning, has come to mean a collection of changes in colour, flavour and texture which lead to the state at which the fruit is acceptable to eat.

Figure VIII.1 shows a ripe (red) fruit cut in a vertical plane, and Fig. VIII.2 an unripe (green) fruit cut in a vertical plane. The increased breakdown of the parenchymatous tissue and the pericarp are obvious in the ripe fruit, which is very fleshy, comparing markedly with the thin, dry stems and leaves of the plant. The plants live in a remarkably dry environment, which originally was the western coastal plain of northern South America. This narrow strip of land, mainly less than 100 miles wide, extends from near the equator in Ecuador, through Peru, to approximately latitude 25 – 30° in Chile. The strip is delineated by the Pacific Ocean to the west and the Andes to the east. The eastern boundary might be a barrier because of the decrease in temperature with height, none of the wild tomatoes possessing any degree of frost resistance.

This part of the western coastal plain of South America, despite its great length, has a fairly uniform climate. But the plain is unusual in that, although it is one of the most arid deserts in the world, the air is cool and the sky is overcast for much of the year. This is attributed to a very cold sea current that flows along the coast and cools the westerly wind which brings moist air from the

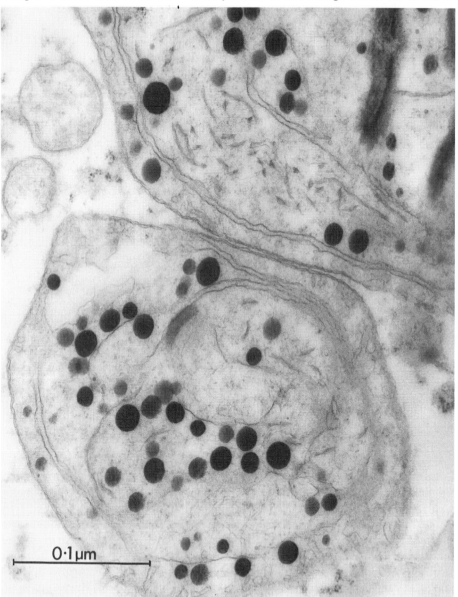

Fig. VIII.1. Section of red fruit showing chloroplast/chromoplasts.

Pacific. When the air reaches the coast it is warmed, and its ability to retain water vapour is increased. However, when the coastal plain receives easterly winds from the Andes, which are dry, they meet the cold moisture-laden winds

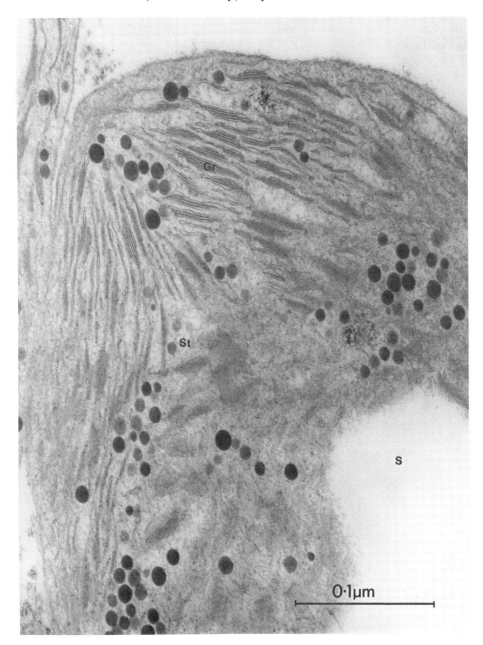

Fig. VIII.2. Section of green fruit showing chloroplasts. Gr = grana, S = stroma, St = starch.

332

from the sea, causing dense persistent fog which obscures the sun and enables vegetation to grow in the desert from April to December. Condensation is extensive, making the surface of the ground quite wet although there is no rain. The typical area of tomato growth is 12° south of the equator at 394 feet above sea level. The humidity ranges between 68 and 95% with virtually no rain, but the fog makes for diffuse radiation. The minimal night temperatures are 15°C and the maximal day temperatures are 19.4°C. The growing day varies from 11.5 to 12.5 hours of daylight. Figure VIII.3 shows the range of air temperature along the western coastal plain of northern South America during the natural growing season.

The temperature change over 2000 miles is remarkably small, and the horizontal lines are the temperature limits of the native habitat. The plants are thus subjected to a very small range of night and day time temperatures (Cooper, 1972).

In summary, the daylength at which these plants grow best and produce well-flavoured fruit is closely controlled at 12 h; similarly, the temperatures at night and day are 11 – 18 and 16 – 24°C, respectively, with a high humidity, 70 – 90%.

1.3 Development of Flavour

As the fruit mature, many changes occur in the composition and flavour of the fruit. When considered in evolutionary terms, it is logical that when the

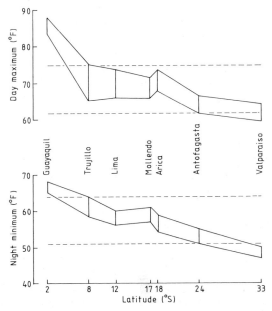

Fig. VIII.3. Range of air temperature along the western coastal plain of northern South America during the natural growing season of tomatoes (Cooper, 1972, reproduced with permission).

seeds are mature and capable of germination then the fruit should be most attractive to animals which would disperse the seeds. However, this is clearly not true with members of the Eriopersicon sub-genera, which have greenish-white fruit, and points to plant selection by man over a long period of time.

By whatever pressures the appearance of ripe tomatoes has been achieved, ripening is now one of the few biochemical processes in plants which bears a similarity to the rapid increases in respiration when animals move quickly. In animals, movement is accompanied by rapid oxidation of sugars, giving a large increase in oxygen uptake and carbon dioxide evolution. In plants, respiration is usually fairly constant and dependent on environmental pressures, such as temperature, light and moisture. But constant respiration is not a characteristic of the group of plants which have fruit known as climacteric ripening fruit. Although the rate of respiration, i.e. the amount of oxygen taken up (or carbon dioxide given out)/unit of tissue in developing tomato fruit, gradually decreases as increase in size slows down, there occurs a point where growth in size has stopped, but there is a sudden, and far-reaching change, in the biochemistry of fruit. Probably the best place to start is the definition of the climacteric in fruit ripening. The development of the modern study of the ripening process stems from work on the physiology of the apple by Kidd and West in the early 1920s. They studied the changes in respiratory activity of Bramley's Seedling, that is the amount of oxygen (O_2) taken up, and the amount of carbon dioxide (CO_2) given out during metabolism of sugars to produce energy; they found that the CO_2 fell to a minimum and then rose more or less rapidly depending on temperature, but started to decline fairly rapidly afterwards (Kidd and West, 1930). They considered the sudden alteration in respiration marked a change from maturation to senescence, and more visible signs of ripening occurred at this time. It was thought to represent a critical phase in the life of the fruit, and was termed the climacteric. A similar rise occurs in the tomato, on or off the vine. The other changes which take place during ripening in the fruit are preceded by the climacteric increase in respiration, and at the same time plant hormones increase.

The coordination of biochemical changes in the tomato seems to rely on plant hormones. Ethylene is produced from methionine by a biosynthetic pathway, shown in Fig. VIII.4 (Yu et al., 1979a,b). The synthesis of 1-aminocylopropane-1-carboxylic acid is controlled by auxin through its effect on the synthesis of a biosynthetic enzyme. Thus the control of enzyme synthesis is at the core of the rapid biochemical ripening process. In all cases, enzyme appearance is the result of a cascade of reactions beginning with activation of a segment of the deoxyribonucleic acid, and production of ribonucleic acid in the nucleus. This is followed by protein synthesis in the cytosol. Plant growth regulators or hormones are thought to control the activation of DNA. Recent reports about tomato ripening have identified several 'ripening related' pieces of DNA, which are obviously important in producing enzymes which control various biochemical mechanism during ripening (Grierson et al., 1986a,b). How the DNA is activated by plant growth regulators, and how these are controlled by time,

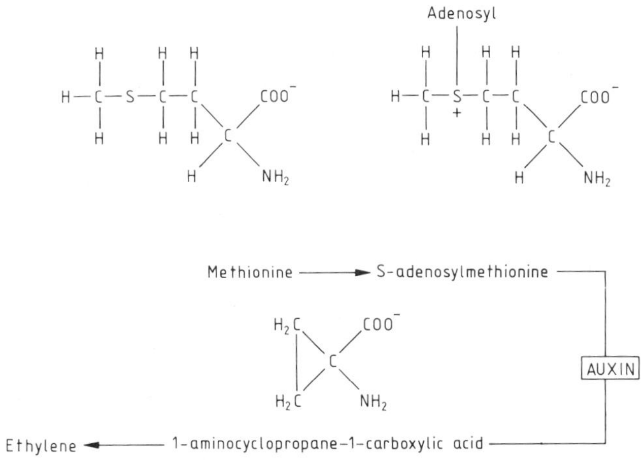

Adenosyl

$$H-\overset{\displaystyle H}{\underset{\displaystyle H}{C}}-S-\overset{\displaystyle H}{\underset{\displaystyle H}{C}}-\overset{\displaystyle H}{\underset{\displaystyle H}{C}}-\overset{C}{\underset{\displaystyle H \quad NH_2}{}}COO^-$$

Methionine ⟶ S-adenosylmethionine

AUXIN

Ethylene ⟵ 1-aminocyclopropane-1-carboxylic acid

Fig. VIII.4. Biosynthetic route from methionine to ethylene.

are very important questions, and the promoters, inducers and repressors of DNA in higher plants will be elucidated in the future by concerted effort in plant molecular biology. Assuming that the evidence clearly points to the formation of enzymes and subsequently flavour compounds being controlled by the activation or repression of the nuclear DNA, we can now turn to how the major flavour components develop during ripening and are affected by various physical parameters.

1.3.1 Sweetness

In green plants, whose nutrition is dependent on addition of carbon dioxide to a five carbon sugar (ribulose bisphosphate), the main carbohydrate which is translocated from one part of a plant to another is the 12 carbon sugar, sucrose.

In tomato plants, sucrose is found in all the parts of the plant except the fruit, and painstaking work by Ho (Walker *et al.*, 1978) has shown that sucrose is rapidly converted by the enzyme invertase into glucose and fructose as soon as it enters the fruit. The sucrose is translocated into the fruit from the main areas of carbon dioxide fixation, and then glucose and fructose are formed. Fruit are therefore a metabolic 'sink', and the metabolic 'source' is the leaves. So much monosaccharide is accumulated by the fruits that 45 – 50% of the dry material is sugar. Only 5 – 8% of the fresh weight of the tomato is dry matter, and half of this is thus sugar. However, the immediate breakdown on the entry into the fruit of sucrose into glucose and fructose by invertase, and the large quantities of these two monosaccharides in the fruit, are not the only methods of developing sweetness. Although it has been shown conclusively that glucose and fructose increase in concentration in the fruit as ripening proceeds (reviewed in Davies and Hobson, 1981), it cannot be assumed that the sucrose entering the fruit is increasing in amount. After many years of research into tomato fruit flavour, it is still possible to find reviews where the rôle of starch in fruit ripening is ignored (Davies and Hobson, 1981). During post-harvest storage, during which light was excluded and fruit could not produce sugars by fixing carbon dioxide, it was shown that many cultivars had built up starch reserves during fruit development, and on the climacteric ripening burst of respiration, starches were broken down (Goodenough and Thomas, 1981). It is therefore important to differentiate carefully the methods of developing full sweetness in tomato fruit flavour.

Commercially it is obviously of vital importance that post-harvest fruit can develop full sweetness if they are shipped slightly underripe (as is often the case). If the full development of sweetness relied only on a source of sucrose being converted to glucose and fructose as the disaccharide entered the fruit from the fruit stalk, then most commercial practices would produce fruit lacking in full flavour. While it is possible to allow for sucrose being imported right up until full ripeness, it has been accepted that sweetness develops fully in fruit ripening off the vine. Goodenough and Thomas (1981) showed that starch was found in all cultivars of fruit that were examined, and furthermore that on storage the starch fell to zero and glucose and fructose increased normally. In some cases, good stoichiometric agreement could be found between starch breakdown and glucose and fructose appearance, but in some cultivars less starch was found than was necessary to give the amount of glucose and fructose recorded (Goodenough and Thomas, 1981). Although it has not been proved by experimental work, one can assume that in some cultivars the starch is more branched than in others, and as such, the methods employed for analysis were unable to successfully break down all the starch to glucose. The estimates for starch were therefore lower than necessary to explain the concentrations of glucose and fructose. This points to a subtle difference in the structure of the amylopectin molecule in the different cultivars.

In summary, the monosaccharides in the fruit could be formed by two routes. First, sucrose enters the fruit from the translocation stream, and is immediately broken down to form glucose and fructose. Secondly, sucrose enters the fruit and is broken down, but the glucose, and possibly some of the fructose isomerised to glucose, is used to build starch. It is more likely that the second route does not occur, and that there is some photosynthesis in green fruit, and starch is formed within the chloroplasts. Several authors have shown how plastids break down to form chromoplasts in ripe fruit (Piechulla et al., 1985; Bathgate et al., 1986). Therefore, photosynthesis could not operate after the orange/green stage of ripeness. Is it feasible that all the starch originates from fruit photosynthesis? Certainly Goodenough (1986) has discussed how a number of mechanisms could result in the post-harvest glucose and fructose being increased by breakdown of starch after the starch has been synthesised from the 'source' sucrose, and mentioned how energetically wasteful this seems. A more feasible method, from the energetic point of view, would be that sucrose from the source provides a high concentration of glucose and fructose in developing fruit, and at the same time photosynthesis in the fruit leads to a store of starch in the plastids. Upon ripening, sucrose is no longer imported, obviously not if the fruit has been removed from the vine, but glucose and fructose from starch appear. This fits with the known facts, and is much more favourable as far as energy conservation for the plant is concerned. However, there is still the observed increase in the enzyme invertase to account for in this scheme. Invertase converts sucrose to glucose and fructose, but possibly also acts upon other polysaccharides. Jeffery *et al.* (1984) were able to show that the invertase activity could be retarded, while increase in starch breakdown and monosaccharide increase were not.

It seems logical that invertase activity is not bound up with starch degradation but with some aspect of cell-wall degradation. Thus, starch is probably degraded to glucose, which is isomerized to fructose, rather than glucose being isomerized to fructose, sucrose synthesised and inverted to give glucose and fructose. The exact proportions of monosaccharide from the source which appear in starch have never been measured, but it is fortunate indeed for commercial exploitation of the fruit that full development of sweetness can occur post-harvest rather than falling dramatically during shipping and storage.

A very useful tool for breeding of fruit may be to find if all starch is formed by photosynthesis, and to increase this in fruit, leading to increased sweetness on ripening.

In 1966, Davies showed that sucrose occurred in greater quantities in other species, and the increased sweetness of small-fruited cultivars, such as Gardeners Delight, may well be due to the influence of the genome of the *pimpinillifolium* types, although sucrose remains at a low concentration in this cultivar (J.N. Scott, personal communication, 1987).

1.3.2 Acidity

The acidity in tomato flavour is produced by accumulation of the two citric acid cycle acids, citric and malic. Very little attempt has been made to explain how the biochemical mechanisms of acid accumulation operate. However, many studies have shown that titratable acidity is invariably significantly correlated with quality (Davies and Hobson, 1981).

The site of citric and malic acid synthesis is the mitochondria. Mitochondria are readily isolated from green fruit, and shown to be fully able to function normally (Hobson, 1969). Very tight coupling, i.e. the uptake of oxygen upon addition of adenosine triphosphate, indicates that the mitochondria must have a full complement of citric acid cycle enzymes, and also a normal electron transport chain, which passes electrons from reduced co-enzymes, via a number of acceptors, to oxygen. All of this takes place within the mitochondria which, even on ripening, seem to remain intact and still are able to use pyruvate to produce oxidative phosphorylation. Normally in plants, the intermediates citrate and malate do not accumulate, and indeed cannot accumulate within the mitochondria, because they will change the physiological pH. It has been shown that many of the enzymes of the citric acid cycle can be identified from intact mitochondria (Hobson, 1974; Jeffery *et al.,* 1984). Although not all enzymes have been detected, the evidence previously mentioned of oxidative phosphorylation, means that they must all be present.

In order to try and explain why the fruit accumulate citrate and malate during fruit development, and then malate decreases at full ripeness whilst citrate increases at first but falls back in concentration at full ripeness, studies have been initiated of the enzymes involved in catabolism and anabolism of the acids. Plants have the usual system of glycolytic and citric acid cycle enzymes, but also seem to make use of a system of malate decarboxylating enzymes. This has led to a phenomenon called the malate effect being identified in fruit, including tomatoes. In the malate effect, a nicotinamide adenine dinucleotide (NAD) or NAD phosphate (NADP) linked enzyme is synthesised in large quantities, and malate is decarboxylated to pyruvate. Carbon dioxide is released as a gas.

In tomatoes it was found that, both on the vine and post-harvest, ripening was associated with a rapid decrease in the enzymes of the citric acid cycle. The evidence was conclusive that only about 60% of enzyme activity was lost, but citrate actually increased at this time and malate decreased. The enzyme malate dehydrogenase is found both within the mitochondria and also in the cytosol. In ripening fruit, the enzyme decreased equally in both areas. However, isolated mitochondria stored at 4°C in an intact state lost only a small percentage of the citric acid cycle enzymes over a 7-day period. In these experiments, a clear pointer exists to active breakdown of citric acid cycle enzymes, but still with intact mitochondria and integrity of the citric acid cycle. Although isocitrate dehydrogenase (NAD linked) decreases in the mitochondria, there is evidence that the cycle continues to turn at full ripeness. How then does citrate accumulate and malate decrease? The answer seems to lie in the increased NADP-linked malate decarboxylating enzyme. This enzyme increases by up to 400% in the cytosol, and presumably pyruvate is produced in large quantities. Citrate synthase is only found inside the mitochondria, and presumably responds to increased flux of pyruvate to produce citrate, which is re-exported to the cytosol rather than being utilized in the mitochondria. The carbon dioxide produced by the 'malate' effect will obviously contribute to the increased respiratory evolution of carbon dioxide at the climacteric. This gives increased credibility to the theory that the climacteric rise is not just increased evolution of carbon dioxide by the electron transport chain, but also by the malate effect (Jeffery et al., 1984, 1986; Goodenough et al., 1985). This has been reviewed recently (Goodenough, 1986).

1.3.3 Colour development

At the time of ripening the chloroplasts are the organelles which undergo the most drastic biochemical differentiation. Although not contributing directly to flavour, the colour change does contribute to overall acceptability. This chloroplast/chromoplast conversion is one example of tissue-specific differentiation of plastids found throughout plant development (Thomson and Whatley, 1980). The ultrastructural changes involve the disappearance of the thylokoid membrane system and the degradation of chlorophyll. Since the light harvesting chlorophyll a/b binding proteins are unstable without chlorophyll, it is assumed that photosynthesis stops at this stage. The early stage of chromoplast formation is characterized by appearance of carotenoids (increased synthesis and/or accumulation). In mature red fruit, the lycopene develops as a large crystal within the chromoplast. The precise control of these ultrastructural changes is, as yet, unknown. Figure VIII.5 shows how the chlorophyll decreases while other pigments increase in the fruit during development. It is interesting to note in this context that the double-stranded DNA of the chloroplasts has been shown to encode rRNAs, tRNAs and a number of chloroplast proteins required for photosynthesis. During light-mediated development of chloroplasts, several plastid-coded proteins, including the large subunit of ribulose bisphosphate carboxylase, a herbicide-binding protein and a number of other photogenes are

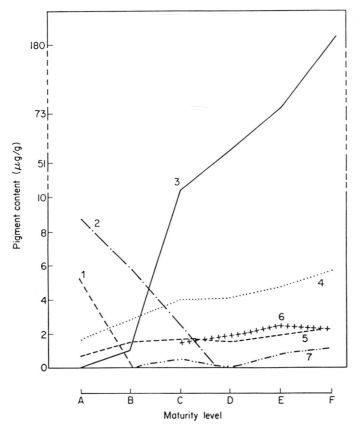

Fig. VIII.5. Changes in the concentration of tomato pigments during maturation (cultivar San Marzano). 1, chlorophyll b; 2, chlorophyll a; 3, lycopene; 4, β-carotene; 5, lutein-5,6-epoxide; 6, lycoxanthin; 7, lycophyll. Maturity levels. A, mature green; B, green-yellow; C, yellow-orange, some trace of green; D, orange-yellow, no trace of green; E, orange-red; F, red (reproduced from 'The Biochemistry of Fruits and their Products' by permission of Academic Press).

continually synthesised. Very little is known about synthesis of these proteins in the chromoplast. Very recently it has been shown that the large subunit of Rubisco is lost during chromoplast development (Bathgate *et al.*, 1985, 1986; Piechulla *et al.*, 1985). Similarly, genes for photosystem I, photosystem II and the stroma decrease. As yet it is not known whether the enzymes involved in the biosynthesis of carotenoid and lycopene are synthesised by the DNA of the chloroplasts or the nucleus.

1.3.4 Textural changes

The changes in the fruit which are able to give full sensory appreciation include textural changes. Although not directly giving rise to flavour, the changes in texture contribute greatly to the release of the flavour components. The unripe fruit contain normal cell walls with cellulose microfibrils, pectin and components of arabinose, rhamnose and galactose as arabin and rhamnogalactan. The structure of plant cell walls has been investigated to a great extent by Albersheim and his co-workers (Talmadge *et al.*, 1973). In order to elucidate the structure of cell walls, Albersheim degraded the sugar residues by enzymic ac-

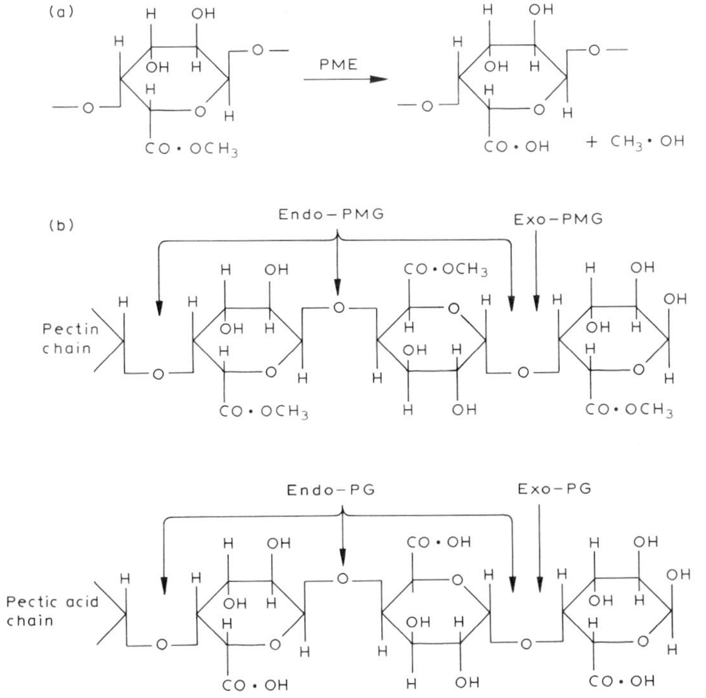

Fig. VIII.6(a). Pectinmethylesterase (PME) removes the methyl group in each monomeric unit of the polymer, pectin.

Fig. VIII.6(b). Enzymes causing degradation by hydrolysis of the polygalacturonate polymer chain (PMG = polymethylgalacturonate, PG = polygalacturonate).

tion. In a very similar way, the cell walls of tomato fruit are degraded by enzymes as the fruit ripen. Pectic material acts as a 'cement' between the primary cellulose wall of cells, and once dissolved the cells will move relative to one another, giving the usual loss of structure found in ripe tomatoes. The enzymes able to break down pectin, and also the structure of pectin, are shown in Fig. VIII.6.

In 1964, Hobson discovered that one of the enzymes able to cleave the pectin material was not present in the unripe green fruit. This enzyme, endopolygalacturonase, is able to break randomly the bonds between adjacent galacturonic acid residues. The enzyme pectinmethylesterase is able to demethylate pectin, and this enzyme is present at all times during fruit development, but increases in concentration during ripening. A third enzyme, known as pectin lyase, is also found in tomato fruit, but does not seem to play a rôle in pectin degradation. The main mediator of breakdown thus seems to be polygalacturonase, which rapidly increases from zero concentration in green mature fruit to a high concentration in mature red fruit. The enzyme seems to exist in three isoenzymic forms, which appear in the fruit at slightly different times. This is the clearest evidence for synthesis of an enzyme specific for ripening. Purified polygalacturonase isoenzymes have been shown to attack tomato fruit cell walls and to solubilise cell walls when applied to segments of unripe tomato tissue (Themmen *et al.*, 1982).

Recently it has been shown using *in vitro* translation and cloned DNA (cDNA) probes, that polygalacturonase mRNA is not detectable in green fruit, but accumulates during ripening along with other 'ripening-related' mRNAs. It has been proposed that these 'ripening-related' mRNAs are the products of a group of genes which code for enzymes important in the ripening process (Grierson *et al.*, 1986a,b).

The loss of firmness is a problem to various retailers and wholesalers, but particularly so in the case of the 'primary products' where tomatoes are used for paste or are canned.

1.3.5 Volatiles

A typical volatile profile from tomatoes has esters, hydrocarbons, long-chain alcohols, and carbonyl compounds in the 285 compounds so far detected. The commonest compounds are *cis*-hex-3-en-1-ol, 2-(2-methylpropyl)thiazole, 3-(methylthio)propan-1-ol and β-ionone. The C_6-aldehydes seem to form an important part of tomato aroma (Guadagni *et al.*, 1972), but as these are absent in rapidly heat-processed cultivars (Kazeniac and Hall, 1978), it seems that a rapid enzymic synthesis is responsible (see later). A long list of aromatic compounds is thus available, but there is little quantitative evidence of the exact contribution to overall aroma.

1.4 Primary Products

It is of great interest that tomatoes are used in so many different processed foods, and still impart a distinctive flavour or give a particular texture. The tomatoes used for processing are grown in the field in southern areas of Europe and the U.S.A. New cultivars have been developed to increase the yield, resistance to disease, and particularly the ability to withstand the stress of mechanical handling. A major factor in development of bushy field cultivars has been the simultaneous ripening of fruit, such that all the area can be harvested at a single time. The fruit have a high resistance to abuse in handling and transport.

A list of 12 modern cultivars which, in 1980, were used in processing is shown in Table VIII.III. Producers can provide themselves with facilities for blending different cultivars to give the most desirable quality features in the product. For example, one or more low viscosity (low consistency) types can be blended with cultivars of higher viscosity to provide a more or less constant final quality in the product. Flavour and colour can also be blended in this way.

The bushy tomato fruit are now moved hydraulically into a factory, and this helps to preserve the flavour by preventing mould and bacterial growth, and also preventing loss of drained weight. The water temperature is kept at $50-75°F$, and storage enables rotten and broken fruit to be taken from the bottom of the tank, immature fruit from the top of the tank, and the bulk of the good fruit in the middle layer of the tank.

TABLE VIII.III

Some modern tomato cultivars

Order of earliness	Cultivar	Soluble solids[a]	Viscosity[b]	Colour[c]	Shape	Weight of fruit (g)	Disease resistance[d]	Transport capability[e]
1	Ventura	3	2	2	Pear	45	F	2
2	Heinz 1706	1	3	1	Pear	40	VF	2
3	Euromech	1	1	1	Square	60	VF	1
4	Petomech	2	1	2	Square	60	VF	1
5	Campbell 28	2	3	2	Round	100	F	2
6	Heinz 2274	1	2	1	Globe	120	VF	1
7	CAL.J	2	1	2	Square	80	VF	1
8	Campbell 35	2	1	1	Round	80	VF	2
9	Heinz	2	3	1	Round	90	FS	3
10	Roma VF	3	2	2	Elongated	50	VF	2
11	Cal-Ace	2	3	1	Square	150	VF	3
12	VF 198	2	2	2	Square	60	VF	2

[a] 1 = Above 5.5, 2 = 5.0–5.5, 3 = Below 5.0.
[b] 1 = High viscosity, 2 = medium viscosity, 3 = low viscosity.
[c] 1 = Over 2.1 (Hunter A/B), 2 = under 2.1.
[d] F = Fusarium, V = verticillium, S = stemphyllium.
[e] 1 = Very good, 2 = good, 3 = poor/fair.

342

1.4.1 Acidification of tomatoes prior to processing

Some of the modern American cultivars tend to possess rather high pH values (3.9–4.9 are typical values) (Sapers *et al.*, 1978), which can make the fruit much more vulnerable to spoilage. Flat sour spoilage in canned tomato becomes more difficult to control as the pH becomes higher than about 4.6. This is now countered by adjustment of pH to 4.1–4.3 using edible organic acids approved for food use, usually citric acid.

In the case of tomato juice canning, no adjustment of acidity is generally necessary, but a full sterilization ensures destruction of all spores. Tomato paste is not made from juice which has received a full sterilization, and some concern is felt over products subsequently made using the paste or diluted paste to form a 'natural-strength' tomato juice for consumption.

1.5 Factory Production of Tomato Juice and Tomato Paste

The production of juice from raw fruit is shown in Fig. VIII.7.

1.5.1 Pulping

As mentioned previously, the enzymes liberated by fracturing ripe tomatoes can rapidly bring about the destruction of the natural tomato pectin. Loss of such pectin causes a lowering of the consistency of the product and, in order

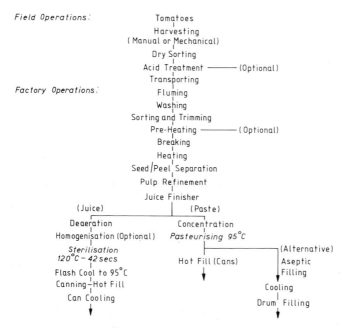

Fig. VIII.7. Flow diagram for production of tomato juice/paste. Reproduced from Schreier, 1981, by permission.

to reduce or prevent this effect, there are systems of heating the macerated tomato pulp during, or very shortly following, the actual crushing of the fruit. This so-called hot-break process is of great significance in imparting good 'body' to the juice or tomato paste. It also aids extraction of seed gum, which helps to stabilise the pulp and reduce syneresis in products made from the pulp.

Some modern cultivars are very resistant to handling damage, and this can be due partly to a greater natural cellulose content and a more fibrous structure which contributes to a greater consistency in the pulp produced from the fruit. Some of the tomato pastes using these cultivars cannot be used in traditional applications, as they are too viscous.

After crushing and heating, the pulp is passed through a two- or three-stage refining process to remove skins and seeds and give the remaining pulp a fine particle size. The refining screen can be as small as 0.4 mm.

The screening systems used for natural juice usually employ 'juice extractors' in which the juice is forced through cylindrical screens by the rotation of helical worms (screw presses). In pulp intended for making tomato paste, it is common practice to use 'cyclone'-type pulpers and finishers, which force the juice through the screens by means of rotating paddles. The system has a higher output than juice extractors, but does tend to beat air into the juice which reduces the flavour by oxidative changes, and loss of vitamin C can be induced.

Recently, completely new machines have been developed which combine the operations of disintegration, seed separation, pulping, finishing and homogenization. The soft tissues and juice pass through the screens, and skins, seeds and other waste are discharged separately.

After the pulp has been refined, it constitutes natural tomato juice, which may be carried directly in consumer packs for distribution to the retail market or in large cans or even bulk containers, for use as an ingredient by soft drink manufacturers. For these purposes, the tomato juice must be heat-treated in plate or tubular heat exchangers, with holding loops giving full sterilising conditions for destruction of flat-sour spores (120°C for 42 s), followed by hot filling into cans immediately after flashing down to not less than 95°C.

If tomato juice is destined for concentration, it does not usually receive this high-temperature treatment, but is passed directly to the evaporation equipment, which is the most important phase in the production of paste.

The preservation of flavour in tomato paste was difficult when open pans were used for evaporation, as caramelisation of sugars with scorched flavours and brown discolouration occurred.

However, the aroma can be changed most radically by enzymic change during pulping, and before sufficient heat has destroyed the enzymes. There appear to be very fast enzymic processes which lead to the composition of the aroma components changing. One such pathway is the degradation of unsaturated compounds to give rise to substances exhibiting 'green-grassy' flavour characteristics (Schreier, 1981). The formation of C_6 aroma constituents during tomato juice production is such that *cis*-hex-3-en-1-ol and hexanal are present, together with *cis*-hex-3-enal, which is the only C_6 aroma compound found in intact fruit. A mechanism for the enzymic formation of C_6-aldehydes in tomato fruit upon the destruction of cell structure involves the enzyme lipoxygenase and a hydroperoxide-cleaving enzyme. Free fatty acids possessing a *cis,cis*-1,4-pentadiene system, such as linoleic or linolenic, are transformed to hydroperoxide by the action of lipoxygenase, favouring the formation of C_9-hydroperoxides, but with C_{13}-hydroperoxides also being formed in the ratio 95:5. Although the C_9-hydroperoxides cannot be further degraded to volatile aldehydes, the C_{13}-hydroperoxides are specifically cleaved to C_6-aldehydes by the action of another enzyme system active in tomatoes. By this mechanism, hexanal originates from linoleic acid, and *cis*-hex-3-enal from linolenic acid. An isomerase transforms *cis*-hex-3-enal to *trans*-hex-2-enal. The reduction of aldehydes to the corresponsing alcohols is catalysed by alcohol dehydrogenase.

The development of other 'secondary' products in tomatoes during juice production includes the carotenoid degradation products methylheptenone, geranylacetone, farnesylacetone and β-ionone. These are much higher in concentration in juice than in intact fruit. These compounds, and the non-enzymic mechanism of production, are discussed by Schreier (1981).

The preservation of flavour was improved when closed 'vacuum pans' were introduced, and these 'boules', as they were called, became standard items of concentration equipment, and many such installations are still in operation, providing an almost continuous flow of good-quality paste from several connected boules.

Continuous evaporators, such as the Manzini 'Titano' systems and the Rossi and Catelli downward forced flow series, have largely superseded the boules in large capacity operations. One plant in the U.S. factory of Tri-Valley growers consists of three DFF evaporators, creating a daily capacity of 675 t of $28-30°$ paste and a consumption of 4000 t of fruit. This gives a seasonal consumption of 200 000 t of fruit. Aseptic tanks and rail cars have been developed to hold this material and transport the paste.

1.6 Canned Tomatoes

The well-known elongated pear-plum shaped fruit is most often used for canning. The San Marzano cultivar is probably the best known internationally. Indigenous to a small district around Salerno in Southern Italy, for many years this cultivar formed the basic raw material of the 'Naples peeled tomato' of Pomadori pelati.

Cultivars of tomatoes suitable for canning must possess a firm flesh and structure with resistance to crushing or collapse during the canning operation, and in subsequent shelf life of the canned product (Goose, 1981). These qualities are shown by the cultivars listed in Table VIII.III.

The traditional method of peeling after scalding was by hand, and sometimes the can filling and peeling can be combined, with the result that tomatoes are extruded out of their skins. Peeling machines are complicated to keep fruit intact for the can.

Lye peeling methods have been successfully applied to tomatoes, and immersions of the fruit in a 17 − 20% solution of caustic soda at 95°C for 30 s satisfactorily peels the fruit.

Steam peeling is generally the method of choice, as it produces less effluent and only a small loss of flavour.

After filling, either by hand or mechanically, the cans are topped up with juice or brine solution. Juice for a covering liquid may be obtained from a pulping plant, and sometimes good quality under-sized fruit are used as a raw product for this purpose. Trimmings and waste from other lines may sometimes be used, but often this produces undesirable off-flavour.

Sometimes brine is added as a carefully metered amount of strong salt solution or as dry tablets.

If the fruit are liable to break down excessively, or if the canned product is to be used as an ingredient in another product, the fruit may benefit by treatment with a 'hardening' (forming) agent. Calcium chloride has been used, but calcium lactate is preferred, as it has a reduced tendency to produce bitter flavours. This material complexes with the pectin and helps retain cell wall rigidity, but it may be banned in some countries.

It is difficult to pack the fruit at elevated temperatures, and cold filling is generally practised. For this reason, it is necessary to ensure a good vacuum in the cans by heating during passage through an exhaust box which is sufficiently large to raise the temperature of the contents and drive out air before the ends are seamed onto the cans. The boxes are water or steam heated, and must achieve a pack temperature of 90°C. Some packers prefer to use steam-flow closing or vacuum seamers to achieve a seal, and this requires very careful headspace control, and overfilling must be avoided. Heat processing of the cans to sterilise the pack at about 103°C is a necessary step in the production, but processing at elevated temperatures is not normally recommended as this is likely to damage the texture and the organoleptic qualities of the product.

1.7 Tomato Powder

The production of dehydrated tomato juice has advanced to a point where it can be considered as a rival to tomato paste for use in some manufactured food products. The further concentration of tomato solids from the 30 − 40% found in tomato pastes to about 97% in the dried powder represents a saving in weight, but little saving in volume.

Tomato powder is a difficult product to handle, and it becomes very tacky at high temperatures because of its thermoplasticity, and it is very hygroscopic, taking up water readily to become sticky and toffee-like.

The use of conventional spray-driers has not proved successful in producing a free-flowing tomato powder, due to the thermoplastic effects.

Foam-mat drying has been used to produce tomato powder. During drying the air bubbles expand under the reduced pressure and produce a fluffy dry material which can be ground to a powder.

In order to produce tomato powder economically which is readily recon-stitutable and of high organoleptic flavour, there have been advances in specialist spray-drying equipment.

The problems of chamber wall temperatures gave rise to the low-temperature spray-drying system with an air outlet temperature of 52°C. In order to provide sufficient drying, the atomised spray falls down a tower 250 feet high and 50 feet in diameter. Paste of 30 – 48% total solids is pumped to the top of the tower and atomised, as it falls it meets an upward draught of hot air dried over silica-gel. Final moisture content is 3% and a very good quality powder is produced.

The large size of the chamber is an economic problem, and smaller vessels are now available where hot air at 145°C is cooled by the paste evaporation to an average of 77 – 80°C. Droplets remain at a low temperature as they settle out and build up in layers to an inch in thickness before falling off and being col-lected. The moisture content is again around 3%.

The retention of pectin in all of these products is vital to the consistency of the juice or puree. If the cultivars have a high resistance to damage and produce a very viscous paste, then the 'hot-break' is sometimes omitted, and the resul-ting breakdown of pectin gives a more liquid type of paste for spray-drying and other applications.

In the 1930s, Italy, France and Hungary were the principal sources of supply of paste to Britain. After the second world war, Portugal joined the paste pro-ducers, followed by Greece and Turkey. Spain built new canneries in the Bada-joz region, and imports of Italian paste fell to only a nominal value. The stan-dard test for acceptance of a paste is the Howard mould count, and the number of mould hyphae identified under a standard microscope is known as 'positive fields'. The setting up of the European Economic Community meant that both Spain and Portugal lost ground to full members of the EEC, such as Italy and Greece. However, all of these countries seem to be competing for a static U.K. market. The canners of peeled tomatoes number more than 150 in southern Ita-ly and perhaps 50 in Spain. The percentage of tomato to water is 36/38% in most cans, and is an indication of the enormous crops of tomatoes canned every year in southern Italy.

2. CUCUMBER AND GHERKIN

2.1 History of Cucurbits

The cucurbits seem to have a dual origin with squashes, pumpkins and gourds evolving from America, while melons and cucumbers came from the tropics of the Old World. A record of the origin of squashes and pumpkins in tropical America is given by Tapley *et al.* (1937).

The controversy over the American origin of the squashes has been paralleled by the discussion of the origin of the Bur Gherkin (*Cucumis anguria*), once known as the West Indian gherkin. Its wild counterparts have not been found in Africa, but that is thought to be of little significance compared with the facts that all its near allies are African, and there are no American relatives and it was found in America only where it could have been introduced by the negroes or through their more-or-less direct agency. So this plant is unquestionably of African origin, and was introduced into the warmer parts of America in the sixteenth century. Both Hooker and de Candolle (quoted in Whitakar and Davies, 1962) agree that it is of African origin, and it has found the conditions for growing in America very much to its liking.

The cucumber (*Cucumis sativus*), or the primitive form from which it is descended, probably originated in Africa, but spread very early on in civilised times to Asia, where it is found wild in northern India and where, according to de Candolle (1882), it has been cultivated for 3000 years. It was grown by the ancient Greeks and Romans. In the Caucasus region there are names for the cucumber that indicate knowledge of the plant before it was known to the Sanskrit, but there is no evidence that it was known to ancient Egypt. It was cultivated throughout Europe in the Middle Ages. Many of these early forms seem to be more rugged and less symmetrical than those grown today, though Mattiolus (1560) and Baukei (1650) (quoted by Sturtevant, 1889) show figures of types very like some of our modern cultivars.

2.2 Fruit Structure

The fruits are inferior berries, indehiscent with the fleshy floral tube adnate to the pericarp. The solid flesh is derived from the placenta.

2.3 *Flavour of* Cucumis

Unlike other members of the cucurbitaceae, the cucumber and gherkin are non-climacteric fruits; respiration gradually declines with age and the commercial maturity of the fruits is just before the end of the growth period. Thus the plants do not have completely mature fruit when harvested. Maturation and senescence are not believed to be at all advanced in the cucumbers used commercially (Wills *et al.*, 1981).

Fruits have been classified into four groups depending on the number of volatile compounds contributing to their aroma (Gormley, 1981). The first category includes fruit which depend on one volatile to produce the characteristic aroma, and cucumbers are a good example of this category. The characteristic odour of cucumbers is given by the unsaturated aldehyde nona-*trans*-2,*cis*-6-dienal, present at about 17 ppm (Takei and Ono, 1939). As well as being the main aldehyde found, it has an odour threshold of only 0.0001 ppm, at least five times lower than that of the compounds *trans*-hex-2-enal and *trans*-non-2-enal. These latter two compounds are also detectable, as is nona-2, 6-dien-1-ol. In fermented and brined cucumbers, only low-boiling saturated carbonyl compounds have been found (Aurand *et al.*, 1965).

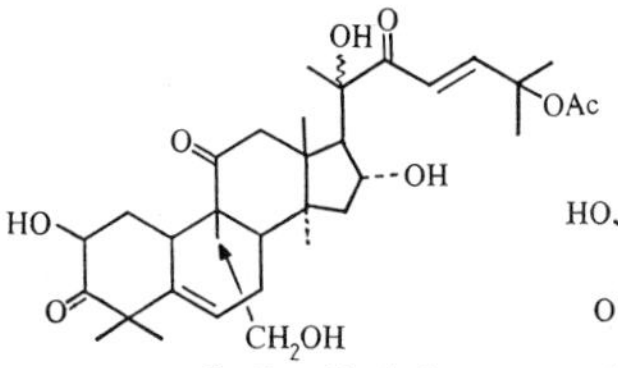

I Cucurbitacin A

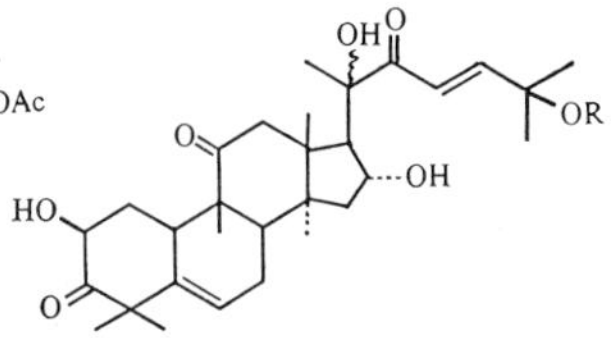

II R = Ac, Cucurbitacin B
III R = H, Cucurbitacin D (Elatericin A)

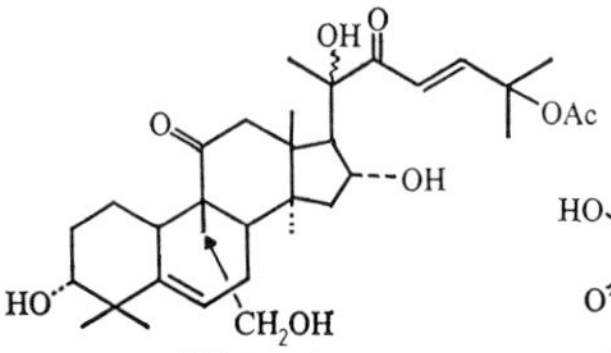

IV Cucurbitacin C

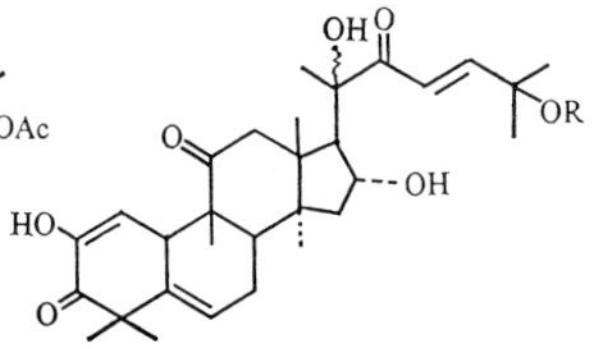

V R = Ac, Cucurbitacin E (Elaterin)
VI R = H, Cucurbitacin I (Elatericin B)

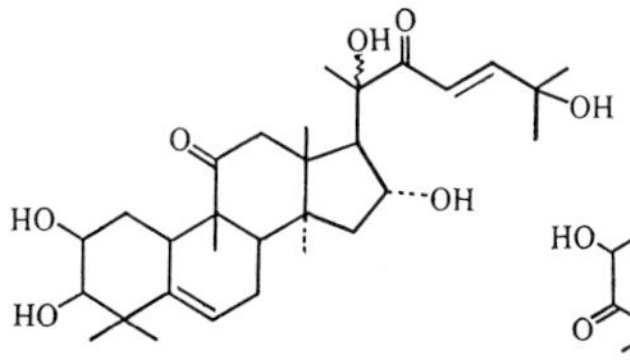

VII Cucurbitacin F

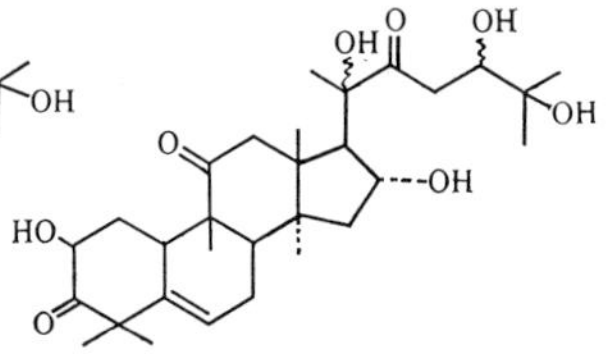

VIII a Cucurbitacin G
VIII b Cucurbitacin H (C24 epimers)

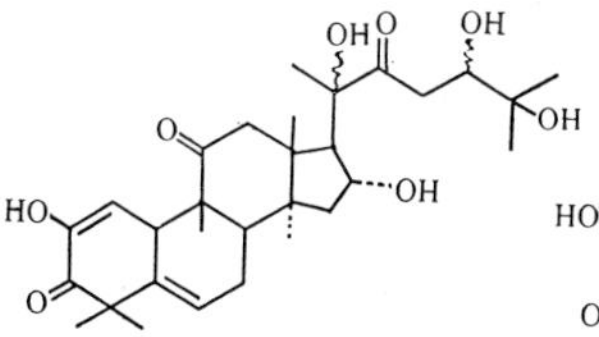

IX a Cucurbitacin J
IX b Cucurbitacin K (C24 epimers)

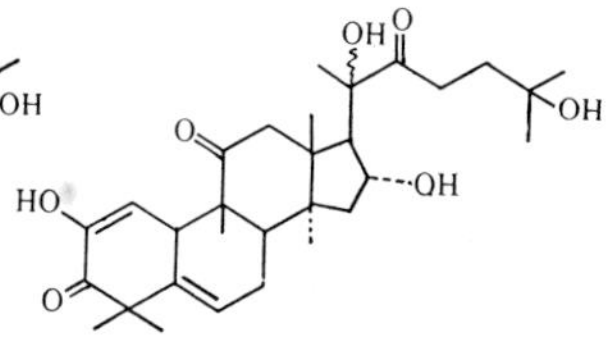

X Cucurbitacin L

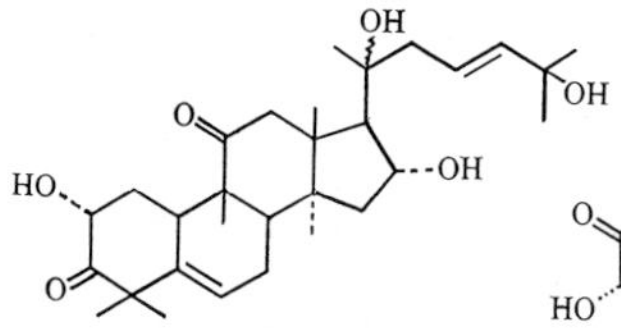

XI 22-Deoxocucurbitacin D

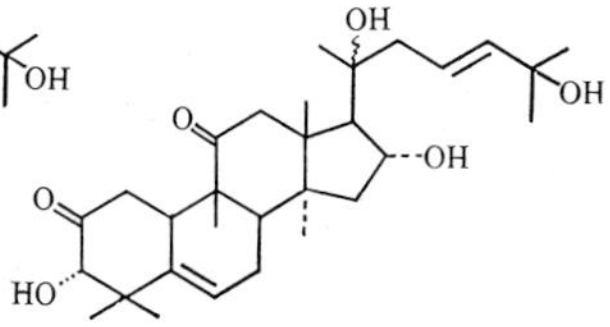

XII 22-Deoxoisocucurbitacin D

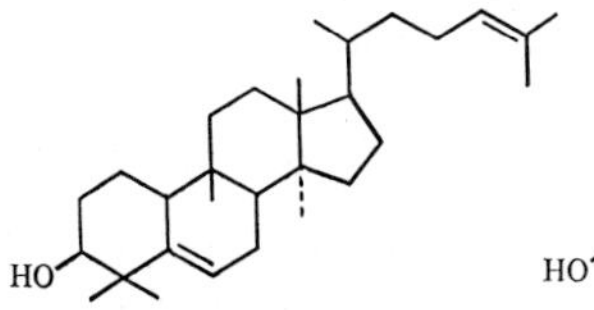

XIII Cucurbita-5,24-dien-3β-ol

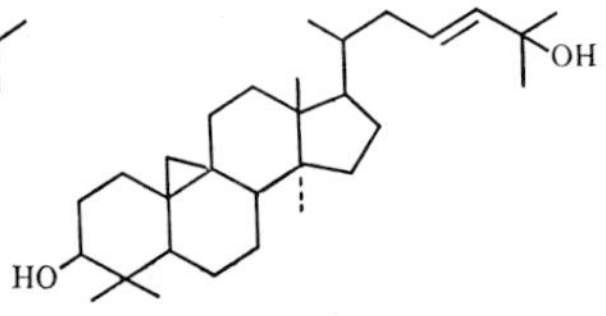

XIV Cycloart-23-en-3β,25-diol

Fig. VIII.8. Structures of the related family of bitterness factors from cucumbers (reproduced, with permission, from Goodwin and Good, 1970).

2.3.1 Bitter factors

One of the most actively studied chemical factors in cucumbers has been the family of related triterpene structures collectively known by the trivial name of cucurbitacins (Lavie *et al.,* 1962). The different cucurbitacins (structures I – XIV, taken from Goodwin and Good, 1970) are shown in Fig. VIII.8. Their distribution and quantitation are discussed extensively by Goodwin and Good (1970). These factors are toxic, and have commercial importance such that fruit free of the bitterness factors were extensively sought. Andeweg and De Bryn (1959) succeeded in isolating a single non-bitter plant in the cultivar improved Long Green. This seedling would have inherited the lack of cucurbitacin formation as a monogenic recessive. All non-bitter cultivars have come from this source.

REFERENCES

Andeweg, J.M. and De Bryn, J.W., 1959. Breeding of non-bitter cucumbers. *Euphytica,* 8: 13 – 20.

Aurand, L.W., Singleton, J.A., Bell, T.A. and Etchells, J.L., 1965. Identification of volatile constituents from pure-culture fermentations of brined cucumber. *J. Food Sci.,* 30: 288 – 295.

Bathgate, B., Purton, M.E., Grierson, D. and Goodenough, P.W. 1985. Plastid changes during the conversion of chloroplasts to chromoplasts in ripening tomatoes. *Planta,* 165: 197 – 204.

Bathgate, B., Goodenough, P.W. and Grierson, D., 1986. Regulation of the expression of the psbA gene in tomato fruit chloroplasts and chromoplasts. *J. Plant Physiol.,* 124: 223 – 233.

Candolle, A. de, 1882. Origine des Plantes Cultivies. Germes Bailliere, Paris, 377 pp.

Cooper, A.J. 1972. The native habitat of the tomato. *Rep. Glasshouse Crops Res. Inst.,* 1971, pp. 123 – 130.

Davies, J.N., 1966. Occurrence of sucrose in the fruit of some species of *Lycopersicon. Nature,* 209: 640 – 641.

Davies, J.N. and Hobson, G.E., 1981. The constituents of tomato fruit – the influence of environment, nutrition and genotype. *CRC Crit. Rev. Food Sci. Nutr.,* 15: 205 – 281.

Druce, G.C., 1914. Report of the Botanical Society and Exchange Club of the British Isles for 1913, p. 433.

Goodenough, P.W., 1986. A review of the rôle of ethylene in biochemical control of ripening in tomato fruit. *Plant growth regulation,* 4: 125 – 137.

Goodenough, P.W. and Thomas, T.H., 1981. Biochemical changes in tomatoes stored in modified gas atmospheres. 1. Sugars and acids. *Ann. Appl. Biol.,* 98: 507 – 515.

Goodenough, P.W., Prosser, I.M. and Young, K., 1985. NADP-linked malic enzyme and malate metabolism in ageing tomato fruit. *Phytochemistry,* 24: 1157 – 116.

Goodwin, T.W. and Good, L.J., 1970. Carotenoids and triterpenoids. In: A.C. Hulme (Editor), The Biochemistry of Fruits and Their Products. Academic Press, London and New York.

Goose, P., 1981. Developments in the production of primary tomato products. *Proc. Inst. Food Sci.,* 14: 15 – 27.

Gormley, T.R., 1981. Aroma in fruit and vegetables. In: P.W. Goodenough and R.K. Atkins (Editors), Quality in Stored and Processed Vegetables and Fruits. Academic Press, London and New York, pp. 35 – 51.

Grierson, D., Tucker, G.A., Keen, J., Pay, J., Bird, C.R. and Schuch, W., 1986a. Sequencing and identification of a cDNA clone for polygalacturonase. *Nucleic Acid Res.,* 14: 8595 – 8603.

Grierson, D., Maunders, M.J., Slater, A., Ray, J., Bird, C.R., Schuch, W., Holdsworth, M.J., Tucker, G.A. and Knapp, J.E., 1986b. Gene expression during tomato ripening. *Philos. Trans. R. Soc. London, Ser. B,* 314: 399 – 419.

Guadagni, D.G., Buttery, R.G. and Venstrom, D.W., 1972. Contribution of some volatile tomato components and their mixtures to fresh tomato aroma. *J. Sci. Food Agric.,* 23: 1445 – 1450.

Hill, J., 1765. *Veg. Syst.,* 9: 32.

Hobson, G.E., 1964. Polygalacturonase in normal and abnormal tomato fruit. *Biochem. J.,* 86: 358 – 365.

Hobson, G.E., 1969. Respiratory activity by miochondria from tomatoes at different stages of ripeness. *Qual. Plant. Mater. Veg.,* 19: 155 – 165.

Hobson, G.E., 1974. Electrophoretic investigation of enzymes from developing *Lycopersicon esculentum* fruit. *Phytochemistry,* 13: 1383 – 1390.

Jeffery, D., Smith, C., Goodenough, P.W., Prosser, I.M. and Grierson, D., 1984. Ethylene independent and ethylene dependent biochemical changes in ripening tomato fruit. *Plant Physiol.,* 74: 32 – 38.

Jeffery, D., Goodenough, P.W. and Weitzman, P.D.J., 1986. Enzyme activities in mitochondria isolated from ripening tomato fruit. *Planta,* 168: 390 – 394.

Kazeniac, S.J. and Hall, R.F., 1978. Processing effects on the sensory acceptability and the volatile composition of tomato juice concentrate. *Proc. Fla. State. Hortic. Sci.* for 1977, 91: 151 – 155.

Kidd, F. and West, C., 1930. Physiology of fruit. I. Changes in respiratory activity in apples at different temperatures during their senescence. *Proc. R. Soc. London,* BIO6: 93 – 109.

Lavie, D., Shvo, Y. and Gottlieb, O.R., 1962. *Annais Ass. Bras. Quim.* 21 Numero espu p. 5.

Linnaeus, C., 1753. Species Plantarum, Day Society ed. pp. 184 – 189.

Miller, P., 1754. The Gardeners Dictionary, 8: 3.

Muller, T.W., 1940. A revision of the genus Lycopersicon. United States Department of Agriculture Miscellaneous Publications, No. 382.

Piechulla, B., Imlay, K.R.C. and Gruissem, W., 1985. Plastid gene expression during fruit ripening in tomato. *Plant Mol. Biol.,* 5: 373 – 384.

Sapers, G.M., Phillips, J.G. and Stoner, A.K., 1978. Tomato acidity and the safety of home canned tomatoes. *Hortic. Sci.,* 13: 187 – 191.

Schreier, P., 1981. Changes of flavour compounds during the processing of fruit juices. In: P.W. Goodenough and R.K. Atkins (Editors), Quality in Stored and Processed Vegetables and Fruit. Academic Press, New York and London, pp. 355 – 371.

Spurr, A.R., 1959. Anatomical aspects of blossom-end rot in the tomato with special reference to calcium nutrition. *Hilgardia,* 28: 269 – 295.

Sturtevant, E.L., 1889. The tomato. *Rep. Maryland Exp. Sta.* for 1888, p. 18.

Takei, S. and Ono, M., 1939. Leaf alcohol. III On the odor of the cucumber. *J. Agric. Chem. Soc. Jpn,* 15: 193 – 195.

Talmadge, K.W., Keegstra, K., Bauer, W.D. and Albersheim, P., 1973. The structure of plant cell walls. I. The macromolecular components of the walls of suspension cultured sycamore cells with a detailed analysis of the pectic polysaccharides. *Plant Physiol.,* 51: 158 – 174.

Tapley, W.T., Enzie, W.D. and Praneseltine, T., 1937. The vegetables of New York. Report of the New York Agricultural Experiment Station for 1935.

Themmen, A.P.N., Tucker, G.A. and Grierson, D., 1982. Degradation of isolated tomato cell walls by purified polygalacturonase *in vitro. Plant Physiol.,* 69: 122 – 124.

Thomson, W.W. and Whatley, J.M., 1980. Development of nongreen plastids. *Annu. Rev. Plant Physiol.,* 31: 375 – 394.

Walker, A.J., Ho, L.C. and Baker, O.A., 1978. Carbon translocation in the tomato: Pathways of carbon metabolism in the fruit. *Ann. Bot.,* 42: 901 – 909.

Whitakar, T.W. and Davies, G.N., 1962. Cucurbits. Botany Cultivation and Utilization. In: N. Poluni (Editor), World Crop Books, Interscience.

Wills, R.H.H., Lee, T.H., Graham, D., McGlasson, W.B. and Hall, E.G., 1981. Postharvest: An Introduction to the Physiology and Handling of Fruits and Vegetables. Granada, London.

Yu, Y.-B., Adams, D.O. and Yang, S.F., 1979a. 1-Aminocyclopropanecarboxylate synthase, a key enzyme in ethylene biosynthesis. *Arch. Biochem. Biophys.,* 198: 280 – 286.

Yu, Y.-B., Adams, D.O. and Yang, S.F., 1979b. Regulation of auxin-induced ethene production in Mung bean hypocotyl: rôle of 1-aminocyclopropane-1-carboxylic acid. *Plant Physiol.,* 63: 589 – 590.

SUBJECT INDEX